D0215965

Biomedical Applications of Synchrotron Infrared Microspectroscopy

RSC Analytical Spectroscopy Monographs

Series Editors:
Neil W. Barnett, *Deakin University, Victoria, Australia*

Advisory Panel:
F. Adams, *Universitaire Instelling Antwerp, Wirijk, Belgium*, M. J. Adams, *RMIT University, Melbourne, Australia*, R. F. Browner, *Georgia Institute of Technology, Atlanta, Georgia, USA*, J. M. Chalmers, *VSConsulting, Stokesley, UK*, B. Chase, *DuPont Central Research, Wilmington, Delaware, USA*, M. S. Cresser, *University of York, UK*, J. Monaghan, *University of Edinburgh, UK*, A. Sanz Medel, *Universidad de Oviedo, Spain*, R. D. Snook, *UMIST, UK*

Titles in the Series:

How to obtain future titles on publication:
A standing order plan is available for this series. A standing order will bring delivery of each new volume immediately on publication.

For further information please contact:
Book Sales Department, Royal Society of Chemistry, Thomas Graham House, Science Park, Milton Road, Cambridge, CB4 0WF, UK
Telephone: +44 (0)1223 420066, Fax: +44 (0)1223 420247,
Email: books@ rsc.org
Visit our website at http://www.rsc.org/Shop/Books/

Biomedical Applications of Synchrotron Infrared Microspectroscopy

Edited by

David Moss
Synchrotron Light Source ANKA, Karlsruhe Research Center, Karlsruhe, Germany

RSCPublishing

RSC Analytical Spectroscopy Monographs No. 11

ISBN: 978-0-85404-154-1
ISSN: 2041-9732

A catalogue record for this book is available from the British Library

Published by The Royal Society of Chemistry,
Thomas Graham House, Science Park, Milton Road,
Cambridge CB4 0WF, UK

Registered Charity Number 207890

For further information see our web site at www.rsc.org

Foreword

It is a pleasure for us to introduce such a splendid collection of articles representing a truly enormous amount of detailed work by many of the leaders of this field, and particularly to acknowledge the editor of this book. We, the readers, owe an enormous debt of gratitude for the magnitude of this work, and for its organization. It is also an honor to have been asked to write this editorial, representing as we do the practitioners of early but small steps, which are no more important than those of our successors in this field. Indeed it has been, and still is, a pleasure to witness the growth, development and deployment of this field, both in the expanding literature and at meetings in the various fields.

The history of infrared synchrotron radiation (IRSR) goes back to the mid 1970's with James Stevenson, Roger Bartlett, working in this area at Tantalus, in Wisconsin, USA, and Pierre Lagarde working on ACO, at Orsay, France. Meanwhile I (GPW) worked in the late 1960's in the infrared using a thermal source and a low throughput high-pressure sample cell, where brightness was a major issue. When I learned about IRSR in 1971, I realized the potential and was then very pleased to hear about the measurements, a few years later. In fact I calculated that 1 mA of stored beam was equivalent in brightness to a typical thermal source, and in those days the synchrotron sources had at least 10 mA in them! Around the same time, Takao Nanba realized the same, and in the early 1980's, working with Bill Duncan and Jack Yarwood, was involved with the commissioning and testing of the first dedicated IRSR beamline on the SRS at Daresbury in the UK. I also visited the SRS at that time and, working with Bill Duncan, wrote a paper with the calculations and applications. Another important milestone in the mid 1980's was the work of Erhard Schweitzer at BESSY-1 in Berlin, Germany, who published a paper pointing out the need to eliminate the synchrotron source noise. It was only after this was addressed in the program at the NSLS, Brookhaven, USA, in the late 1980's to early 1990's, with much work done as part of Carol Hirschmugl's PhD thesis, that we could proceed.

For the applications critical to the theme of this book, Phil Brierley was the next key person, because he put me in touch with John Reffner in 1993 at the ICOFTS meeting in Calgary. John at the time was working on IR microscopes, which need exactly the IRSR source. Quickly, working with Larry Carr and Paul Dumas, the first microscope was loaned to the NSLS and installed in September 1993 using a sewer pipe to eliminate water vapor.

The above is a brief history of the development of the tools for diagnostics – the applications of these tools are the subject of this book, and all the key players have contributed their own knowledge in the chapters that follow. The list of facilities contained in this book illustrates how quickly the field has blossomed around the world. The continuous improvement in instrumentation, and in applications of IRSR, is continuing apace at this time.

This collection of chapters narrates the history of the development and application of infrared microspectroscopy from a large number of viewpoints, and covering history, philosophy, as well as key applications. This has been a slightly unusual field, where progress has profited deeply from a collaborative involvement in instrumentation and technology by the research scientists. And this field, like few others, is characterized by a broad world-wide spirit of support, collaboration and camaraderie. The spirit that 'a rising tide floats all boats' has been instilled into the community from the outset by the early pioneers. Collaboration rather than competition has allowed a rapid dispersal and development, which unfolds numerous times in the papers, and is testimony to the dedication of everyone involved.

As to the future, no satisfaction is higher than that of sharing in the discovery of something in science that reveals something new. This is the critically important criterion that drives this field, and must remain the focus for all of us if we are to make real contributions to understanding how nature works. In this as in any field, a scientist must keep fully aware of all the available tools to solve the particular problem at hand, and of complementarity and competition among these tools. Responsible development of the field, then, will focus on ensuring that we continue to strive to provide unique data for a wider range of applications. To do so will require developments in each of the areas covered by the various chapters – higher performance in spatial, spectral, detection and temporal resolution. Above all, it will involve research scientists from physics, chemistry, materials science, biology, medicine and optics, working closely to help each other, just as we have all enjoyed up to this point.

Gwyn P. Williams
Newport News, Virginia

Foreword

What more is there to say, after being introduced as the "Grand Old Man" in the field of biomedical FT-IR spectroscopy by the editor, David Moss?

I don't know about grand, but old must imply having been around for a long time, a criterion I certainly meet. In the Editor's Preface, my young colleague and friend, who has a way with words, walks us through the scope and strategy of this multi-author book with superb eloquence. What then is there left for the grand old man to add?

Well, for me the journey began in the late 1950s with a PhD thesis on infrared spectroscopic studies of natural products; at that time the expression bio-molecules had not yet been coined. To borrow the phrase "standing on the shoulders of giants" from a true grand old man, Sir Isaac Newton, the single most significant event in my early career as a spectroscopist was my good fortune to attend the VIIth European Congress on Molecular Spectroscopy, held in Budapest in 1963. The "three wise men from Europe", Harold (Tommy) Thompson, Reinhard Mecke and Jean Lecomte were there, as were other grandmasters of molecular spectroscopy from all over the world: Herzberg, Pimentel, Jones, Miller, Sheppard, Lord, Stammreich, Shimanouchi, Longuett-Higgins, Hadzi, Zerbi, Lippincott, Morcillo, Krimm and many others; the list reads like a *Who's who in spectroscopy*. Interestingly, it was not only their science and notability, but their personality, dedication and commitment that spurred on the budding spectroscopist in me.

Leading up to 1963 there had been few attempts to apply infrared spectroscopy to the study of complex biomolecules, a fact largely due to the limitations imposed by the conventional dispersive infrared spectrometers in existence at that time. In fact, in those early years it was Raman spectroscopy that had the upper hand in its constant rivalry with infrared spectroscopy. However, with the advent of FT-IR spectrometers in the 1970s and 1980s, this all changed and numerous groups from around the globe turned their attention to the world of biomolecules; a complex realm in which individual biomolecules conform to all known physical and chemical laws, yet where a collection of the same biomolecules in a living environment reveals new, exciting properties not displayed by their individual, inanimate counterparts. The idea of using infrared spectroscopy to explore the properties of living organisms is as outlandish as it is compelling. The attraction, however, has proven irresistible to many practitioners of infrared spectroscopy, and has led to what is known today as biomedical infrared spectroscopy.

Combining infrared synchrotron radiation with a microscope was only a further, logical step in the evolution of this field. As my distinguished colleague Gwyn Williams reveals in his opening remarks, the beginnings of IR synchrotron radiation go back to the mid 1970s and were then amalgamated with biomedical infrared microspectroscopy in the early 1990s. I remain highly grateful to Gwyn, who shortly after having the first microscope installed at the National Synchrotron Light Source at Brookhaven in 1993, co-supervised and coached one of my PhD students, Lin-P'ing Choo, in our early study of *in situ* characterization of beta-amyloid in Alzheimer's diseased tissue by synchrotron FT-IR micro-spectroscopy. The chapters in this book, while not exhaustive, are representative of this emerging field, which has clearly come a long way, as attested by the contributions in this book, and numerous articles published in various spectroscopic and non-spectroscopic journals.

I also fully endorse the editor's comments regarding the need to bridge the fundamentally different thinking and language used by MDs and PhDs and to close the gap that has long separated the two solitudes of infrared spectroscopy and medicine. I too share a vision of inexpensive IR analyzers at point-of-care places, be it hospitals or private medical offices. Based on today's advances in IR spectroscopic imaging I dare go even further, envisaging a new field of "functional IR imaging", analogous to "functional NMR imaging", used today not only by radiologists but also by psychologists. What we need is a generation of young spectroscopists not afraid to tackle the seemingly "impossible".

Last, but not least, I also agree with you, David, that with so much emphasis on clinical applications, we should not lose sight of the fundamental aspects of this research. In the mind of my mentor and scientific godfather, Dr Gerhard Herzberg, the difference between fundamental or curiosity-oriented research, as he called it, and applied research is fluid, since sooner or later good basic research will yield useful applications. As recent history has shown, this may not always be obvious; Bell Research Labs initially refused to patent the laser, deeming it an esoteric invention that would never, ever lead to practical applications.

Now that the marriage between infrared spectroscopy and medicine seems to have been consummated, we are looking forward to many offspring.

Henry H. Mantsch
Winnipeg

Preface

This book is the product of the EU-funded research project "Diagnostic Applications of Synchrotron Infrared Microspectroscopy" (DASIM), which ran from 2005 to 2008 with the participation of nearly 70 synchrotron physicists, spectroscopists, biologists and clinicians from nine European countries.

The project had its origins in a truly seminal time for the field of IR microspectroscopy using synchrotron light, in Europe especially. At the time of the future project consortium's first meeting, at Frascati on 26 February 2003, there were only two synchrotron IR microscopy beamlines available to users in Europe: at LURE in Orsay, France, run by Paul Dumas, and at the SRS in Daresbury, UK, run by Mark Tobin. At this time Europe was lagging badly behind the USA, where the NSLS in Brookhaven alone was already running six IR beamlines. But the situation was poised to change: my own beamline at ANKA in Karlsruhe, Germany, designed and built by Yves-Laurent Mathis (who had learned the trade as a PhD student at LURE), opened for general user operation just three days after the Frascati meeting, and that was only the start of a period of unprecedented growth. By the end of the DASIM project, there were further IR microscopy beamlines in operation at BESSY in Berlin, at DAFNE in Frascati and ELETTRA in Trieste, at the Swiss Light Source in Villigen, at MAX in Lund, and at the joint European facility ESRF in Grenoble. Simultaneously, France and the UK had upgraded with new state-of-the-art facilities at SOLEIL in St. Aubin and DIAMOND in Didcot. These developments were also paralleled in other parts of the world, with new beamlines appearing in the USA and Japan as well as in Canada, Australia, Japan, China, Singapore and Taiwan as new entrants to the field. And there's not yet much sign that the tide is ebbing: still further beamlines are under discussion or in planning in Spain, China, Thailand, India, Jordan and Brazil.

This explosive growth in the provision of experimental facilities was driven by the anticipated demand in many fields of science, and the potential for infrared spectroscopy in biology and medicine was seen as one of the most promising areas of application. The synchrotrons here are just part of a larger

RSC Analytical Spectroscopy Monographs No. 11
Biomedical Applications of Synchrotron Infrared Microspectroscopy
Edited by David Moss
© Royal Society of Chemistry 2011
Published by the Royal Society of Chemistry, www.rsc.org

research field: Simultaneously with the pioneering work of Gwyn Williams, Larry Carr and Paul Dumas at NSLS on the use of synchrotron light for IR microscopy, the spectroscopy community led by Henry Mantsch, Max Diem, Don McNaughton, Dieter Naumann and Michel Manfait (and others too numerous to mention) was starting to use their conventional benchtop instruments for investigating the vibrational spectra of tissues and cells, revealing correlations that showed great promise in particular for practical applications in disease diagnosis. With its high brilliance advantage, translating into the ability to obtain high quality spectra from a measurement area smaller than a human cell, the potential contribution of the synchrotrons to this field was clear. Perhaps Gwyn Williams' most inspired contribution to the field was his early recognition of the need for a biologist on the infrared beamline staff: Lisa Miller's tireless work soon attracted the attention of synchrotron facilities all over the world to biomedical IR microscopy as a rapidly accelerating train that they urgently needed to board.

The endeavour is striking in its extreme level of interdisciplinarity – the typical cancer surgeon's ignorance of relativistic charged particles is probably only exceeded by the typical accelerator physicist's ignorance of tumour biology! Dialogue between the diverse communities is thus essential, and by 2003 all of the European synchrotrons running or planning IR microscopy beamlines had made contacts with local collaboration partners to explore biomedical applications. The DASIM project was conceived to network this research effort on the European scale, providing a forum for international cross-fertilization of ideas and multidisciplinary expertise, facilitating access to synchrotrons for researchers from EU countries without their own facility, and promoting methodological validation by adding a multicentric quality control perspective. By including all European synchrotron facilities with an infrared microscopy beamline as well as spectroscopists, biologists and clinicians, together with its international associates from the USA, Canada, Australia, India and China, the project brought together essentially all recognized leaders in the field.

Our strategy for the dissemination of results was an important criterion in the European Commission's assessment of the DASIM project application, and we presented the production of a multi-author textbook after the end of the project as our most important measure towards this aim. The book is intended to cover all aspects of the field with particular emphasis on an understandable presentation of necessary background knowledge, in language that all the disparate experts involved can understand. It seeks to demystify the subject both for clinicians and biologists who find synchrotron physics difficult to understand and for physicists who find medical/biological terminology incomprehensible. As the subtitle "A Practical Approach" implies, the book is intended to go beyond the mere presentation of research results in order to serve as a practical "how to do it" guide for those working in or wishing to enter the field of biomedical synchrotron IR microspectroscopy and imaging. Section 1, "Fundamentals", explains the basics of synchrotrons and spectroscopy as well as the needs of clinicians and biologists with respect to these technologies. Section 2, "Technical Aspects", goes into depth on optical issues, sample preparation, study design and data analysis. Section 3, "Case Studies", brings together these elements through practical examples. Since the field of biomedical synchrotron infrared microspectroscopy is too young to have a "Grand Old Man" to write the foreword, I'm delighted and honoured that Henry Mantsch and Gwyn Williams, the Grand Old Men in the fields of biomedical FTIR spectroscopy and synchrotron IR microspectroscopy respectively, have agreed to share the task.

The observant will have noticed that I broadened the scope of the book's title by using "Biomedical applications" rather than keeping the "Diagnostic applications" of the DASIM project's title. The word "biomedical" can have various meanings, but in this case I meant it as a contraction of "biological and medical", intended to reflect that synchrotron infrared microspectroscopy has great potential as a tool for fundamental cell biology research as well as in the practical application of clinical diagnostics. Indeed, the former is the direction that I would prefer to emphasize. An important aspect of the DASIM project was our discussion of the role that synchrotrons should be playing in the wider field of biomedical IR spectroscopy, and our conclusion was that routine diagnostic work was not the likely direction. Not that the transport of patient samples to a synchrotron facility for analysis would be an inconceivable scenario: if patients' lives depended on it, that would surely be done. However, there is general agreement amongst those working in the field that a comparably inexpensive IR analyzer in the hospital's own pathology laboratory could do the job perfectly well, and in view of the progress in this research field I personally am convinced that such analyzers will indeed emerge in future – as soon as the clinical trials have demonstrated improved survival rates for patients diagnosed with the help of IR spectroscopy. For the synchrotrons with their expensive "cherry on the cake" capabilities in IR microspectroscopy, the DASIM consortium's view was that the rational role is at the cutting edge of fundamental research, for example in studies of single living cells. Such studies at synchrotrons will certainly contribute fundamental knowledge that will go into the design of dedicated benchtop analyzers, as already shown in the work on understanding scattering effects by Peter Gardner and his colleagues, presented in Chapter 8, and therein lies the continuing link between synchrotrons and diagnostics.

But providing a tool for advancing fundamental knowledge in cell biology must also be an aim, and indeed it is an aim dear to the heart of many researchers. Most of the pioneers of tissue and cell spectroscopy entered the field from their previous work on the spectroscopy of purified biological molecules *in vitro*, and thus were accustomed to interpreting their spectra at the level of individual atoms in the sample, making significant contributions to the science of protein structure and dynamics. In contrast, spectral features can be used for classifying tissue samples in the pattern recognition sense with no requirement at all for understanding how those spectral features arise. That has an obvious usefulness for practical clinical diagnosis, but to the enquiring scientific mind there's something missing. Does that signpost the future direction? Well, there is a logic to justify that guess: this is still a very young field, and in all areas of science the phenomenological observations need to be collected before the work of understanding them can begin. But the work scientists choose to do is also dictated by what they can obtain funding for. Sadly, we live in a world where considerations of practical utility have come to dominate most countries' research funding policy. There seems to be neither an awareness of the history that tells us that practical usefulness out of scientific research is nearly always unpredictable, nor a preparedness to face the question of what future generations of applied scientists will be applying if our generation doesn't do any pure science.

Be that as it may, the vision of a book dedicated in equal measures to fundamental biological research and clinical applications didn't come out: the reader will find that this book is mostly about the latter. For the reasons discussed above, it's probably just too early for that book as envisioned: I'll look forward to reading it sometime in the future. And in the meantime I'll keep "biomedical" instead of "diagnostic" in my title, even if only as a signpost to the way ahead.

Editing a book on this scale is a major undertaking, and I'd like to express my thanks to my colleagues Yves-Laurent Mathis, Biliana Gasharova, Michael Süpfle, Jochen Bürck and Siegmar Roth for taking on far more than their fair share of the everyday workload over the past few months; to John Chalmers, veteran editor of ten books, for his invaluable advice and assistance to a novice in the editing game; to Merlin Fox of RSC Publishers for his support and for his highly elastic deadlines; to all of the chapter authors, both those who made my life easier by submitting on time, and those who exceeded their deadline by geological timescales (at least your contributions were worth the wait); to Giuseppe Bellisola for taking the time and trouble to write a chapter that regrettably couldn't be accommodated, and which I hope soon to see in print elsewhere; and, leaving the best until last, to my darling wife Anna for accepting that her place on my lap was constantly occupied by my MacBook, and for always knowing when I needed a cup of tea.

David Moss
Ettlingen, Germany

Contents

Section 1: Fundamentals

RSC Analytical Spectroscopy Monographs No. 11
Biomedical Applications of Synchrotron Infrared Microspectroscopy
Edited by David Moss
© Royal Society of Chemistry 2011
Published by the Royal Society of Chemistry, www.rsc.org

Section 2: Technical Aspects

Section 3: Case Studies

Dedication

By a tragic irony, during my work on this book largely about spectroscopy as a tool for cancer diagnosis, my dear mother-in-law Hilde Böttcher herself was diagnosed with incurable cancer.

My wife and I decided to take care of her at home, to give her the comfort of her loving family in her final months. We brought her bed to our house, rearranging the living room furniture to accommodate her, and much of the editing work was done on my laptop at her bedside, where she took great interest in my progress.

Hilde died on 9th May 2010, two weeks after I'd completed submission of the manuscript.

Caring for Hilde in our home was made possible by the impressively calm and competent support of Sister Angelika, our community oncological nurse. At her request, I'd like to dedicate this book to the following thought: however far medical technology advances in the future, let's never forget that medicine is about individual people who need to be treated with dignity, respect and love.

Section 1
Fundamentals

CHAPTER 1

Vibrational Spectroscopy: What Does the Clinician Need?

SHEILA E. FISHER,[a,*] ANDREW T HARRIS,[b] NITISH KHANNA[c] AND JOSEP SULE-SUSO[d]

[a] Clinical Research Fellow, Section of Experimental Therapeutics, University of Leeds, Room 6.01, Clinical Sciences Building, St James's University Hospital, Leeds, LS9 7JT, UK and Hon Senior Research Fellow, School of Health Studies, University of Bradford, UK; [b] Cancer-Research UK Research Training Fellow, Oral Biology, Leeds Dental Institute, University of Leeds, UK; [c] Specialist Registrar in Medical Microbiology, Western Infirmary, Glasgow, Scotland, UK; [d] Associate Specialist and Senior Lecturer in Oncology, Cancer Centre, University Hospital of North Staffordshire and Keele University, Stoke-on-Trent, UK

1.1 Introduction

Modern medicine demands rapid, consistent and reliable techniques for population screening, clinical and laboratory diagnosis, prediction of treatment outcomes and to guide the use of ever more expensive therapeutic agents. This chapter will explore clinical need using examples of major diseases and highlight the potential of vibrational spectroscopy to play a part in future clinical management.

To achieve success it is vitally important that basic scientists, spectroscopists, biologists and clinicians are able to communicate effectively and understand each other's requirements and challenges. The advancement of multi-disciplinary working has been a key feature of the Diagnostic Applications

RSC Analytical Spectroscopy Monographs No. 11
Biomedical Applications of Synchrotron Infrared Microspectroscopy
Edited by David Moss
© Royal Society of Chemistry 2011
Published by the Royal Society of Chemistry, www.rsc.org

of Synchrotron Infrared Microspectroscopy (DASIM) initiative, a Specific Support Action funded by the European Union to bring these groups together. During this period, the work of the group has encompassed, in addition to synchrotron based work, the wider aspects of diagnostic vibrational technology and the range of diseases and disorders to which it can be applied. In 2005, when the network began its work, although existing networks between biologists, clinicians and synchrotron scientists were making substantial progress in IR microspectroscopy of cells and tissues leading to the identification of spectroscopic biomarkers of potential diagnostic relevance, the European dimension was missing in these efforts because very few countries had synchrotron IR microspectroscopy facilities. Global networking was established between scientists but the technology had made limited impact on clinical practice and was poorly understood by doctors and other health professionals. During the life of the network, collaborations have been established which cross disciplines and new technology and techniques are constantly being developed and improved, potentially bringing the power and resolution previously offered only by synchrotron sources to the hospital laboratory, the clinic and to the patient. Understanding of the strengths and weaknesses of different spectral modalities, *e.g.* infrared (IR), Raman and fluorescence, and exploration as to the place of each in biomedical work is an emerging theme which continues to advance apace. At international level, teams are now working together to set parameters in terms of harvesting of clinical material, storage, preparation for imaging, and preprocessing of data, all of which are essential given the changes which happen in biological samples as soon as they are separated from their host tissue and blood supply and also the tremendous complexity of the systems and biochemical processes to be imaged. Since disease may change the biochemical composition, not only of cells and tissues but also of blood and other body fluids, the potential to use these as 'biomarkers' of disease processes is an important area for clinically based research. To map disease related changes it is necessary to carry out spectroscopic measurements at a microscopic level, matched to the size of cells or subcellular structures such as the nuclei and major organelles.

Clinically based research on vibrational techniques has focused on IR and Raman spectroscopy. The scientific basis of this approach relies on measurement of the natural vibrational frequencies of the atomic bonds in molecules. These frequencies depend on the masses of the atoms involved in the vibrational motion (*i.e.* on their elemental and isotopic identity), on the strengths of the bonds, and on the resting bond lengths and angles – in other words, on all the parameters that constitute the structure of the molecule.[1] For this reason, IR spectroscopy is a powerful technique for the identification, quantification and structural analysis of small molecules, and has been established for many decades as an indispensible tool in organic chemistry, polymer chemistry, pharmaceuticals, forensic science, and many other areas. Thus very high resolution material can be imaged at subcellular level and beyond to allow detailed understanding of biological processes.

Resolutions as small as 7 μm×7 μm×2 μm can be achieved by bench top IR machines and 0.3 μm×0.3 μm×0.5 μm by Raman,[2] allowing the level of exploration required for understanding of clinically important spectral changes. Images can be acquired by transmission or reflectance modes and increasing use is being made of confocal techniques. The resolution offered by Raman and its freedom from difficulties imaging aqueous based preparations or environments makes it a promising modality for biological imaging, both *'in vitro'* in the laboratory setting and *'in vivo'* for non-invasive diagnosis in the clinic. Its spatial resolution permits detection of subcellular components (mitochondria, nucleoli, condensed chromatin) in cells, and opens new avenues of monitoring cellular processes without the use of stains or marker molecules, using the inherent spectral properties of molecular constituents. Evolving techniques are increasing the depth of imaging possible *'in vivo'*, which is critically important for clinical utility.

However, the complexity and variation in these processes are a challenge. Alterations in cell function may be a product of changes in biochemical pathways but are likely to represent changes in magnitude or amplification of cell pathways, requiring quantitative measurements of molecules. Identification of specific 'biomarkers' of disease is likely to be possible only in a limited range of conditions. Ranges of quantitative change with cross over between disease state and normality are likely to prove a challenge when making careful assessment of any proposed clinical instruments in a well conducted clinical trial to establish sensitivity and specificity. These are defined as follows:

$$\text{Sensitivity} = \frac{\text{number of true positives}}{\text{number of true positives} + \text{number of false negatives}} \quad (1.1)$$

A sensitivity of 100% recognises all people with the condition.

$$\text{Specificity} = \frac{\text{number of true negatives}}{\text{number of true negatives} + \text{number of false positives}} \quad (1.2)$$

A specificity of 100% means that the test recognises all healthy people as healthy.

This kind of data is imperative for quality assurance and to give confidence to doctors who may be considering use of vibrational technology based devices. Taking the current promise into the clinical environment will require robust scientific and clinical collaborations.

This chapter will consider clinical applications which might be met by vibrational spectroscopy, using cancer, infective diseases and vascular surgery as examples and gives a brief overview. This is by no means exhaustive but suffices to illustrate the potential applications of the technology, which are addressed in greater detail in the appropriate following chapters.

1.2 Vibrational Spectroscopy in Cancer

1.2.1 Introduction

Cancer represents a global health challenge. In 2002 there were 10.9 million new cases of cancer globally, with 6.7 million deaths.[3] It was estimated at this time that 24.6 million people were living with cancer, within 3 years of their diagnosis. These figures will undoubtedly be underestimates as remote populations will be under recorded. Breast cancer is the most prevalent cancer with lung cancer having the highest mortality. The World Health Organization states that by 2020 the rates of cancer will have doubled.[4]

Therapeutics research has embraced the concept of 'designer' drugs, based on exploration of biochemical pathways, giving considerable impetus to the biomedical disciplines of genomics, epigenetics, proteomics and metabolomics, all of which look in complementary ways to understand disease processes at molecular level. The result is that by 2006, approximately 2000 new chemotherapeutic agents were in the pathway towards clinical use.[5] This explosion in science and therapeutics provides challenges both in management of individual patients and health economics. There is considerable pressure through the media to offer new therapies, often with little consideration of side effects and lack of efficacy. In the UK, the management of funding decisions has become the work of the National Institute for Health and Clinical Excellence (NICE). Their remit is to take decisions on risk versus benefit and cost per quality adjusted life year (QALY), a health economics based measure, allowing comparison between patient groups and treatments, in a system which will always have cost pressures.

The areas to which vibrational spectroscopy could contribute to clinical practice include:

- Screening – Who is likely to develop the disease? How reliable is the prediction? Can this be achieved without invasive methods? Can inexpensive devices be produced for use in population screening, ideally at venues close to home?
- Diagnosis – Is the disease present? How advanced is it? Can vibrational methods be combined with other technologies?
- Intraoperative monitoring – Has the cancer been fully removed?
- Prediction of response to therapy – Will the proposed treatment work? Is there any evidence of residual cancer?
- Follow-up – Has the disease returned?

These clinical challenges require some common approaches in terms of recognition of abnormal cell pathways, cell morphology and tissue characteristics but also important individual features. In the next section, we will briefly examine each of these needs in turn.

1.2.2 Screening, Early Diagnosis and Surveillance

Screening is the process by which those who have the disease or, in some conditions, are likely to suffer it in the future, can be identified with the aim of maximising cure and reducing morbidity to the lowest possible level. Well known examples include mammographic screening for breast cancer and the cervical smear programme. The expenditure involved in bringing patients into the service and analysing their results can only be justified economically (on cost per patient diagnosed) and ethically (in terms of the anxiety and morbidity of invasive investigation when screening suggests disease but subsequently this is not confirmed by more intense investigation [false negative result]), when the disease is sufficiently prevalent and the test sufficiently accurate. No screening test achieves 100% sensitivity and specificity and some patients will be falsely reassured by screening tests. The requirement is for a test which will be reliable, with high sensitivity and specificity, easy to administer and which is acceptable to patients, and is ideally achieved by a non-invasive modality.

Some screening tools are aimed at visible areas, mainly skin where differentiation between melanoma and non-melanoma is topical. Moncrieff *et al.* (2002) and Claridge *et al.* (2003) have used colour of lesions and matched this to spectral characteristics.[6,7] The oral mucous membrane is another area of interest as up to one third of patients who subsequently develop cancer go through a phase where visibly identifiable changes are present. However clinical prediction as to which changes are likely to transform to cancer is difficult. Histological assessment, where a piece of tissue is surgically removed from the patient, usually under a local anaesthetic, processed, stained and examined under a microscope, can help but there remains a lack of concordance amongst even specialist pathologists as to which lesions are of concern. Sankaranarayanan *et al.* (2007) in a large population study in India, where oral cancer is most prevalent, showed that a screening programme can result in early diagnosis, with a reduction in morbidity and mortality.[8]

There are two advances which could make a major impact on this area of medicine. One is a non-invasive technique which could accurately predict the potential for malignant transformation by '*in vivo*' mucosal surveillance. Work has been done in a number of premalignant entities; for example, Kendall *et al.* (2003) have studied Barrett's oesophagus and cervical dysplasia with promising results.[9] The cervical cancer screening programme has been in operation for many years, and depends on smears of exfoliated cells, examined using staining patterns and morphological parameters. Suspect cases then require a formal operation, known as cone biopsy, to allow definitive staging. A reliable imaging tool could eliminate this procedure with less delay and less discomfort to the patient. Utzinger *et al.* (2001) suggest that a simple algorithm based on two specific intensity ratios, taken from 13 patients '*in vivo*', can discriminate between high-grade squamous dysplasia and other pathologies, misclassifying only one sample.[10] Teh *et al.* (2009) have achieved a sensitivity of 90.5% and specificity of 90.9% in the diagnosis of gastric dysplasia using Raman spectroscopy.[11]

Given the prevalence of cancer and the improvements in survival achieved at some anatomical sites, it is not surprising that more and more people face this diagnosis more than once in their lives. Whether there is a general predisposition to cancer as a generic behavioural change in cells or whether susceptibility is site specific is something which will only be determined by extensive epidemiological work. Certainly for some sites, of which head and neck cancer is a good example, the chance of a second primary cancer is of the order of 15%.[12] The justification for longer term follow up of these patients is to identify and intervene early if a second cancer is found. The site can be anywhere in the upper aerodigestive tract as the entire mucosal surface is subject to incremental cellular changes which ultimately lead to cancer. Multifocal disease is also observed. The challenges here are those described for screening, to determine which anatomical area is at immediate risk. Current practice involves regular examination, either in the clinic or under sedation or full anaesthesia. Where there is suspicion of new disease, patients are scanned using computed tomography (CT) or magnetic resonance imaging (MRI). The personal cost of this surveillance is anxiety and a requirement for regular hospital visits. The cost to healthcare budgets is considerable, especially where the cancer is common and surveillance requires formal admission. Lung and colorectal cancer both require formal endoscopy and biopsy at intervals during the surveillance period. To test by a non-invasive probe or to develop a 'surrogate' test using a body fluid would result in a considerable health gain. Candidate systems include breath for lung, faeces for colorectal, and urine for kidney and bladder surveillance. Of wider impact is the possibility of markers in serum or saliva.[13]

A major potential for vibrational techniques is to screen body fluids, such as plasma, saliva and urine, or breath. If a reliable vibrational spectroscopy dataset can be developed such tests could be offered from a community setting, especially important in areas where the population is scattered, such as Canada or parts of Australia, or where economic costs make hospital based screening impossible, such as rural India. Madhuri *et al.* (2003) compared native fluorescence spectra derived from plasma in subjects with oral cancer or liver disease and healthy subjects, showing good discrimination.[14] Harris *et al.* (2009) compared 20 plasma samples from patients with respiratory disease with 20 from patients with head and neck cancer, thus ensuring a similar population in terms of gender, age, smoking, alcohol consumption and the likely presence of inflammatory mediators.[15] They used Principal Component Analysis (PCA), Linear Discriminant Analysis (LDA) and a 'trained' evolutionary algorithm, achieving sensitivity and specificity levels of 75%.

The detection of lung cancer through breath analysis has become recently an area of high interest. Sulé-Suso *et al.* (2009) showed variations in the amount of volatile compounds released by different lung cells (epithelial, fibroblast, and cancer).[16] Further work is required to assess whether volatiles released by '*in vitro*' growing cells could also be detected in breath and used in clinical practice for lung cancer diagnosis.

1.2.3 Therapy

Treatment for cancer involves chemotherapy (systemic drug treatment), radiotherapy, which can be external beam or by insertion of radioactive wires directly into the cancer (brachytherapy), and/or surgery. Treatment choice depends on the type of cancer, where it is in the body, how extensive it is and its likely natural history. The experience of a cancer patient can be very variable with some cancers, such as breast, having a very long natural history, often years from diagnosis to death even in advanced disease and others, such as head and neck cancer, having a rapid course with death in less than a year being common for those whose disease cannot be controlled. Each modality of treatment comes at a price, both in terms of individual experience and health economics. Chemotherapy targets the cancer, but also affects the parts of the body where cells divide rapidly. Extreme fatigue, lethargy, anaemia, oral mucositis, vomiting and diarrhoea are common. Hair loss used to be one of the most obvious signs that someone was having chemotherapy. Many new agents spare the patient this side effect but skin reactions are common and are often considered to correlate with the degree of response to the treatment.

The toxicity profile of radiotherapy is similar. Early side effects include those listed above. Late effects include fibrosis, scarring, stiffness and reduction in blood supply to the tissue in the treatment field which extends beyond the site(s) directly involved with the cancer. Complications can be devastating, for instance radiotherapy to the pelvis can cause 'frozen pelvis' with marked functional disturbance. Fistulation between, for example, bowel and major blood vessels can occur with fatal sequelae. In head and neck cancer, treatment followed by late fibrosis can impair mouth opening, speech and swallowing, with unacceptable psychosocial and functional consequences.

Progress is being made in these areas by constant searches for new and more specific cancer drugs. As noted in the introduction, IR has a good record in the pharmaceutical environment and if cancer characteristics could be matched to therapeutic agent, morbidity might be minimised and efficacy maximised. If we knew which patients would, or would not, respond to a given treatment, we could be more positive in our therapies and, one hopes, spare those patients for whom radical but ineffective therapy simply takes away their final period of reasonable health. For radiotherapy, careful imaging and volume prediction has resulted in advanced computer guided treatment plans which maximise the dose to the cancer whilst sparing as much surrounding tissue as possible. Some aspects of response relate to hypoxia. Ability to image during treatment, looking at this kind of response, or even at the residual cancer burden, might allow further refinement. Recent advances in Raman imaging might make this feasible in the future.

For those cancers where surgery is the modality of choice, the most important aspects of efficacy are achieving full excision of the cancer, together with
any related premalignant tissue. In oral cancer, close or involved margins have
an adverse impact on both locoregional recurrence (return or persistence of the
cancer at the primary site and/or the lymphatics which drain the area) and
survival.[17] For some cancer sites a radical resection is possible without marked
morbidity. For all sites, there is eventually an anatomical structure which limits
further surgery. For sites such as the head and neck, the more radical the
surgery, the greater is the postoperative morbidity and this has an impact on
key functions like eating, speech and swallowing. Cancer infiltrates tissue at
microscopic level, so the boundaries of the cancer are not visible to the treating
surgeon. It is usual practice to remove the cancer with a 'safety margin' of
normal tissue. Figure 1.1 shows the invasive front of a cancer. This can
be 'pushing' in character, and the true boundary relates closely to palpable
disease (cancer feels firm). Where the disease spreads as islands of cells, as in the
example shown, palpation will not be a reliable guide to surgical margins.

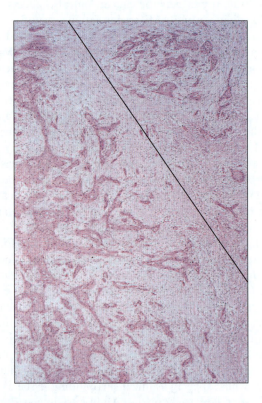

Figure 1.1   Histological stained section of an oral squamous cell carcinoma. The line
shown, if followed by a surgeon removing the cancer, would appear not to
cross any active disease. However, islands of cancer cells are present
beyond the excision margin, which would lead to recurrence of the disease.

There will also be problems with the method many surgeons use to try to check margins, frozen sections. In this approach, small pieces of tissue from the planned margins are snap frozen, stained and examined by a pathologist. If islands of cancer remain beyond the margins sampled, recurrence is likely. In head and neck cancer, even where margins are reported, after full and careful examination, as free of cancer, 10% of patients will suffer a locoregional relapse of their disease. Salvage after recurrence or residual disease is unlikely. A tool to immediately assess margins would be invaluable in surgical practice. Challenges are the ability to image to the necessary depth, to pick up microscopic deposits of a few cells only and to function in an environment contaminated with blood and saliva.

In addition to the primary cancer, spread can take place, either through the lymphatic system, which in health drains tissue fluids and filters pathogens in structures called lymph nodes, or through the blood stream. Spread to lymph nodes which drain the primary cancer site is often the first sign of systemic disease and removal of the lymph nodes often forms part of radical surgical treatment with curative intent. As with the primary site, this operation comes at a price in terms of morbidity but leaving involved nodes *in situ* will result in locoregional treatment failure. Attempts have been made to minimise the need for surgery using techniques which image involved nodes. One which has undergone clinical trials is sentinel node biopsy where a dye and/or radioactive tracer is injected into the primary cancer and the nodes become apparent by simple inspection and/or radioimaging.[18] This assessment is time sensitive if the correct nodes are to be identified and has yet to gain widespread clinical acceptance. As noted above, imaging by a vibrational spectroscopy probe may perhaps become feasible in the future.

Once the cancer has been removed, it is the task of the pathologist to determine if the cancer has been fully removed. If one imagines the volume and surface area of the specimen, the time required to do a systematic margin check at microscopic level can be appreciated. The time to similarly fully examine a specimen containing lymph nodes is a similar challenge. If 'sampling' is practised, looking only at selected areas, the risk of understaging (underestimating the extent of spread of) the cancer exists and the patient is denied the option of adjuvant radiotherapy or chemoradiotherapy which would increase their chance of locoregional disease control and overall survival. The need to examine the whole specimen carefully is illustrated by the presence of very small deposits of cancer outside the capsule which forms the boundary of lymph nodes (Figure 1.2). Such disease, which can be very small in volume and which can occur in lymph nodes less than 3 mm in diameter, has a marked adverse effect on disease control and survival.[19,20] This is an area where clinically important advances are certainly feasible. One advantage of vibrational techniques is that the threshold for identification of a meaningful difference is a function of the processing of data. A screening tool through which sections were passed could hypothetically be programmed for high sensitivity, selecting out those sections which the pathologist should carefully examine. Some authorities have suggested that automated diagnosis might be feasible in the

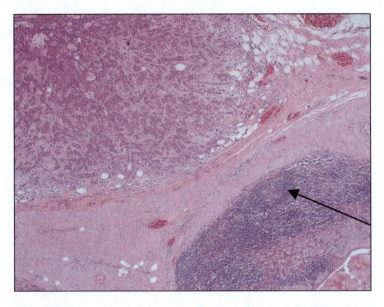

Figure 1.2   This is a stained histological section showing a lymph node. Beyond the
boundary of the lymph node is a separate island of cancer (arrowed).

future but the analysis of cancer and treatment recommendations rely on a
complex interpretation of the disease and its patterns of spread. A tool to help
pathologists focus their skills most effectively is feasible; this should be a primary aim of research towards the clinical setting.

There is a considerable literature on the differentiation of cancer and non-
cancer so we have limited our description to examples of work in this area.
Tobin *et al.* (2004) used synchrotron IR to differentiate oral cancer tissue, using
air dried tissue sections, from the surrounding normal tissue using PCA and
LDA.[21] Sections were subsequently stained and good correlation was achieved;
indeed, in one section where visually the tissue appeared to be 'non-cancer' but
for which the spectra fitted the 'cancer' profile, after staining it became clear
that there were cancer cells in the area from which the spectra had been
obtained. The possibility of obtaining IR spectra from cells which have already
been stained opens up a new avenue in the study of pathology samples with
Fourier Transform Infra-Red (FTIR) spectroscopy.[22]

For some cancers the prognosis (outcome) is defined by tissue characteristics.
This is true in prostate cancer, one of the most common cancers in men, and the
aggressiveness is decided using a rating system known as Gleason grading. The
higher the Gleason grade, the poorer the prognosis will be. Baker *et al.* (2008,
2009) have reported high correlation between FTIR and Gleason grades in a
series of prostate cancer patients.[23,24] Other groups have concentrated on
attempting to understand the biological basis for the spectral changes seen, as
typified by the studies of Anastassopoulou.[25]

In addition to controlling the locoregional disease, cancer causes mortality by distant spread, a process known as metastasis. Metastasis occurs primarily at sites with a high blood flow, typically lung, bone, brain, liver and kidneys. Once spread has occurred, it is highly likely that metastases will prove to be multiple so a systemic therapy is needed. Chemotherapy is given either through a catheter inserted into a vein or, increasingly in modern practice, orally. Much effort has been placed into development of new and more specific agents to target cancers and reduce effects on normal tissue. Despite that aim, reduction in cancer size comes at a price, usually fatigue, anaemia, nausea, and vomiting and diarrhoea. There remains a small but significant mortality. If it were to prove possible to predict which patients will respond to which agent, this would be a significant health gain.

In summary there is good and increasing evidence for the efficacy of vibrational spectroscopy techniques in cancer screening and diagnosis. The next phase will be validation of these techniques in carefully designed studies of sufficient statistical power to prove efficacy in clinical practice.

1.3   Vascular Disease

1.3.1   Introduction

This part of the chapter reviews the present stage of optical spectroscopy in the diagnosis of vascular pathology. Ischaemic heart disease (IHD) and peripheral vascular disease constitute a massive burden to the western world.[26-29] Approximately 250 000 people per year die in Britain of cardiac disease.[30] Of those the preponderance are males, usually over 50 years of age. However, with genetic and increased lifestyle risk factors many younger males, and females, are also suffering from this disease.

1.3.2   Pathophysiology

Both these disease patterns originate from the same source, *i.e.* atheromatous plaques within arterial vessel walls. In order to understand this pathology it is necessary to be familiar with the anatomical structure of arteries. The arteries are composed of three layers: an intima, which lines the interior of the blood vessel, the media, a structural support for the vessel, and then the adventitia, surrounding the outer surface of the vessel (Figure 1.3). The constituents of these layers are endothelial cells, smooth muscle cells, and collagen and proteoglycan elements which make up the extracellular matrix.[31] There are three types of artery: the large or elastic artery, of which the aorta is an example, medium sized vessels, which include branches of the aorta such as the coronary arteries, and then the small arteries (usually less than 2 mm in diameter). Atherosclerosis affects only the large and medium sized vessels.[31]

Atherosclerosis literally means *hardening of the arteries*. This is caused by smooth muscle proliferation within the intimal layer. These smooth muscle cells

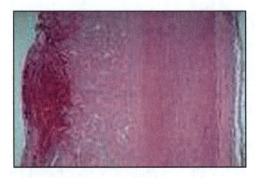

Figure 1.3 Histological section through an atheromatous vessel.

originate in the media layer, but migrate into the intima in response to damage
to the intima from hypertension, smoking and other toxins.[32,33] Lipid accumulation occurs at the site, with macrophage infiltration. The macrophages
engulf the accumulating lipids and form foam cells, which give the macroscopic
appearance of fatty streaks, forming a plaque.[31] The plaque grows and reduces
the size of the vessel lumen; as this plaque enlarges it will decrease lumen size.
The plaque is covered by a fibrotic cap, which if it ruptures may cause acute
thrombus formation which will result in vessel occlusion and ischaemia distally.

   At present assessment of patients with atheromatous lesions within the
cardiac or peripheral arterial system is by X-ray angiography (injection of a
radio-opaque dye and imaging the flow by sequential X-ray films); this can
quantify the severity of the vessel stenosis.[34] Patients may also undergo other
forms of assessment such as external ultrasound scanning, with spiral computer
tomography or magnetic resonance angiography as options. These investigative
modalities can all provide assessment of vessel stenosis and whether calcification is present, but they do not possess the ability to detect the chemical
composition of the atherosclerotic plaque.

   It is well recognised that lesion composition is a very important aspect of this
disease.[35–37] Plaques can be divided into stable or unstable, based on their
chemical composition not their area or volume. Unstable plaques have lipid
pools which can rupture creating a thrombus cascade, causing occlusion of the
vessel.[38] This is thought to be the predominant pathophysiology for acute
myocardial infarctions, and these patients are thought to have a poorer prognosis. However, densely calcified plaques are slow growing in nature, and are
inherently stable, thus producing progressive clinical symptoms giving a prior
warning of future events.[39,40]

   Cardiac risk factors cannot predict how disease will progress, and as these
important chemical changes occur at an asymptomatic phase, there is a need to
delineate underlying chemical structure at this early stage, thus allowing optimal planning of treatment. At present the majority of patients present after a
clinical vascular event, such as a myocardial infarction (MI, 'heart attack') or
cerebrovascular accident (CVA, 'stroke'). At this stage, permanent morbidity

may well be present and reversal of disability is unlikely to be achieved. Intravascular ultrasound can provide more detailed anatomy of plaques, as can magnetic resonance imaging, but they cannot detail chemical composition.[41,42] To be able to enumerate the compounds present in the atherosclerotic plaque, with quantities present, would provide the ability to intervene early in those patients at high risk of plaque rupture and prevent clinical vascular events and their attendant morbidity and mortality.

1.3.3 Vibrational Spectroscopy in Vascular Disease

Spectroscopy, more specifically Raman spectroscopy, has the ability to give '*in vivo*' recordings of chemical compositions of atheromatous plaques.

The initial work in this field consisted of '*in vitro*' work carried out on human vessels. Early work centred on fluorescence spectroscopy. Bartorelli and co-workers (1991) and Richards-Kortum and colleagues (1989) studied the use of spectra in the ultraviolet (UV) light region on diseased arterial tissue.[43,44] They independently demonstrated the ability of this technique to distinguish between atheromatous and normal tissue. However, recent work has centred on the use of Raman as this has an ability to distinguish different chemical compositions.[45-56] Raman spectra are unique to compounds (which produce 'fingerprints' as stated above) whereas fluorescence spectra are limited in their differences.

Buschman and colleagues (2001) utilised Raman spectroscopy to differentiate normal versus diseased arterial walls.[57] They described a technique for Raman spectroscopy which could potentially be performed *in situ*. From 16 patients, hearts were obtained from explantation during transplant procedures; 200 samples of human coronary artery tissue were gathered. Of these 16 patients, 7 were awaiting transplant due to dilated cardiomyopathy, and the rest were on the organ waiting list for severe ischaemic heart disease. The methodology included the arterial samples being washed then immediately frozen in liquid nitrogen. The samples were placed into two categories. The first consisted of 113 samples which were used in the original data collection. They were thawed and underwent spectroscopic evaluation, formalin (10%) fixed and underwent histological evaluation, in which the two pathologists were blinded to the Raman results. The remaining 87 samples were evaluated against the first set to give comparable data.

Their results indicated that non-atherosclerotic tissue, non-calcified atherosclerotic plaque and calcified atherosclerotic plaque all give different Raman spectra, and hence this technique can differentiate these three states of a vessel wall. Microscopic Raman spectra were obtained to study the chief constituents of the vessels under examination: elastic laminae, collagen fibres, smooth muscle cells, fat cells, foam cells, necrotic cores, cholesterol crystals, β-carotene containing crystals, and calcium mineralisations. The results indicate different levels of each depending on which arterial sample was being observed. For example, calcified atherosclerotic plaques contained a lot more foci of calcium mineralisation than normal tissue but fewer fat cells than normal tissue, whereas atherosclerotic tissue contained much higher concentrations of foam cells than normal tissue, as would be expected. This paper gives a thorough, detailed methodology and indicates the potential for the clinical use of Raman spectroscopy in atheromatous disease. All techniques have limitations which must be understood and addressed, such as the difficulty in obtaining clear distinctions between cellular constituents such as foam cells and cholesterol crystals, which gave similar spectra. These spectra also are very close to those of smooth muscle cells, collagen fibres and elastic laminae.

The ability of Raman spectroscopy to distinguish diseased from normal vessel wall is also supported by Nogueira *et al.* (2005),[58] utilising FTIR spectroscopy, and studying carotid artery samples from post-mortem examinations. Seventy-five vessel samples were obtained from 22 subjects. Once spectral results had been obtained the sample was marked with India ink and then formalin treated and underwent histopathological study. The results indicate that calcified plaques have very different spectra, but those of atherosclerotic and non-atherosclerotic tissues are very similar. Their study highlights another problem faced; the initial diseased state is in between the intima and media layers, possibly being just too deep for spectral analysis when an endovascular approach is used, acquiring the spectra through the vessel wall. Hence atheromatous vessels produce spectra similar to 'normal' tissue.

Römer and co-workers (1998) sought to evaluate the ability of Raman spectroscopy to correlate with histopathological assessment of human coronary arteries.[59] One hundred and sixty-five coronary artery samples were obtained from explanted recipients' hearts. Their report aimed not only to show an ability to distinguish between 'normal' and diseased tissue but to quantify the chemical composition of diseased vessels. Again calcium rich lesions produce different spectra, with diseased and normal tissue having slight differences which when analysed through mathematical models show a clear distinction. Their data suggest cell constituents differ in diseased states and can be quantified by spectroscopy, although subject to errors as previously mentioned.

In order to progress from the laboratory to a clinical setting work was undertaken on a probe which could be utilized for *in situ* vessel analysis. Motz *et al.* (2004) undertook work on a Raman probe.[60] There are many difficulties when designing such an instrument; for example, it must be long enough and flexible to access remote organs; have a diameter tolerable to the vessel; contain optical filters strong enough to reduce the background 'noise' from the fibres themselves. Background 'noise' can often mask the signal of the tissue. The laser must have a low enough exposure to be unable to cause any damage to the vessel or organs in the beam. The authors produced a probe which underwent evaluation in simulation tests with experimental models and '*in vitro*' analysis. Other authors have also described optical probes for Raman spectroscopy '*in vivo*' work, and outlined the design problems to overcome.[61–66]

Motz and colleagues (2006) used their Raman probe '*in vivo*' approximately two years later.[67] Spectra were taken from 14 femoral bypass and 6 carotid endarterectomy operations. The sites from which spectra were taken were marked with a suture and a small biopsy was taken for histopathological analysis. However, sutures could not be used for the carotid endarterectomy samples, and so estimates were made of where these samples were taken. The field was bloodless due to saline flushes being used prior to sampling. The authors state that the results obtained correlate well with the pathology report, indicating that spectra could distinguish the presence of markers of atheromatous disease.

The other practical issue is that all lights have to be turned off in theatre to reduce background 'noise'; this may not be appropriate or even safe in all circumstances.

The promise centres on the ability of Raman probes to image the chemical composition of plaques and thus these techniques may be able to predict which patients are at risk of acute arterial occlusion, an aim which would have considerable benefit to patients.

1.4 Microbiology and Infective Disease

Infective disease remains one of the main healthcare burdens faced by both the developing and the developed world. Control of infective outbreaks proves challenging in terms of:

- Identification of the causal organism
- Determination of the correct therapy
- Recognition of biological change, *e.g.* a mutation which renders the organism resistant to a therapy which has been reliably used to contain it.

Pathogenic (capable of causing disease) organisms include bacteria, viruses, and yeasts.

Very often it is the most vulnerable, the very young, infirm, or those individuals suffering other conditions, who are at greatest risk of infections. The EPIC study (*European Prevalence of Infection in Intensive Care*) of 1417 intensive care units (ICUs) in 17 different countries found that almost 45% of patients on these units had a microbiologically confirmed infection.[68] At present, septic patients (broadly defined as patients with either a confirmed infection, or highly suspected infection with clinical signs of a systemic inflammatory response syndrome), whether on ICUs or other hospital wards, are initially treated with empirical broad spectrum (essentially, best guess) antimicrobial therapy until microbiology data on culture results and sensitivities are available, usually one to two days later.[69]

Despite the use of broad spectrum antimicrobial therapy, 10–30% of patients still receive inappropriate antibiotics, which may impact on mortality.[70,71] It is in the ICU patients where appropriate antimicrobial therapy can lead to a 30–60% reduction in mortality rates.[72] Furthermore, despite adequate antimicrobial therapy, mortality rates amongst patients with microbiologically confirmed infections are statistically greater when compared to patients with no infection.[73] These infections can be community or hospital acquired (nosocomial); however it is estimated that nosocomial infections are 5 to 10 times greater in ICUs than on general medical wards.[74-77] Patients in ICUs are at the highest risk, due to immunosuppression, inserted lines, catheters, mechanical ventilators, and multiple illnesses.

The most clinically important disadvantage of using broad spectrum antibiotics empirically is the emergence of drug resistance.[78] Concerns have focused on methicillin-resistant *Staphylococcus aureus* (MRSA),[79,80] vancomycin-resistant enterococci (VRE),[81-83] extended spectrum β-lactamase producing Gram-negative bacilli,[84] multi-drug resistant *Mycobacterium tuberculosis*,[85] and resistant *Candida* species.[86] In the USA and Japan there have also been reports of isolation of *S. aureus* with a reduced susceptibility to vancomycin.[87]

Any drug has the potential for adverse reactions, ranging from simple diarrhoea to pseudomembraneous colitis, or possible anaphylactic shock. It is apparent that the use of unnecessary medication can be detrimental to a patient. However, due to the inability to quickly determine the nature of an organism, an empirical approach is utilised to prevent a patient clinically deteriorating prior to the availability of microbiological data. Furthermore, there is a large body of evidence suggesting that the initial use of antimicrobial agents to which the identified pathogens are resistant increases the risk of hospital mortality.[88-96]

At present, after harvesting, the biological material (usually a blood culture, aspirate or swab) is taken to the laboratory where it is manually 'plated out' on dishes of nutrient material until colonies of the organism can be grown. These are then characterised, using simple morphological characteristics, *e.g.* Gram staining. Colonies are treated with candidate antibiotics and the zone around the antibiotic (zone of inhibition) is measured. From this, a therapy which is likely to work in clinical practice can be derived. All of this is labour intensive and takes time. It is often as long as 48 hours before a definitive recommendation can be made to the treating physician.

In order to attain a faster diagnosis and begin targeted antibiotic therapy, studies have focused on the use of the polymerase chain reaction (PCR).[97,98] Belgrader and colleagues (1999) investigated a high speed PCR technique to detect bacteria in seven minutes.[99] PCR requires a degree of sample preparation prior to analysis. In addition, due to the extremely high sensitivity of PCR, contamination from non-template DNA present in the laboratory environment is problematic. Although automated microbial detection systems frequently identify common organisms, a number of organisms have proven to be difficult to identify consistently. Evidence exists that enterococci,[100] pneumococci,[101] and certain β-lactam resistant Enterobacteriaceae[102] pose problems for certain automated systems. Furthermore, these systems can misidentify rare organisms, leading to inadequate or inappropriate antimicrobial therapy.[103,104] Bacterial detection systems requiring minimal specimen processing and with the ability to detect small numbers of organisms accurately and consistently would be highly advantageous.

Maquelin *et al.* (2003) reported a clinical trial of vibrational spectroscopy.[105] Firstly, a reference library was created of bacterial and fungal pathogens. The reference strains were seeded in blood samples from healthy volunteers. These blood samples were then inoculated into standard blood culture bottles appropriate for the organism. Once the blood cultures were flagged as positive via a BACTEC blood culture system, the material was incubated on a culture medium for six hours and then underwent Raman spectroscopy, from the culture medium. For the IR detection, the biomass from the blood culture bottle was incubated overnight and then transferred to detection plates. References were then obtained for the 106 different strains of bacterial species used and for 34 yeast strains. Recordings were taken until the system was calibrated through a mathematical model to identify the various pathogens. Next, the prospective data collection began from blood culture bottles from two separate hospitals over three- to four-month periods. These were incubated for six to eight hours and then underwent spectroscopic evaluation. Raman correctly identified 92% of the samples as compared to conventional phenotypic identification. The IR method correctly identified 98% of the organisms. These are very promising results, as the phenotypic mechanism system currently used in practice cannot guarantee 100% accuracy.

More recently Ibelings *et al.* (2005) studied the use of Raman alone on clinical samples.[106] They initially constructed a Raman reference library for 93 strains of 10 *Candida* species. Once Raman recordings had been taken the differences in spectra were detected using mathematical models. They tested this reference data set using a collection of peritoneal fluid samples, obtained from patients suffering from peritonitis. In total, 88 fluid samples were collected from 45 patients. The samples were cultured for 48 hours and underwent routine microbiological testing and Raman spectral readings: 29 samples were positive for *Candida,* and, 26 were correctly identified by Raman. This study does show promise for Raman in clinical practice but does not prove that it is any quicker than conventional methods, one of the main reasons for its consideration in the clinical setting.

Low resolution Raman spectroscopy has been shown to be a cost-effective model of spectroscopy and benefits from being a portable device that can easily couple with optical fibres. This raises the possibility of direct point of care testing, where the system of microbial detection can be available in the ICU itself, therefore allowing for rapid sample analysis and microbial identification using a reference library. This would ease the burden of samples processed in microbiology laboratories and reduce costs involved in specimen transport. Mello *et al.* (2005) managed to correctly identify a variety of pathogens associated with gastroenteritis indicating promise for future progress in this area.[107] Harz *et al.* (2009) recently reported direct analysis of clinically relevant single bacterial cells (*Neisseria meningitidis*) from cerebrospinal fluid using micro-Raman spectroscopy.[108] Bacterial meningitis is a disease for which urgent and effective therapy must be given and, if the promise of this initial communication is supported by larger studies, the use of Raman would represent a significant clinical advance.

In addition to basic organism identification, Raman spectrometry has been shown to be able to identify certain resistant strains, therefore providing vital information on treatment. Sockalingum *et al.* (1997) managed to identify *Pseudomonas aeruginosa* isolates with varying degrees of resistance to impenem.[109] Although the technique does not elucidate mechanisms of antibiotic resistance, Raman spectroscopy identifies the indirect biochemical changes that occur when an organism exhibits resistant traits towards a particular antibiotic. In addition, Hosseini and colleagues (2003) developed a non-invasive and non-contact method of measuring antimicrobial concentrations in the eye using Raman spectroscopy, therefore raising the possibility of using this method to assess the pharmacokinetics of antimicrobials in normally inaccessible areas.[110] In a significant breakthrough Amiali *et al.* (2007) explored the use of IR spectroscopy in MRSA detection.[111] Using mathematical modelling of their spectral results, they report 97% accuracy for MRSA within one hour. This compares to current techniques giving results in approximately three days.

The current literature purports that vibrational spectroscopy will hold a place in the clinical setting. Accuracy rates are reported to be greater than 90% with Raman or IR, and a combination of these could confer higher certainty. However, there are concerns regarding the capability of spectroscopy to detect polymicrobial infections, where the ability to identify the organisms involved is of paramount importance. Abscesses frequently contain multiple pathogenic organisms that need to be targeted if adjunctive antimicrobial chemotherapy is to be of any use. One method that has been developed to allow manipulation of a sample without direct physical contact is through the use of optical tweezers. Raman tweezers focus a near-IR laser on a sample. The tweezers fix the cell within an optical trap from which it may be manipulated, therefore allowing analysis of an individual cell.[112] The continued interesting work on bacterial taxonomy through spectroscopy and its clinical application as discussed by Beekes, Naumann and co-workers (2007)[113] reinforces the potential for future applications in microbiology and the classification of pathogenic organisms.

1.5 Conclusions

In this chapter we have placed studies using vibrational spectroscopy in a clinical context, indicating the potential for the future and for strong basic science/clinical alliances to be made to ensure that the promise seen in this area is translated through to clinical practice.

References

1. Diagnostic Applications of Synchrotron Infrared Microspectroscopy. EU SSA. Available at http://www.dasim.eu.
2. M. Romeo and M. Diem. Spectral diagnosis workshop, Boston, 2007. Available at http://www.spectraldiagnosis.com.

3. D. M. Parkin, F. Bray, J. Ferlay and P. Pisani, Global Cancer Statistics 2002, *CA Cancer J. Clin.*, 2005, **55**, 74–108.

4. L. Eaton, World cancer rates set to double by 2020, *Br. Med. J.*, 2003, **326**, 728.

5. R. Rosen, A. Smith and A. Harrison. Future Trends and Challenges for Cancer Services in England: a review of literature and policy. King's Fund ISBN 978 1 85717 549 3, 2006.

6. M. Moncrieff, S. Cotton, E. Claridge and P. Hall, Spectrophotometric intracutaneous analysis: a new technique for imaging pigmented skin lesions, *Br. J. Dermatol.*, 2002, **146**, 448–457.

7. E. Claridge, S. Cotton, P. Hall and M. Moncrieff, From colour to tissue histology: Physics-based interpretation of images of pigmented skin lesions, *Med. Image Analysis*, 2003, **7**, 489–502.

8. R. Sankaranarayanan, K. Ramadas and G. Thomas *et al.*, Oral visual screening reduces oral cancer mortality: evidence from a randomised controlled trial, *Oral Oncol. Suppl.*, 2007, **2**(1), 120.

9. C. Kendall, N. Stone, N. Shepherd, K. Geboes, B. Warren, R. Bennett and H. Barr, Raman spectroscopy, a potential tool for the objective identification and classification of neoplasia in Barrett's oesophagus, *J. Pathol.*, 2003, **200**, 602–609.

10. U. Utzinger, D. L. Heintzelman, A. Mahadev-Jansen, A. Malpica, M. Follen and R. Richards-Kortum, Near-infrared Raman Spectroscopy for in-vivo detection of cervical precancers, *Appl. Spectrosc.*, 2001, **55**(8), 955–959.

11. S. K. Teh, W. Zheng, K. Y. Ho, M. Teh, K. G. Yeoh and Z. W. Huang, Near-infrared Raman spectroscopy for gastric precancer diagnosis, *J. Raman Spectrosc.*, 2009, **40**(8), 908–914.

12. J. S. Cooper, T. F. Pajak, P. Rubin, L. Tupchong, L. W. Brady, G. E. Laramore, V. A. Marcial, L. W. Davis, J. D. Cox and S. A. Leibel, Second malignancies in patients who have head and neck cancer: incidence, effect on survival and implications based on the RTOG experience, *Int. J. Radiat. Oncol. Biol. Phys.*, 1989, **17**, 449–456.

13. T. Shpitzer, G. Bahar, R. Feinmesser and R. M. Nagler, A comprehensive salivary analysis for oral cancer diagnosis, *J. Cancer Res. Clin. Oncol.*, 2007, **133**(9), 613–617.

14. S. Madhuri, N. Vengadesan, P. Aruna, D. Koteeswaran, P. Venkatesan and S. Ganesan, Native fluorescence spectroscopy of blood plasma in the characterisation of oral malignancy, *Photochem. and Photobiol.*, 2003, **78**(2), 197–204.

15. A. T. Harris, A. Lungari, C. J. Needham, S. Smith, M. A. Lones, S. E. Fisher, X. B. Yang, N. Cooper, J. Kirkham, A. Smith, D. Martin-Hirsh and A. S. High, Potential for Raman spectroscopy to provide cancer screening using a peripheral blood sample, *Head & Neck Oncol.*, 2009, **1**, 34, http://headandneckoncology.org/content/1/1/34.

16. J. Sulé-Suso, A. Pysanenko, P. Španěl and D. Smith, Generation of acetaldehyde and carbon dioxide by malignant and non-malignant lung cell lines *in vitro* as measured by SIFT-MS, *Analyst*, 2009, **134**, 2419–2425.

17. D. N. Sutton, J. S. Brown, S. N. Rogers, E. D. Vaughan and J. A. Woolgar, The prognostic implications of the surgical margin in oral squamous cell carcinoma, *Int. J. Oral & Maxillofac. Surg.*, 2003, **32**, 30–34.

18. L. W. Alkureishi, G. L. Ross, D. G. MacDonald, T. Shoaib, H. Grey, G. Robertson and D. S. Soutar, Sentinel node in head and neck cancer: use of size criterion to upstage the N0 neck in head and neck squamous carcinoma, *Head & Neck*, 2007, **29**(2), 95–103.

19. J. N. Myers, J. S. Greenberg, V. Mo and D. Roberts, Extra-capsular spread: a significant predictor of treatment failure in patients with squamous cell carcinoma of the tongue, *Cancer*, 2001, **92**(12), 3030–3036.

20. J. Jose, J. W. Moor, A. P. Coatesworth, C. Johnston and K. A. MacLennan, Soft-tissue deposits in neck dissections of patients with head and neck squamous cell carcinoma-prospective analysis of prevalence, survival and its implications, *Arch. Otolaryngol., – Head & Neck Surg.*, 2004, **130**(2), 157–160.

21. M. J. Tobin, M. A. Chesters, J. M. Chalmers, F. J. M. Rutten, S. E. Fisher, I. M. Symonds, A. Hitchcock, R. Allibone and S. Dias-Gunase-kara, Infrared microscopy of epithelial cells in whole tissues and in tissue culture using synchrotron radiation, *Faraday Discussion, Royal Society of Chemists.*, 2004, **126**, p. 27–39.

22. J. Pijanka, G. D. Sockalingum, A. Kohler, Y. Yang, F. Draux, P. Dumas, C. Sandt, G. Parkes, D. G . van Pittius, G. Douce, V. Untereiner and J. Sulé-Suso, Synchrotron based FTIR spectra of stained single cells. Towards a clinical application in pathology. Laboratory Investigation (DOI:10.1038/labinvest.2010.8).

23. M. J. Baker, E. Gazi, M. D. Brown, J. H. Shanks, P. Gardner and N. W. Clarke, FTIR-based spectroscopic analysis in the identification of clini-cally aggressive prostate cancer, *Br. J. Cancer*, 2008, **99**, 1859–1866.

24. M. J. Baker, E. Gazi, M. D. Brown, J. H. Shanks, N. W. Clarke and P. Gardner, Investigating FTIR based histopathology for the diagnosis of prostate cancer, *J. Biophot.*, 2009, **2**(1), 104–113.

25. J. Anastassopoulou, E. Boukaki, C. Conti, P. Ferraris, E. Giorgini, C. Rubini, S. Sabbatini, T. Theophanides and G. Tosi, Microimaging FT-IR spectroscopy on pathological breast tissues, *Vib. Spectrosc.*, 2009, **51**, 270–275.

26. R. de Paula Jr. Alderico and S. Sokki, Raman Spectroscopy for diagnosis of atherosclerosis: a rapid analysis using neural networks, *Med. Eng. Phys.*, 2005, **27**, 237–244.

27. C. J. L. Murray and A. D. Lopez, *The global burden of disease: A comprehensive assessment of mortality and disability from diseases, injuries, and risk factors in 1990 and projected to 2020*, Harvard University Press, Cambridge, MA, 1996.

28. S. Leeder, H. Raymond and H. Greenberg *et al.*, *A race against time: The challenge of cardiovascular disease in developing economies*, Earth Institute at Columbia University, New York, 2004.

29. P. Kumar and M. Clark, *Textbook; Clinical medicine*, 4th Edition, W. B. Saunders, 1999.
30. British Heart Foundation Statistics Website. www.heartstats.org.
31. R. S. Cotran, V. Kumar and T. Collins, *Textbook: Robbins: Pathologic Basis of Disease*, 6th Edition, W. B. Saunders Company, 1999.
32. M. G. Davies and P. O. Hagen, Pathology of intimal hyperplasia, *Br. J. Surg.*, 1994, **81**, 1254.
33. E. Allaire and A. W. Clowes, The intimal hyperplastic response, *Ann. Thoracic Surg.*, 1997, **64**, S38.
34. E. M. Tuzcu, B. Berkalp and A. C. De Franco *et al.*, The dilemma of diagnosing coronary artery calcification: angiography versus intravascular ultrasound, *J. Am. Coll. Cardiol.*, 1996, **27**, 832–838.
35. J. R. Guyton and K. F. Klemp, Development of the lipid-rich core in human atherosclerosis, *Arterioscler. Thromb.*, 1996, **16**, 4–11.
36. H. M. Loree, B. J. Tobias, L. J. Gibson, R. D. Kamm, D. M. Small and R. T. Lee, Mechanical properties of model atherosclerotic lesion lipid pools, *Arterioscler. Thromb.*, 1994, **14**, 230–234.
37. D. M. Small, GL Duff memorial lecture. Progression and regression of atherosclerotic lesions. Insights from lipid physical biochemistry, *Arteriosclerosis*, 1988, **8**, 103–129.
38. V. J. Dzau, Pathophysiology of atherosclerosis and plaque complications, *Am. Heart J.*, 1994, **128**, 1300–1304.
39. P. Libby, Molecular basis of the acute coronary syndromes, *Circulation*, 1995, **91**, 2844–2850.
40. R. J. G. Peters, W. E. M. Kok and M. G. Havenith *et al.*, Histopathological validation of intracoronary ultrasound imaging, *J. Am. Soc. Echocardiogr.*, 1994, **7**, 230–241.
41. J. F. Toussaint, G. M. LaMuraglia, J. F. Southern, V. Fuster and H. L. Kantor, Magnetic resonance images lipid, fibrous, calcified, hemorrhagic, and thrombotic components of human atherosclerosis *in vivo*, *Circulation*, 1996, **94**, 932–938.
42. L. Wexler, B. Brundage, J. Crouse, R. Detrano, V. Fuster, J. Maddahi, J. Rumberger, W. Stanford, R. White and K. Taubert, Coronary artery calcification: pathophysiology, epidemiology, imaging methods, and clinical applications: a statement for health professionals from the American Heart Association, *Circulation*, 1996, **94**, 1175–1192.
43. A. L. Bartorelli, M. B. Leon and Y. Almagor *et al.*, *In vivo* human atherosclerotic plaque recognition by laser-excited fluorescence spectroscopy, *J. Am. Coll. Cardiol.*, 1991, **17**, 160B–168B.
44. R. Richards-Kortum, A. Mehta and G. Hayes *et al.*, Spectral diagnosis of atherosclerosis using an optical fiber laser catheter, *Prog. Cardiol.*, 1989, **118**(2), 381–391.
45. P. Weinmann, M. Jouan, Q. D. Nguyen, B. Lacroix, C. Groiselle, J. P. Bonte and G. Luc, Quantitative analysis of cholesterol and cholesterol esters in human atherosclerotic plaques using near-infrared Raman spectroscopy, *Arteriosclerosis*, 1998, **140**, 81–88.

46. G. Deinum, D. Rodriguez and T. J. Römer *et al.*, Histological classification of Raman spectra of human coronary artery atherosclerosis using principle component analysis, *Appl. Spectrosc.*, 1999, **53**(8), 938–942.
47. J. F. Brennan, T. J. Römer and R. S. Lees *et al.*, Determination of human artery composition by Raman spectroscopy, *Circulation*, 1997, **96**(1), 99–105.
48. L. Silveira Jr., S. Sathaiah, R. A. Zángaro, M. T. Pacheco, M. C. Chavantes and C. A. Pasqualucci, Correlation between near-infrared Raman spectroscopy and the histological analysis of atherosclerosis in human coronary arteries, *Lasers Surg. Med.*, 2002, **30**(4), 290–297.
49. R. Manoharan, Y. Wang and M. S. Feld *et al.*, Histochemical analysis of biological tissues using Raman spectroscopy, *Spectrochim. Acta. Part A*, 1996, **52**, 215–249.
50. G. O. Angheloiu, J. T. Arendt and M. G. Müller *et al.*, Intrinsic fluorescence and diffuse reflectance spectroscopy identify superficial foam cells in coronary plaques prone to erosion. Arterioscl., *Thromb., Vasc. Biol.*, 2006, **26**, 1594.
51. R. Manoharan, J. J. Baraga, M. S. Feld and R. P. Rava, Quantitative histochemical analysis of human artery using Raman spectroscopy, *J. Photochem. Photobiol. B.*, 1992, **16**, 211–233.
52. R. Manoharan, J. J. Baraga, R. P. Rava, R. R. Dasari, M. Fitzmaurice and M. Feld, Biochemical analysis and mapping of atherosclerotic human artery using FT-IR microspectroscopy, *Atherosclerosis*, 1993, **103**, 181–193.
53. L. Silveira Jr., S. Sathaiah, R. A. Zâhgaro., M. T. Pacheco, M. C. Chavantes and C. A. Pasqualucci, Near infra-red Raman spectroscopy of human coronary arteries: histopathological classification based on mahalanobis distance, *J. Clin. Laser Med. Surg.*, 2003, **21**(4), 203–208.
54. E. U. Otero, S. Sathaiah and L. Silveira Jr. *et al.*, Raman spectroscopy for diagnosis of calcification in human heart valves, *Spectroscopy*, 2004, **18**(1), 75–84.
55. S. W. E. Van de Poll, T. J. Römer, G. J. Puppels and Laarse Avandev, Raman spectroscopy of atherosclerosis, *J. Cardiovasc. Risk.*, 2002, **9**, 255–261.
56. J. P. Salenius, J. F. Brennan and A. Miller *et al.*, Biochemical composition of human peripheral arteries examined with near-infrared Raman spectroscopy, *J. Vasc. Surg.*, 1998, **27**, 710–719.
57. H. P. Buschman, J. T. Motz, G. Deinum, T. J. Römer, M. Fitzmaurice, J. R. Kramer, A van der Laarse and M. S. Feld, Diagnosis of human coronary atherosclerosis by morphology-based Raman spectroscopy, *Cardiovasc. Pathol.*, 2001, **10**, 59–68.
58. G. V. Nogueira, L. Silveira Jr., A. A. Martin, R. A. Zangaro, M. T. Pacheco, M. C. Chawartes and C. A. Pasqualucci., Raman spectroscopy study of atherosclerosis in human carotid artery, *J. Biomed. Optics.*, 2005, **10**(3), 0311171–7.
59. T. J. Römer, J. F. Brennan III, T. C. Bakker Schut, R. Wotthis, R. C Varden, H. Oogen, J. J. Emeis, Avandev Laarse, A. V. Binschke and S. J.

Puppels, Raman spectroscopy for quantifying cholesterol in intact coronary artery wall, *Atherosclerosis*, 1998, **141**, 117–124.

60. J. T. Motz, M. Hunter, L. H. Galindo, J. A. Gardecki, J. R. Kramev, R. R. Dasai and M. S. Feld, Optical fibre probe for biomedical Raman spectroscopy, *Appl. Optics.*, 2004, **43**(3), 542–554.

61. M. Shim and B. Wilsom, Development of an *in vivo* Raman spectroscopic system for diagnostic applications, *J. Raman Spectrosc.*, 1997, **28**, 131–142.

62. B. Shim, B. Wilson, E. Marple and M. Wach, Study of fibre-optic probes for *in vivo* medical Raman spectroscopy, *Appl. Spectrosc.*, 1999, **53**, 619–627.

63. J. Y. Ma and S. Y. Li, Optical-fibre Raman probe with low-background interference by spatial optimization, *Appl. Spectrosc.*, 1994, **48**, 1529–1531.

64. T. F. Cooney, H. T. Skinner and S. M. Angel, Comparative study of some fibre-optic remote probe designs. 1. Model for liquids and transparent solids, *Appl. Spectrosc.*, 1996, **50**, 836–848.

65. T. F. Cooney, H. T. Skinner and S. M. Angel, Comparative study of some fibre-optic remote Raman probe designs. 2. Tests of single-fibre, lensed and flat- and bevel-tip multi-fibre probes, *Appl. Spectrosc.*, 1996, **50**, 849–860.

66. U. Utzinger and R. Richards-Kortum, Fibre optic probes for biomedical spectroscopy, *J. Biomed. Opt.*, 2003, **8**, 121–147.

67. J. T. Motz, M. Fitzmaurice, A. Miller, S. J. Gandhi, A. S. Haka, L. H. Salindo, R. R. Desari, J. R. Kramev and M. S. Feld, *In vivo* Raman spectral pathology of human atherosclerosis and vulnerable plaque, *J. Biomed. Optics.*, 2006, **11**(2), 0210031–9.

68. A. Vincent, J. L. Bihari and D. J. Suter *et al.*, The prevalence of nosocomial infection in intensive care units in Europe results of the European Prevalence of Infection in Intensive Care (EPIC) Study, *J. Am. Med. Assoc.*, 1995, **274**, 639–644.

69. K. Maquelin, C. Kirschner, L.-P. Choo-Smith, N. van den Braak, H. P. Endtz, D. Naumann and G. J. Puppels, Identification of medically relevant microorganisms by vibrational spectroscopy, *J. Microbiol. Meth.*, 2002, **51**, 255–271.

70. M. H. Kollef, Inadequate antimicrobial treatment: an important determinant of outcome for hospitalized patients, *Clin. Infect. Dis.*, 2000, **31**(Suppl. 4), S131–S138.

71. A. P. Wheeler and G. R. Bernard, Treating patients with severe sepsis, *N. Eng. J. Med.*, 1999, **340**, 207–214.

72. E. H. Ibrahim, G. Sherman, S. Ward, V. J Fraser and M. H Kolleff, The influence of antimicrobial treatment of bloodstream infections on patient outcomes in the ICU setting, *Chest*, 2000, **118**, 146–155.

73. S. Osmon, D. Warren, M. Sondra, B. S. Seiler, W. Shannon, V. J. Fraser and M. H. Kollef, The Influence of Infection on Hospital Mortality for Patients Requiring > 48 h of Intensive Care, *Chest*, 2003, **124**, 1021–1029.

74. L. G. Donowitz, R. P. Wenzel and J. W. Hoyt, High risk of hospital acquired infection in the ICU patient, *Crit. Care Med.*, 1982, **10**, 355–357.

75. P. H. Chandraseekar, J. A. Kruse and M. F. Matthews, Nosocomial infection among patients in different types of intensive care units at a city hospital, *Crit. Care Med.*, 1986, **14**, 508–510.

76. R. B. Brown, D. Hosmer, H. C. Chen, D. Teres, M. Sards, S. Bradley, E. Opitz, D. Szwedzinslci and D. Opalenik, A comparison of infections in different ICUs within the same hospital, *Crit. Care Med.*, 1985, **13**, 472–476.

77. R. L. Brawley, D. J. Weber, G. P. Samsa and W. A Rutala, Multiple nosocomial infections: an incidence study, *Am. J. Epidemiol.*, 1989, **130**, 769–780.

78. D. J. Weber, P. D. Raasch and W. A. Rutala, Nosocomial infections in the ICU: The growing importance of antibiotic-resistant pathogens, *Chest*, 1999, **115**(3), 34S–41S.

79. J. M. Boyce, Methicillin-resistant Staphylococcus aureus: detection, epidemiology, and control measures, *Infect. Dis. Clin. North Am.*, 1989, **3**, 901–913.

80. M. E. Mulligan, H. C. Standiford and C. A. Kauffman, Methicillin-resistant Staphylococcus aureus: a consensus review of the microbiology, pathogenesis, and epidemiology with implications for prevention and management, *Am. J. Med.*, 1993, **94**, 313–328.

81. B. E. Murray, Vancomycin-resistant enterococci, *Am. J. Med.*, 1997, **102**, 284–293.

82. G. M. Eliopoulos, Vancomycin-resistant enterococci: mechanism and clinical relevance, *Infect. Dis. Clin. North Am.*, 1997, **11**, 851–865.

83. J. M. Boyce, Vancomycin-resistant enterococcus: detection, epidemiology, and control measures, *Infect. Dis. Clin. North Am.*, 1997, **11**, 367–384.

84. D.L. Paterson, W.C. Ko and S. Mohapatra et al. Klebsiella pneumoniae bacteremia: impact of extended spectrum beta-lactamase (ESBL) production in a global study of 216 patients (abstract). 37th Interscience Conference on Antimicrobial Agents and Chemotherapy; Sept. 28-Oct. 1; Toronto, Ontario, Canada,1997.

85. CDC. Guidelines for preventing the transmission of Mycobacterium tuberculosis in health-care facilities, 1994. Morb. Mort. Week. Rep.; 43(RR-13):1–132.

86. T. C. White, K. A. Marr and R. A. Bowden, Clinical, cellular, and molecular factors that contribute to antifungal drug resistance, *Clin. Microbiol. Rev.*, 1998, **11**, 382–402.

87. CDC. Update, Staphylococcus aureus with reduced susceptibility to vancomycin – United States. *Morb. Mort. Week. Rep.*, 1997, **46**, 813–815.

88. M. H. Kollef and S. Ward, The influence of mini-BAL cultures on patient outcomes: implications for the antibiotic management of ventilator-associated pneumonia, *Chest*, 1998, **113**, 412–420.

89. F. Alvarez-Lerma, Modification of empiric antibiotic treatment in patients with pneumonia acquired in the intensive care unit. ICU-Acquired Pneumonia Study Group, *Intens. Care Med.*, 1996, **22**, 387–394.

90. C. M. Luna, P. Vujacich, M. S. Niederman, C. Vay, C. Gherardi, J. Materaad and E. C. Jolly, Impact of BAL data on the therapy and outcome of ventilator-associated pneumonia, *Chest*, 1997, **111**, 676–685.
91. J. Rello, M. Gallego and D. Mariscal *et al.*, The value of routine microbial investigation in ventilator-associated pneumonia, *Am. J. Respir. Crit. Care Med.*, 1997, **156**, 196–200.
92. P. Montravers, R. Gauzit, C. Muller, J. P. Maruse, A. Fichelle and J. M. Desmcents, Emergence of antibiotic-resistant bacteria in cases of peritonitis after intra-abdominal surgery affects the efficacy of empirical antimicrobial therapy, *Clin. Infect. Dis.*, 1996, **23**, 486–494.
93. J. W. Chow, D. M. Fine, D. M. Shlaes, J. P. Quinn, D. C. Hooper, M. P. Johnson, R. Ramphal, M. M. Wagener, D. K. Miyashiro and V. L. Yu, Enterobacter bacteremia: clinical features and emergence of antibiotic resistance during therapy, *Ann. Intern. Med.*, 1991, **115**, 585–590.
94. J. Romero-Vivas, M. Rubio, C. Fernandez and J. J. Picazo, Mortality associated with nosocomial bacteremia due to methicillin-resistant Staphylococcus aureus, *Clin. Infect. Dis.*, 1995, **21**, 1417–1423.
95. V. C. Chang, C. C. Huang and S. T. Wang *et al.*, Risk factors of complications requiring neurosurgical intervention in infants with bacterial meningitis, *Pediatr. Neurol.*, 1997, **17**, 144–149.
96. C. H. Heath, D. I. Groveand and D. F. Looke, Delay in appropriate therapy of Legionella pneumonia associated with increased mortality, *Eur. J. Clin. Microbiol. Infect. Dis.*, 1996, **15**, 286–290.
97. D. M. Neimeyer, Polymerase chain reaction: a link to the future, *Mil. Med.*, 1998, **163**(4), 226–228.
98. Y. W. Tang, G. W. Procop and D. H. Pershing, Molecular diagnostics of infectious diseases, *Clin. Chem.*, 1997, **43**(11), 2021–2038.
99. P. Belgrader, W. Benett and D. Hadley *et al.*, Infectious Disease: PCR detection of bacteria in seven minutes, *Science*, 1999, **284**(5413), 449–450.
100. R. Leclercq, M.-H. Nicolas-Chanoine, P. Nordmann, A. Philippon, P. Marchais, A. Buu-Ho, H. Chardon, H. Dabernat, F. Doucet-Populaire, C. Grasmick, P. Legrand, C. Muller-Serieys, J. Nguyen, M.-C. Ploy, M.-E. Reverdy, M. Weber and R.J. Courcol, Multicenter evaluation of an automated system using selected bacteria that harbor challenging and clinically relevant mechanisms of resistance to antibiotics, **20**(9), 2001, 626–635.
101. D. M. Livermore, M. Struelens, J. Amorim, F. Baquero, J. Bille, R. Canton, S. Henning, S. Gatermann, A. Marchese, H. Mittermayer, C. Nonhoff, K. J. Oakton, F. Praplan, H. Ramos, G. C. Schito, J. Van Eldere, J. Verhaegen, J. Verhoef and M. R. Visser, Multicentre evaluation of the VITEK 2 Advanced Expert System for interpretive reading of antimicrobial resistance tests, *J. Antimicrob. Chemother.*, 2002, **49**, 289–300.
102. C. C. Sanders, M. Peyret, E. S. Moland, C. Shubert, K. S. Thomson, J. M. Boeufgras and W. E. Sanders Jr., Ability of the VITEK 2 advanced expert

system to identify beta-lactam phenotypes in isolates of Enterobacter-iaceae and Pseudomonas aeruginosa, *J. Clin. Microbiol.*, 2000, **38**(2), 570–4.

103. A.-M. Freydiere, R. Guinet and P. Boiron, Yeast identification in the clinical microbiology laboratory: phenotypical methods, *Med. Mycol.*, 2001, **39**, 9–33.

104. K. H. Schadow, D. K. Giger and C. C. Sanders, Failure of the Vitek AutoMicrobic system to detect beta-lactam resistance in Aeromonas species, *Am. J. Clin. Pathol.*, 1993, **100**(3), 308–10.

105. K. Maquelin, C. Kirschner and L.-P. Choo-Smith, N. A. Ngo-Thi, T. van Vreeswijk, M. Stammler, H. P. Endtz and H. A. Bruining, Prospective study of the performance of vibrational spectroscopies for rapid identification of bacterial and fungal pathogens recovered from blood cultures, 2003, 41(1):324–329.

106. M. S. Ibelings, K. Maquelin, H. Ph. Endtz, H. A. Bruining and G. J. Puppels, Rapid identification of Candida spp. in peritonitis patients by Raman spectroscopy, *Clin. Microbiol. Infect.*, 2005, **11**, 353–358.

107. C. Mello, D. Ribeiro, F. Novaes and R. J. Poppi, Rapid differentiation among bacteria that cause gastroenteritis by use of low-resolution Raman spectroscopy and PLS discriminant analysis. *Analytical & Bioanalytical Chemistry*, 2005, **383**(4), 701–706.

108. C. Harz, P. Rosch and J. Popp, Vibrational spectroscopy – a powerful tool for the rapid identification of microbial cells at the single cell level, *Cytometry Part A: J. Int. Soc. Anal. Cytol.*, 2009, **75**(2), 104–113.

109. G. D. Sockalingum, W. Bouheda, P. Pina, P. Allouch, R. Mandray, R. Labia, J. M. Millot and M. Manfait, ATR-FTIR spectroscopic investigation of imipenem susceptible and resistant Pseudomonas aeruginosa isogenic strains, *Biochem. Biophys. Res. Commun.*, 1997, **232**, 240–256.

110. K. Hosseini, F. H. M. Jongsma, F. Hendrikse and M. Motamedi, Non-invasive monitoring of commonly used intra-ocular drugs against endopthalmitis by Raman spectroscopy, *Lasers Surg. Med.*, 2003, **32**(4), 265–270.

111. N. M. Amiali, M. R. Mulvey, J. Sedman, A. E. Simor and A. A. Ismail, Epidemiological typing of methicillin-resistant Staphylococcus aureus strains by Fourier Transform infrared spectroscopy, *J. Microbiol. Methods*, 2007, **69**, 146–153.

112. P. J. Lambert, A. G. Whitman, O. F. Dyson and S. M. Akula, Raman spectroscopy: the gateway into tomorrow's virology, *Virol. J.*, 2006, **3**(51), 1–8.

113. M. Beekes, P. Lasch and D. Naumann, Analytical applications of Fourier transform-infrared (FT-IR) spectroscopy in microbiology and prion research, *Vet. Microbiol.*, 2007, **123**(4), 305–319.

CHAPTER 2

Mid-infrared Spectroscopy: The Basics

JOHN M. CHALMERS

VS Consulting, Stokesley, UK and School of Chemistry, University of Nottingham, UK

2.1 Introduction

The main purpose of this chapter is to give a *beginner's guide* for those readers not familiar with mid-infrared (IR) spectroscopy. It will cover the basics, describing concisely what mid-IR radiation is, how it may be used, both qualitatively and quantitatively, the basic sampling techniques and practices undertaken when using mid-IR spectroscopy, and the basics of the instrumentation most commonly used today, *i.e.*, the Fourier transform infrared (FT-IR) spectrometer. This chapter will then move on to discuss the technique of FT-IR microscopy, and why, in particular, one might wish to use mid-IR radiation emanating from a synchrotron as a source for undertaking studies by FT-IR microscopy. The chapter will conclude with some considerations of data analysis techniques and how mid-IR spectroscopy relates to both near- and far-IR spectroscopy and how it is the complement of Raman spectroscopy (see Chapter 4 in this book for more information on Raman spectroscopy).

If you are experienced with mid-IR spectroscopy then you may feel there is no real need to 'read on' with this chapter, but please persist, as it provides a useful background to the subsequent chapters within this book.

RSC Analytical Spectroscopy Monographs No. 11
Biomedical Applications of Synchrotron Infrared Microspectroscopy
Edited by David Moss
© Royal Society of Chemistry 2011
Published by the Royal Society of Chemistry, www.rsc.org

2.2 Mid-infrared Radiation and Mid-infrared Spectroscopy

2.2.1 Electromagnetic Radiation: What is it?

The colours of a rainbow that we perceive – red, orange, yellow, green, blue, indigo, and violet – are components that go towards making up the visible spectrum, which is a sub-part of the electromagnetic spectrum. The electromagnetic spectrum is an energy continuum that extends from the relatively low energy radio and television wave region through the visible spectrum region to the high-energy region of its spectrum where the likes of X-rays and γ-rays occur.[1] Electromagnetic radiation, as implied by its name, can be considered as two orthogonal oscillating fields, one electric and one magnetic, which operate at right angles (orthogonal) to each other, see Figure 2.1.

The interaction of electromagnetic radiation with molecules for vibrational spectroscopy is considered through the interaction of the electrical component with respect to molecular vibrations. These waves are a consequence of the motion caused by a stream of particles, which have no mass; the particles are called *photons*. The highest energy photons have the shortest wavelength and highest frequency. All electromagnetic radiation travels at the speed of light, c, in a vacuum. A property that differentiates the colours in the visible spectrum, and indeed separates energy elements in other ranges of the electromagnetic spectrum, is their wavelength or frequency (or more fundamentally their energy); these two properties are interrelated, since the wavelength, λ, in vacuum is related to the frequency, v, by:

$$\lambda = \frac{c}{v} \tag{2.1}$$

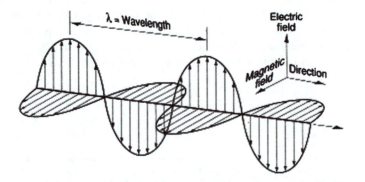

Figure 2.1 Schematic of electromagnetic radiation showing the orthogonal electrical and magnetic fields, the direction of propagation of the wave, and a measure of the wave's wavelength.

The reciprocal of the wavelength is called the wavenumber, $\hat{v}$, that is the number of waves in a unit length; it has the unit cm^{-1}. Thus:

$$\lambda = \frac{c}{v} = \frac{1}{\hat{v}} \tag{2.2}$$

The wavelength range covered by the visible spectrum extends from about 770 nm (red) to 380 nm (violet) (*i.e.*, from about 7.7×10^{-6} m to 3.8×10^{-6} m). (Radio wavelengths can be of the order of 10^3 m; X-rays have wavelengths in the region of 10^{-10} m.) Mid-infrared radiation occurs over the range 2.5 μm ($4000\,cm^{-1}$) to 25 μm ($400\,cm^{-1}$). Electromagnetic radiation can be expressed in equivalent units of wavelength, frequency and/or energy; the favoured choice in any particular region depends both on the field of study and the convenience of the size of the unit. For instance, in the visible region, a wavelength of 400 nm (0.4 μm; 4×10^{-9} m) is equivalent to: a wavenumber of $25000\,cm^{-1}$; a frequency of *ca.* 7.5×10^{14} Hz [750 terahertz (THz) or sometimes expressed as $7.5\times10^{14}\,s^{-1}$ (historically, 7.5×10^{14} cycles per second)]; an energy of about 5×10^{-19} J (Joules) or about 3.1 eV (electron volts).

2.2.2 Mid-infrared Radiation: What is it?

Extending beyond the long wavelength (lower frequency) region of the visible spectrum region lies the infrared ('*beyond red*') region of the electromagnetic spectrum. This may be divided up into three regions, known in increasing wavelength coverage as the near-, mid- and far-IR regions; these cover the wavelength ranges of about 780 nm to 2500 nm (2.5 μm), 2.5 μm to 25 μm, and 25 μm to about 1000 μm, respectively. In the context of this chapter, the reason for this distinction will become apparent later, when the molecular vibrations of organic and bio-molecules are discussed. Friedrich Wilhelm Herschel (Sir William Herschel) discovered near-IR (NIR) radiation in 1800. Using a blackened thermometer bulb to measure dispersed solar radiation, he discovered the existence of radiation (thermal) beyond the long-wavelength (red) region of the visible spectrum, which was invisible to the human eye. (It is far-IR radiation that we strictly feel as heat from sources such as the sun, a radiator, *etc.*).

So the mid-IR region, the principle region of the electromagnetic spectrum with which we are concerned in this chapter, and indeed in this book, spans the wavelength range 2.5 μm to 25 μm. For (convenient) reasons, which will become apparent when we discuss both modern instrumentation (that is the FT-IR spectrometer) and qualitative analysis (interpretative mid-IR spectroscopy), nowadays spectroscopists and analysts concerned with interpreting mid-IR spectra usually refer to the mid-IR region as spanning the range $4000\,cm^{-1}$ to $400\,cm^{-1}$, where the unit 'cm^{-1}' known as the *wavenumber* is related to the wavelength (expressed in cm) by eqn (2.2), *vide supra*. For example, wavelengths of 2.5 μm, 10 μm and 25 μm are equivalent to $4000\,cm^{-1}$, $1000\,cm^{-1}$ and $400\,cm^{-1}$, respectively.

2.2.3   What is Mid-infrared Spectroscopy?

Mid-infrared spectroscopy is the interrogation of the molecular structure of materials, in our case biological molecules, by mid-IR radiation. Molecules are not static entities; the bonds that connect the atoms comprising a molecule are in continuous movement. Molecules can translate, that is move in any one of three dimensions, they can rotate about three orthogonal axes, and, important in our case, they undergo molecular vibrations, that is each internal (intra-) molecular bond or bond between molecules (inter-), *e.g.*, a hydrogen-bonded pair, within a molecule, or associated pair, can be considered to undertake a series of molecular vibrations; the number of absorption bands occurring in a mid-IR spectrum arising from these molecular vibrations is considered in section 2.7. While these intra- (and inter-) molecular vibrations are not independent, they can be envisaged as such as a bond stretch or a bond deformation, which can be, for example, a bond twist, a bond wag, a bond scissoring motion, or a bond rocking mode. To illustrate this point, let us consider a hypothetically isolated methylene, $-CH_2$ group, see Figure 2.2.

The series of *fundamental* (normal modes) vibrations shown in Figure 2.2 is known as a set of fundamental *group* vibrations (frequencies) for the $-CH_2$ group. For this molecular group, they are shown to comprise two stretching vibrations and four deformation vibrations. The two stretching vibrations, a

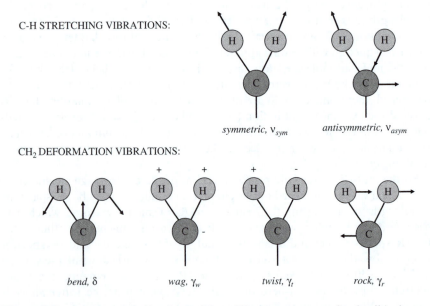

Figure 2.2   Fundamental vibrations of the $-CH_2$ group (see text for details). Arrows indicate respective motions of the atoms; $+$ and $-$ indicate opposite motions perpendicular to the plane of the molecular group.

symmetric and an antisymmetric vibration, may be envisaged as those in which the two H atoms move in unison in relation to the C atom or they move in opposite directions in relation to the C atom, respectively; stretching vibrations are commonly assigned the symbol v, and hence these are notated as v_{sym} and v_{asym}, respectively. (This view of CH stretching modes is over-simplistic, since the stretching modes are not restricted solely to movement by the H atoms, but also involve movement of the C atom (and indeed all other atoms of the molecule of which it is a part.) However, for many vibrational modes such as the CH stretching modes, only a few atoms have large displacements and the rest of the molecule is almost stationary.) The fundamental intra-molecular deformation vibrations of a $-CH_2$ group comprise descriptively a bend (δ), wag (γ_ω), twist (γ_t) and rock (γ_r); in Figure 2.2, the plus and minus signs indicate atoms moving oppositely in and out of the plane of the page. While it must be remembered that the whole molecule to which such a $-CH_2$ unit may be attached is vibrating, for particular vibrations such as these, most of the vibration is localised and they therefore tend to undergo their molecular vibrations within a narrow frequency range, which happens to coincide within a narrow frequency range (well-defined wavenumber region) of the mid-IR spectrum region. Consequently, when these molecules are interrogated using mid-IR radiation, by, for example, placing them into the beam of an IR spectrometer, they will, providing they satisfy certain well-established theoretical conditions (*vide infra*), absorb the mid-IR radiation that occurs at a frequency that coincides with that of the molecular vibration, thereby giving rise to a characteristic absorption band; this is the basis of *mid-IR spectroscopy* in which the loss of IR radiation transmitted through a sample is measured across the spectrum. A plot (spectrum) is recorded of absorption bands versus mid-IR wavenumber (see, for example, Figure 2.4(e) below). These largely localised vibrations are therefore termed a series of *group* vibrations, and each occurs therefore within a narrow well-documented (correlated) wavenumber region; in this case, for example, for the two stretching vibrations, between about $2950 \, cm^{-1}$ and $2860 \, cm^{-1}$, the antisymmetric vibration is of higher frequency, occurring at higher wavenumber. While it is not the purpose of this chapter (nor this book) to provide an informative text on the assignments of bands within a mid-IR spectrum, we will see in subsequent chapters frequent mention of, for example, the Amide I and Amide II bands, which are very characteristic of proteins and peptides; these occur at about $1650 \, cm^{-1}$ and $1550 \, cm^{-1}$, respectively; the former is attributed largely to the $C=O$ stretching (sometimes notated as $vC=O$) associated with a secondary amide, the $-C(=O)N(H)-$ grouping, while the latter (for a *trans* secondary amide) is considered a combined mode involving the motions of both N–H bending and C–N stretching. The exact position at which a fundamental band occurs within its characteristic mid-IR spectrum, and the shape it has, depend, *inter alia*, on the influence of its neighbouring atoms in the molecular structure, the molecular conformation, ordering and configuration, its environment, temperature, pressure, *etc.*; within the context of this book, this can, for example, be influenced by the state of aggressiveness a cell or tissue sample has experienced.

Not every fundamental vibration of a molecule gives rise to an absorption band within its characteristic spectrum (*vide infra*); to absorb mid-IR radiation there must be a change of dipole moment during the molecular vibration, and the magnitude of this dipole moment change determines the intensity of an absorption band. So, while vC=O bands are very strong in a mid-IR spectrum, more symmetric vibrations such as vC=C can be very weakly absorbing or even non-absorbing.

Between $4000 \, \text{cm}^{-1}$ and $1500 \, \text{cm}^{-1}$ (excluding overtone, combination and Fermi resonance bands),[2,3] a mid-IR spectrum can be sub-divided into three characteristic regions, this is largely a consequence of the characteristic vibrational frequencies due to the masses of the atoms involved in the group frequency vibration and the force constant of the bonds linking these atoms together. These approximate regions are: $3600–2500 \, \text{cm}^{-1}$, which is associated with X–H stretching vibrations, where X is C, O, N or S (P and Si, being heavier, give rise to X–H stretching vibrations between 2500 and $2100 \, \text{cm}^{-1}$); $2500–2000 \, \text{cm}^{-1}$ is the region in which triple (*e.g.*, –C≡N) and cumulative double bonds (*e.g.*, –N=C=S–) occur; between about 2000 and $1500 \, \text{cm}^{-1}$, the fundamental stretching vibrations of double bonds occur (*e.g.*, C=O, C=C, C=N). The region between 1500 and $400 \, \text{cm}^{-1}$ is known as the *fingerprint region,* so-called because, for instance, quite similar molecules can give different band patterns, sometimes subtle, in this region reflecting, for example, their polymorphic form, degree of crystallinity, or in our case, for instance, the effects of malignancy or keratinisation of a tissue or cell sample. These attributes are key to the value and success of mid-IR spectroscopy in being able to fingerprint, identify and/or classify materials.

2.2.4   Quantitative Mid-infrared Spectroscopy: The Basics

The traditional means of measuring the mid-IR spectrum of a sample using an IR spectrometer has been by a transmission measurement, in which the intensity of mid-IR radiation passing through the sample, presented in some appropriate form, and reaching the detector is compared to the intensity of mid-IR radiation reaching the detector in the absence of the sample. If these two values are I and I_0, respectively, then the % of radiation transmitted, $\%T$, by the sample at a given wavenumber, $\hat{v}$, can be represented as:

$$\%T_{(\hat{v})} = \frac{I_{(\hat{v})}}{I_{0(\hat{v})}} \times 100 \tag{2.3}$$

Transmittance, T, has values between 0 and 1, so the transmittance of the sample at a particular wavenumber, $\hat{v}$, is given by $\frac{I_{(\hat{v})}}{I_{0(\hat{v})}}$.

The more a sample absorbs the mid-IR radiation, the lower the value of T. The transmittance of a pure sample of pathlength (thickness) l, where l is in cm, is expressed as:

$$T_{(\tilde{v})} = \frac{I_{(\tilde{v})}}{I_{0(\tilde{v})}} = e^{-\alpha_{(\tilde{v})}l} \tag{2.4}$$

where $\alpha_{(\tilde{v})}$ is the *linear absorbance coefficient* (cm^{-1}) at wavenumber $\hat{v}$.[4]

The amount of mid-IR radiation absorbed, A, by the pure sample at $\hat{v}$ is expressed as:

$$A_{(\tilde{v})} = \log_{10}\left(\frac{1}{T_{(\tilde{v})}}\right) = \log_{10}\frac{I_{0(\tilde{v})}}{I_{(\tilde{v})}} \tag{2.5}$$

A is known as the *absorbance*. For a pure sample:

$$A_{(\tilde{v})} = a_{(\tilde{v})}lc \tag{2.6}$$

where: c is concentration of the sample, and $c \times l$ represents the relative number of absorbing molecules in the mid-IR beam; $a_{(\tilde{v})}$ is the *absorptivity* at wavenumber $\hat{v}$ and has units concentration/pathlength.

Equation 2.6 is a form of the basic law underlying quantitative analysis by mid-IR spectroscopy; it is known as the Lambert–Beer Law and commonly referred to simply as *Beer's Law*. Providing this linear relationship holds for all molecules within the mid-IR beam, that is there are no intermolecular interactions between molecules, then for a mixture Beer's Law can be applied in an additive form. For an *n*-component mixture, then the measured absorbance A' for all the i components present is given by:

$$A' = \sum_{i=1}^{n} a_i l c_i \tag{2.7}$$

Because of the linear relationship between the absorbance A and concentration c, quantitative analyses are always carried out on absorbance spectra (not transmission spectra), whether they are simple univariate determinations or involve multivariate analyses. Multivariate analyses, used to statistically analyse or classify data that involve more than one component (*e.g.*, different states of cancer aggressiveness) over a wide spectral range (*i.e.*, many wavenumbers), is that which is commonly practised in samples for medical diagnostics, such as those considered in this book.

2.3 Mid-infrared Spectroscopy Instrumentation

2.3.1 What is FT-IR? Why FT-IR?

FT-IR (or FTIR) is an abbreviation for Fourier transform IR. The terminology came into common usage in the 1970s to distinguish FT-IR mid-IR spectrometers from what had been mostly conventionally used until then to record mid-IR spectra; these were dispersive mid-IR spectrometers, the basics of which are more easily conceived. In a dispersive spectrometer, mid-IR radiation from a broadband (wide wavelength range continuum) source is dispersed (narrow wavelength elements are sequentially selected by such as a prism or grating); the dispersed elements of radiation are in turn focused in sequence onto a single-element detector; when a sample is placed in the path of this scanned mid-IR beam it produces its characteristic mid-IR absorption spectrum. Some of the 'magic' of FT-IR spectrometers that saw them replace dispersive spectrometers as standard instruments is that they can be very much more sensitive, they produce much higher signal-to-noise ratio (SNR) spectra, and that the wavenumber axes of the spectra that they record are far more reproducible and precise. The high sensitivity and SNR available with FT-IR mid-IR spectrometers enable their use to analyse easily much smaller samples (see section below on FT-IR microscopy), and their wavenumber repeatability advantage is essential to multivariate analyses procedures.

The basics of an FT-IR spectrometer design can be perceived in a Michelson interferometer (see schematic in Figure 2.3). A Michelson interferometer comprises five basic elements: a source, three optical elements, and a detector. The optical elements are a fixed mirror, a moving (or scanning) mirror and a beam-splitter. Its principle of operation (see Figure 2.3) is most easily understood using a single wavelength, λ_1, (monochromatic) source, the collimated radiation from which is incident on the beam-splitter, which then splits the transmitted beam 50 : 50 onto each of the fixed mirror and the moving mirror. On returning to the beam-splitter, the partially transmitted beams recombine and *interfere*. At what is known as *zero path difference* (ZPD), that is when the distances between the beam-splitter and the two mirrors are equal, there is no pathlength difference (*optical retardation*) between the two beams so that when they interfere they do so *constructively*. When the distance travelled by the two beams is from equal pathlength to $\lambda_1/4$ then this introduces a pathlength difference between the two beams of $\lambda_1/2$, which results in so-called *destructive interference*. Destructive interference is maximum at each $\lambda_1/2$ pathlength difference (*i.e.*, at $\lambda_1/2$, $3\lambda_1/2$, $5\lambda_1/2$, *etc.*); at each integral pathlength difference (*i.e.*, λ_1, $2\lambda_1$, $3\lambda_1$, *etc.*) the beams will undergo constructive interference. As the moving mirror scans at constant velocity, the detector records a sinusoidal variation (cosine function) of intensity, *the interferogram*, of the monochromatic radiation, which is sampled digitally (*vide infra*). When this sinusoidal variation, in terms of the displacement domain (distance of the moving mirror) or time, is Fourier transformed (a mathematical transformation) the resultant is a spectrum (in the frequency domain); in the case of monochromatic radiation, this is a single line of wavelength λ_1 (wavenumber $\tilde{v}_1$).

In an FT-IR spectrometer, the source is, of course, a broadband (polychromatic, continuum) source, thus the interferogram recorded can be considered as a summation of all the cosine functions of the individual wavelengths present in the source; these will only all be in phase at the position of ZPD, and give rise to a strong interferogram *centre-burst* (see Figures 2.3(b) and 2.4). The

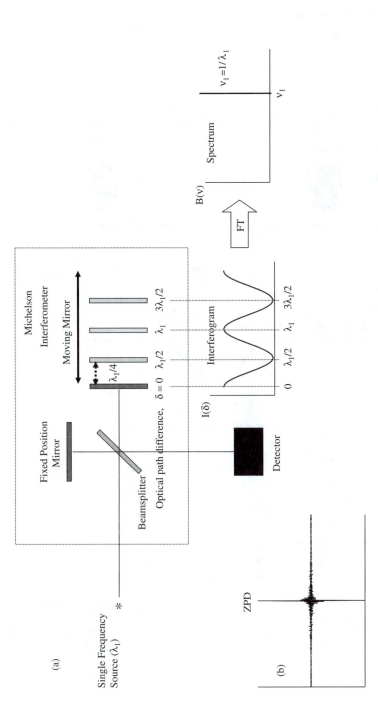

Figure 2.3 (a) Schematic of a basic Michelson interferometer and the production of a single wavelength (monochromatic radiation) cosine function interferogram, the Fourier transform of which is a single line of wavelength λ_1. (b) Example interferogram from a broadband source, showing the strong centre-burst – the point at which the cosine interferograms of all the wavelengths present are in phase. Adapted from reference [25]; copyright Wiley-VCH Verlag GmbH & Co. KGaA. Reproduced with permission.

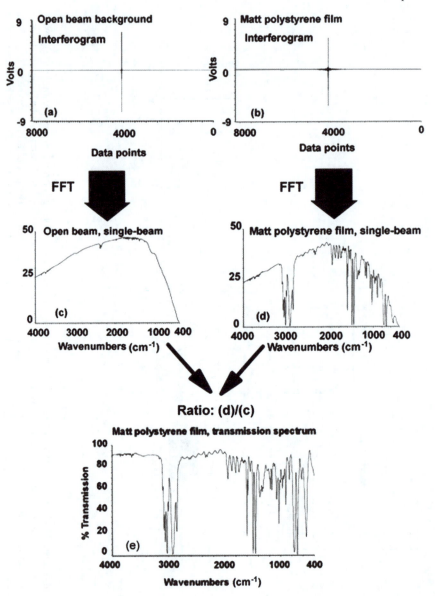

Figure 2.4   Example of the processes involved in producing an FT-IR transmission spectrum from a matt polystyrene film. FFT, fast Fourier transform algorithm. Reproduced by permission of The Royal Society of Chemistry from *Industrial Analysis with Vibrational Spectroscopy*, J. M. Chalmers & G. Dent (1997).

Fourier transform of such an interferogram produces a mid-IR single-beam spectrum.

The normal practical process of producing a mid-IR spectrum using an FT-IR spectrometer is to record two interferograms, each of which is then converted (Fourier transformed) into a single-beam spectrum. One of these interferograms (usually recorded first) is recorded without the sample being imposed in the radiation beam path, so it represents a convolution of the energy profile of the mid-IR beam and characteristics of the optical components, including the detector. This single-beam spectrum is referred to as the *single-beam background* spectrum. The second interferogram is recorded with the sample placed appropriately in the interrogating mid-IR beam path; it is then Fourier transformed into a single-beam spectrum. The ratio of these two single-beam spectra yields the mid-IR transmission spectrum of the sample, which can then be converted into an absorption spectrum [eqn (2.5)]. This process is summarised pictorially in Figure 2.4.

For a continuum source, an interferogram may be represented as:

$$I(\delta) = \int_{-\infty}^{+\infty} B(\tilde{v}) \cos 2\pi\tilde{v} \, d\tilde{v} \qquad (2.8)$$

and its spectrum is given by:

$$B(\tilde{v}) = \int_{-\infty}^{+\infty} I(\tilde{v}) \cos 2\pi\delta \, d\delta \qquad (2.9)$$

Equations (2.8) and (2.9) are two halves of a cosine *Fourier transform pair*: $I(\delta)$ is the intensity of mid-IR radiation reaching the detector, where δ is the optical retardation (*vide supra*); $B(\tilde{v})$ represents the single-beam spectrum as a function of $\tilde{v}$ calculated by computing the cosine Fourier transform of $I(\delta)$. Equations (2.8) and (2.9) can be seen to imply that for infinitely high spectral resolution the moving mirror within the interferometer (FT-IR spectrometer) must travel a distance from + infinity to − infinity. Clearly, the moving mirror can only travel a certain limited distance, and this determines the spectral resolution of the recorded spectrum, that is how well resolved closely spaced absorption bands might appear. As an example, Figure 2.5 shows two spectral lines of equal intensity. To be able to distinguish between these two lines in a composite spectrum, the interferogram of this composite spectrum must generate an optical retardation that produces at least the first maximum of their beat pattern. In practice, an interferogram comprises a set of equally spaced digitised data points collected during the scanning of the moving mirror of the interferometer, hence an FT-IR spectrum consists of a set of equally spaced digitised data points. The wavenumber interval between these equally spaced

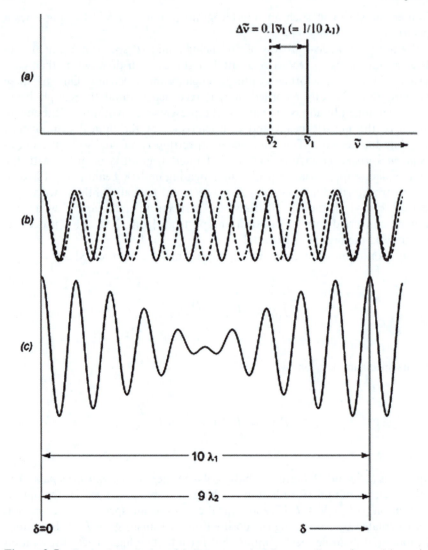

Figure 2.5 Spectral resolution: (a) Spectrum showing two lines of equal intensity; these occur at wavenumber $\tilde{v}_1$ (solid line) and wavenumber $\tilde{v}_2$ (dashed line); they are separated by $0.1\tilde{v}$. (b) The interferogram of each spectral line is shown individually as a solid and a dashed line, respectively. (c) The resulting interferogram from both spectral lines, with the first maximum of the beat signal at $10/\tilde{v}_1$; to resolve these two spectral lines, it is necessary to generate an optical retardation of at least this value. Reproduced from reference [4] by permission of John Wiley & Sons, Inc., Hoboken; © 2007.

data points defines the spectral resolution according to a well-established criterion, the *Nyquist criterion*.[4]

A mid-IR spectrum from such as a tissue or cell sample (or indeed most condensed-phase samples) will be recorded typically at a spectral resolution of

$4\,cm^{-1}$ or $8\,cm^{-1}$. The single scan cycle to record a mid-IR spectrum at these spectral resolutions with today's FT-IR spectrometers is rapid, and of the order of one second or less. By taking advantage of the wavenumber scale reproducibility of FT-IR spectrometers, this is put to good advantage by recording multiple scans from a sample in a reasonable timescale and averaging these to obtain a mean representative spectrum with a high signal-to-noise ratio.

It is outside the purpose of this book to go deeply into the details of the various optical designs of commercial FT-IR spectrometer interferometers and their peculiarities and nuances, the means of sampling data, the way interferograms are processed to obtain mid-IR spectra, *etc.* There are many books and reference works that cover these aspects; the interested reader is recommended particularly to consult reference 4. Suffice it to repeat here what was mentioned above.

- Mid-IR FT-IR spectrometers can be very sensitive, and they can be used to produce rapidly high SNR mid-IR spectra.
- Mid-IR FT-IR spectrometers have a distinct (energy) throughput advantage over other spectrometers operating with a broadband source.
- Very important, too, is that the wavenumber axis of spectra generated with a mid-IR FT-IR spectrometer is very precise and reproducible.

2.3.2 FT-IR Microscopy

Early in the application of modern mid-IR FT-IR spectrometers it became readily appreciated that their sensitivity could be utilised to analyse very small samples or interrogate localised areas of larger samples. This was particularly so when the FT-IR spectrometer was equipped with a highly sensitive, liquid nitrogen cooled mercury cadmium telluride (MCT) detector. This recognition of the throughput and sensitivity advantages very quickly led to the development of systems in which an FT-IR spectrometer is interfaced to an optical microscope, which is equipped with a liquid nitrogen cooled MCT detector.[5] The designs of these systems, in which the optical and mid-IR paths are collinear through the sample, allow one to visually inspect (and microphotograph, video capture) a sample located on the microscope sampling stage, then an area of interest can be delineated by a set of masks (apertures) located in a remote focal plane, and its mid-IR spectrum recorded. (The single-beam background spectrum will be recorded through the apertured area but without the sample being present.)

An example of the optics associated with a commercial FT-IR microscope system is shown in Figure 2.6. In such systems, one is able to record a single-point spectrum, or an automated x–y microscope stage can be programmed so that either: (i) a linear series of mid-IR spectra is recorded from a defined aperture area, or (ii) a two-dimensional array of spectra is recorded from sequentially adjacent (or partially overlapping) areas delineated by the defined apertured area, or (iii) a sequence of single-point spectra may be recorded using the defined aperture area from pre-determined positions on the sample. In these

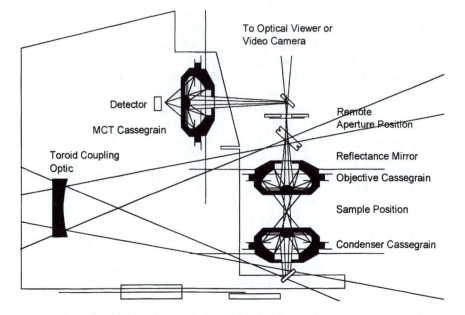

Figure 2.6   FT-IR microscope: Optical schematic of a microscope used in conjunction
with an FT-IR spectrometer; the position of the toroidal coupling optic
can be switched to allow for the recording of a spectrum in either the
transmission mode or the reflection mode. Reproduced by permission of
PerkinElmer Life and Analytical Sciences.

automated processes, the sample on the microscope stage is moved (translated)
under the defined aperture. The different sampling techniques (transmission
and reflection) are considered below in section 2.4. From these multi-spectra
datasets, grey-scale or false colour maps may be generated that highlight,
through relative absorption intensity differences, differences in the chemical or
morphological form of the overall sample area investigated. Alternatively, these
datasets might represent the input data for a multivariate classification analysis
such as assigning particular areas on the sample with a state of cancer
aggressiveness.

2.3.3   Why Synchrotron-sourced Mid-infrared FT-IR Microspectroscopy?

(See also Chapter 3.) As the apertured sample area interrogated using an FT-IR
microscope system gets smaller, so does the relative SNR of the recorded
spectrum. Also, as one works with an aperture with a linear dimension that
approaches the wavelength of mid-IR radiation, in our case 2.5 μm to 25 μm,
then the recorded spectra become increasingly distorted due to diffraction
effects (which increase with increasing wavelength), and as a consequence
become not fit for their analytical purpose. The continuum of mid-IR radiation

emanating from a synchrotron offers some distinct advantages when it is used as the source for an FT-mid-IR microscopy study of an apertured small sample area. The mid-IR radiation emanating from a synchrotron does so in a well-defined beam that has very low divergence, so that its photons are contained within a narrow transverse area; this makes a synchrotron a very much *brighter* mid-IR source than that used conventionally as a source for a mid-IR FT-IR spectrometer. In the mid-IR region, the radiation from a synchrotron, although brighter, is less intense than that from a conventional mid-IR source such as a globar (heated silicon carbide rod); it is more intense in the far-IR region, and this advantage is put to good use for many far-IR studies. Figure 2.7 compares the calculated IR radiation emanating from a synchrotron source, in terms of both power (intensity) and brightness, with that typically from a (black body) thermal source.[6] Because of its lower power in the mid-IR region, the brightness only becomes advantageous to use with mid-IR spectroscopy measurements through apertures of about 40 µm diameter or less.[7] Figure 2.8 compares the IR signal passing through various aperture sizes using a synchrotron source with a globar source, from which it can be seen that very little light passes through a 10 µm square aperture when a globar source is used.[8] The mid-IR beam size at the sampling position within an FT-IR microscope system using synchrotron-sourced mid-IR radiation on a recent generation synchrotron may be as small as about 10 µm × 10 µm (see, for example, Figure 2.9),[9] and therefore well suited to FT-IR microscopy studies on small samples, such as a single cell, or of localised areas on a sample, such as a sectioned biopsy tissue sample.

Figure 2.10 compares the mid-IR spectra recorded on the same FT-IR microscopy set-up from a 10 µm × 10 µm area on a tissue sample using either a conventional globar source or the synchrotron radiation as a source (in this case the SRS at Daresbury, UK). The SNR improvement in using the synchrotron as a source of mid-IR radiation is clearly evident.[10,11]

2.4 Mid–infrared Spectroscopy: Sampling Techniques and Practices

As with the rest of this chapter, we will only consider here those sampling techniques and practices that are relevant to the studies contained within this book.

The methods of presenting samples such as a tissue or isolated single cell for study in an FT-IR microscope have to date been predominantly confined to transmission and, the so-called, transflection sampling techniques. The latter is actually a reflection–absorption technique (*vide infra*). Of increasing recent interest is use of the so-called ATR sampling technique for the analysis of tissue samples. ATR is an abbreviation for attenuated total reflection and is an internal reflection spectroscopy technique. On the horizon are perhaps near-field techniques. Each of these will now be considered in turn.

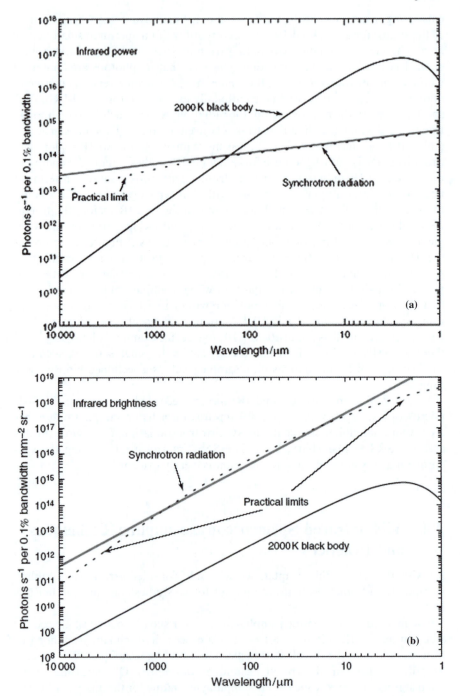

Figure 2.7 Calculated performance comparison between a synchrotron radiation (1 A) source and a thermal source: (a) power; (b) brightness. Reproduced from reference [6] with permission of John Wiley & Sons Ltd., Chichester; © 2002.

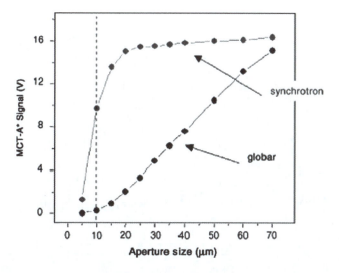

Figure 2.8   Infrared signal intensity recorded using a confocal IR microscope with a single-point detector through various apertures using a synchrotron or globar source.[8] Copyright 2005, with permission from Elsevier.

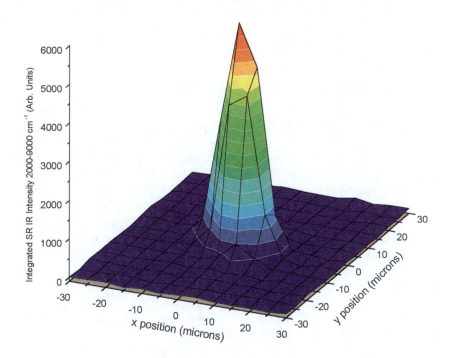

Figure 2.9   Plot demonstrating the small spot size that can be achieved using synchrotron-sourced mid-infrared radiation. The plot represents the integrated signal intensity from 2000–9000 cm^{-1} through a 10 µm pinhole scanned on a microscope stage in an FT-IR spectrometer. Reproduced from reference [9] by kind permission of the Advanced Light Source (ALS), Berkeley Laboratory.

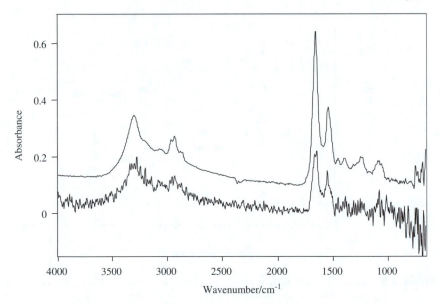

Figure 2.10   Comparison of SRS (upper) and Globar-source (lower) FT-IR spectra,
8 cm^{-1} spectral resolution, 1024 scans, from a 10 μm×10 μm area on a
tissue section. The upper spectrum has been offset for clarity. Reproduced from reference [10] with permission of John Wiley & Sons, Ltd.,
Chichester; © 2008.

2.4.1   Transmission Sampling Technique

Transmission sampling, as its name implies, is associated with measuring the
mid-IR radiation transmitted by an appropriate sample, thereby recording its
absorption characteristics. To record a mid-IR transmission spectrum appropriate to a qualitative analysis or for classification purposes from a tissue
sample, for example, requires that the specimen thickness be about 5–10 μm.
Any thicker, and some of the absorption bands may effectively saturate, that is
be too strong to be considered for proper analysis; any thinner, and the SNR of
its spectrum is likely to be poor. Ideally, the absorbance intensity of the
strongest band, the so-called Amide I band (*vide supra*) recorded from a gly-
cogen-free/low concentration tissue sample, should have an intensity between
0.7 and 1.0 absorbance units. The required sample area must completely fill the
diameter of the interrogating mid-IR beam (or be appropriately masked) so
that at the focal plane on the sampling stage of the FT-IR microscope the only
radiation passing through the sample and reaching the detector is that which
has been transmitted through the sample area of interest.

The preparation of a tissue specimen of appropriate thickness from such as a
biopsy sample (see also Chapter 5) usually follows standard histology practices,
particularly since it is a common practice to microtome two parallel adjacent
sections of near equivalent thickness. One of these sections is stained in a
conventional way, most commonly with haematoxylin and eosin (H&E), for

conventional histopathological examination and diagnosis; the adjacent section is usually left unstained and used for the mid-IR microspectroscopic measurement.[12] A visual comparison between these two sections then aids in selecting the most appropriate region(s) of the unstained section for study during method development. An alternative method of preparing thin sections is to cryo-microtome thin sections from a biopsy sample that has been rapidly frozen; this lessens the chances of possible changes being introduced into cell structure and morphology during the traditional embedding, fixation and deparaffinization processes that accompany conventional H&E staining protocols.

Glass microscope slides are totally opaque to almost all mid-IR radiation, so the most common substrate on which to support an unstained tissue section for a mid-IR spectroscopic examination is a thin polished window of CaF_2 or BaF_2 (of typical thickness *ca.* 0.5 mm).

Individual cells may also be deposited onto a CaF_2 window support, but perhaps the more practical and most appropriate (in terms of its match to cytological practices) method nowadays is to use a *low-e glass slide* and record a transflection spectrum, see next section.

2.4.2 Transflection Sampling Technique

(See also Chapter 5.) The, so-named, transflection technique can be considered essentially a reflection–absorption technique undertaken at an acute angle of incidence (as near normal incidence as practical is optimum), see schematic in Figure 2.11. It is commonly employed to record mid-IR spectra using an FT-IR microscope from both tissue and cell samples. The most commonly used substrate today for a transflection measurement is known as a 'MirrIR low-e microscope slide', often simply referred to as a 'low-e slide', produced by Kevley Technologies (Chesterland, OH, USA). Low-e slides are produced with dimensions very similar (25 mm × 75 mm) to those of a standard glass microscope slide thereby making them very convenient for mounting on the microscope stage of an FT-IR microscope system. Although the slides are made of

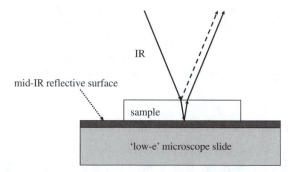

Figure 2.11 Schematic of the transflection sampling technique. The dashed arrow shows the weak specular reflection component; see text for details.

glass, they are coated with a thin layer of a proprietary silver compound, which makes them almost completely reflective to mid-IR radiation, while still remaining nearly totally transparent to visible light. The slides also have the advantage of being chemically inert to common organic solvents, and are stable to 400 °C. (These slides are less costly than CaF_2 and BaF_2 substrates.)

In a transflection mid-IR spectroscopic measurement of a tissue sample, a microtomed section (about 3–8 μm thickness) of the tissue sample is supported flat on a low-e slide mounted in the microscope stage of an FT-IR microscope. The sample can then be examined visually in transmission and an appropriate region selected for study; in the recording of its mid-IR spectrum, the mid-IR radiation passes through the section, is reflected back by the coating on the low-e slide and then passes through the sample again, thereby effectively producing a mid-IR spectrum that is closely equivalent to a mid-IR transmission spectrum recorded from a specimen of double the section thickness. However, super-imposed on this 'double-transmission' spectrum will be a very much weaker front surface (specular) reflection spectrum, which, because of its dependence on refractive index changes associated with absorption bands (*anomalous dispersion*), has features at the positions of absorption bands that have a first-derivative like appearance, and these relatively weak features may well distort slightly and shift the peak maximum of absorption bands, compared to those evident in a 'pure' transmission spectrum. Notwithstanding this, so long as the experimental method has consistency, these effects have been demonstrated by many studies to be of minimal consequence in terms of the required classification procedures being developed for assigning mid-IR spectral profiles to states of cancer aggressiveness.

Preparing cytological samples in a form suitable for recording a mid-IR spectrum from an individual cell requires that the cells are spread out and well separated on an appropriate substrate; because of these requirements, transflection (low-e) microscope slides have proven a very effective substrate onto which cytological samples can be prepared. Exfoliated cells can be deposited as sparse monolayers from suspensions onto a low-e slide using commercially available apparatus developed originally for cytology. Two such systems are the CytoSpin® (Thermo Shandon, Pittsburgh, PA, USA) and the ThinPrep® method (Cytyc Corporation, Marlborough, MA, USA). The methodologies for preparing specimens for a mid-IR spectroscopic examination associated with each of these have been outlined in a recent publication by Diem *et al.*[13] This reference also discusses issues that relate to each methodology and cell fixation; in addition, the authors discuss the use of low-e slides as substrates, onto which cells may be cultured, and the processes and sub-strate and sample issues as they relate to such specimen preparations.

2.4.3   Attenuated Total Reflection (ATR) Technique

Attenuated total reflection (ATR) spectroscopy is an internal reflection spectroscopy sampling technique. It is limited to sampling the surface layer of a sample, to a depth of between about 0.3 and 3 μm, but offers the advantage of not requiring a thin section to be prepared from such as a tissue sample. This

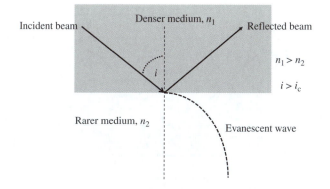

Figure 2.12 Schematic of internal reflection spectroscopy: The IR beam is transmitted through the denser medium of refractive index, n_1, and internally reflected at the boundary with the medium of lower refractive index, n_2, providing it is incident on this boundary at an angle greater than the critical angle, i_c. The dashed curve represents schematically the intensity of the evanescent wave. Reproduced from *Qualitative and Quantitative Analysis of Plastics, Polymers and Rubbers by Vibrational Spectroscopy*, J. M. Chalmers and N. J. Everall, in *Vibrational Spectroscopy of Polymers: Principles and Practices*, eds N. J. Everall, J. M. Chalmers and P. R. Griffiths, John Wiley & Sons Ltd., Chichester (2007).

'surface layer' mid-IR spectrum is generated by placing the surface of a sample into intimate optical contact with a higher refractive index optical element (*e.g.*, a prism, or other geometric shape), which is transparent (or semi-transparent) to mid-IR radiation. Mid-IR radiation passing through this internal reflection element (IRE), also known as an ATR element (or crystal), will be totally internally reflected at its boundary in contact with a (non-absorbing) medium, providing that its angle of incidence at the boundary is greater than the critical angle, and the medium is of lower refractive index.

A schematic of internal reflection is shown in Figure 2.12. The so-called *evanescent wave* (decaying wave) penetrates the surface of the sample, and will be *attenuated* by the mid-IR absorption characteristics of the lower refractive index medium. For the purposes of this chapter, the ATR approach may envisaged as shown in Figure 2.13, in which the mid-IR beam 'penetrates' the surface layer of the sample to a depth, dp, providing the angle of incidence, α, of the mid-IR radiation is greater than the critical angle, α_c; $\sin \alpha_c$ is equal to the ratio of the refractive indices n_2/n_1, where $n_1 > n_2$, and n_1 is the refractive index of the ATR element and n_2 is the refractive index of the sample. The depth of penetration, dp, is given by (the so-called *Harrick equation*)[14,15]:

$$d_p = \frac{\lambda}{2\pi n_1 \left(\sin^2 \alpha - n_{21}^2 \right)^{1/2}} \tag{2.10}$$

where λ is the wavelength of the mid-IR radiation.

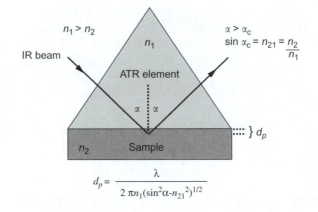

Figure 2.13   Schematic representation of an ATR prism showing the principle of the internal reflection sampling technique. See text for details.

It can be seen by looking at eqn (2.10) that dp has dependency on λ, so that the 'depth of surface layer probed' increases with increasing wavelength; a consequence of this is that, compared with a spectrum recorded in transmission (or transflection), the relative band absorption intensities will increase with increasing wavelength. A procedure known as an '*ATR correction*',[4] or similar, available with most FT-IR spectrometers, involves essentially dividing ATR spectral intensities by their corresponding wavelengths, thereby generating a spectrum, for comparative purposes, in which bands throughout the spectrum are more similar in relative intensity to those recorded from a transmission spectrum of the sample.

Also evident from eqn (2.10) is that the higher the refractive index of the ATR element, the shallower the dp. Another consequence is that the higher the angle of incidence of the mid-IR radiation, the shallower becomes dp, but as all ATR objectives supplied currently with commercial FT-IR microscope systems are fixed angle of incidence (usually nominally 45°), then this variable cannot be utilised in such systems. However, it is worth noting that because of anomalous dispersion[i,16] the shapes of bands in an ATR spectrum are also slightly different and their band maximum positions may be shifted slightly to lower wavenumber. (The closer the angle of incidence is to the critical angle, the greater is the distortion of strong bands in ATR spectra.) An algorithm is available within the software of, at least, one commercial FT-IR spectrometer that also corrects for this band distortion.[16]

---

[i]This behaviour occurs in the vicinity of an absorption band. It is a consequence of the fact that in the vicinity of an absorption band a sample's refractive index decreases markedly with decreasing wavenumber (increasing wavelength) on the high wavenumber side of an absorption band. It then increases rapidly through the absorption band centre, and then decreases markedly on the low wavenumber side of the band.[16]

2.4.3.1 ATR Microspectroscopy

The internal reflection elements (IRE) used in commercial ATR objectives available for use with FT-IR spectrometers are made either from ZnSe, Si, Ge or Type IIa diamond. The refractive indices (at *ca.* 1000 cm^{-1} and 25 °C) of these are 2.4 (ZnSe and Type IIa diamond), 3.42 (Si) and 4.0 (Ge). (A disadvantage with the Ge element is that it is not transparent to visible light, so precludes one being able to view through it a sample on the microscope stage.) In non-absorbing regions, the refractive index of most organic materials is about 1.5, and if we assume this value and insert it for n_2 in eqn (2.10) along with corresponding values for n_1 for the internal reflection elements and an incidence angle α of 45°, then the resultant *dp* versus wavenumber variations determined are shown in Figure 2.14. One can see, for example, that at 1000 cm^{-1}, depending on the IRE material, the *dp* ranges from about 0.6 to 2.0 μm, and that over the mid-IR spectral range *dp* varies from about 0.3 to 1.6 μm for an n_1 of 4.0 but is significantly greater for an n_1 of 2.4, for which it varies from about 0.5 to 5 μm.

As implied in section 2.3.3 above, the lateral spatial resolution, that is, that at which a mid-IR spectrum, which is free from diffraction distorting effects, can be recorded in transmission (or transflection) using an FT-IR microscope is approximately equal to the wavelength, that is, about 10 μm at 1000 cm^{-1}. Because of the higher (than air) refractive index of an ATR objective IRE, and consequent refraction (focusing) of the mid-IR beam, they have the potential for delivering higher (improved) lateral spatial resolution, *e.g.*, for Type IIa diamond this would be an improvement of about 2.4×, making it about 4.2 μm at 1000 cm^{-1}. This capability is of particular interest for such as intra-cellular studies.

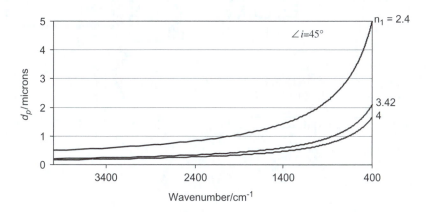

Figure 2.14 Plots of depth of penetration, d_p, versus wavenumber in internal reflection spectroscopy for ATR elements of ZnSe or Type IIa diamond ($n_2 = 2.4$), Si ($n_2 = 3.42$) and Ge ($n_2 = 4.0$) for $n_1 = 1.5$ and an angle of incidence of 45°. Reproduced from reference [16] with permission of John Wiley & Sons, Ltd., Chichester; © 2007.

2.4.4 Near-field FT-IR Microscopy

Conventional optical microscopy, whether visible or IR, is strictly limited in lateral spatial resolution due to the diffraction of radiation; this is particularly restrictive in mid-IR microspectroscopy, where high spectral contrast and purity is needed for high quality characterisation and classification purposes. And, while visibly one may be able to resolve structures in size of about one half of the wavelength of the light, in a mid-IR spectrum, a spectrum undistorted (or nearly so) by diffraction effects can only be recorded at a spatial resolution of about 10 μm, particularly when masking apertures are used to define a sampling area. One solution to overcoming the wavelength limitation is to use a near-field technique. Many FT-IR near-field approaches use an evanescent field generated at the tip of an appropriate probe.[17] One approach that has been reported recently with mid-IR radiation emanating from a synchrotron as its source uses a near-field thermal probe to detect a mid-IR absorption spectrum.[18] Figure 2.15 shows how during its evaluation it was interfaced via an optical interface (side-arm port) to an FT-IR spectrometer microscope installed above the synchrotron beam line at the SRS at Daresbury in the UK.

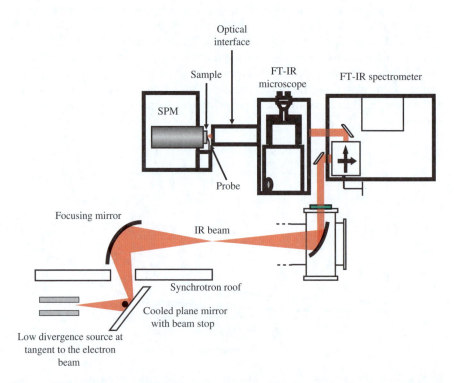

Figure 2.15   Schematic of the near-field photothermal IR microspectroscopy (PTMS) set-up at the SRS beamline Daresbury, UK. SPM, scanning probe microscope. Reproduced from reference [18] with permission of IOP Publishing Ltd. © 2002.

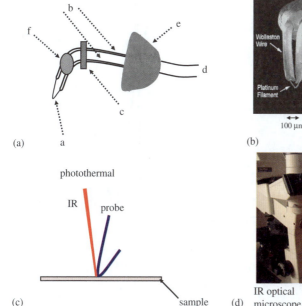

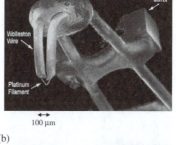

(a)

(b)

(c)

(d) IR optical microscope   Optical interface

Figure 2.16 (a) Schematic of Wollaston wire photothermal IR microspectroscopy (PTMS) thermal probe (a, Pt/10% Rh core, 5 μm diameter, exposed length 50 μm; b, Wollaston wires, 75 μm diameter; c, mirror (part of force feedback system for measuring deflection of Wollaston wires, when probe is being used for imaging); d, electrical connections; e, washer; f, strengthening bead); (b) photomicrograph of Wollaston thermal probe;[19] (c) schematic of 'sampling' technique; (d) photo of installation at Daresbury SRS, UK (see Figure 2.15).

One of the probes, similar to that used for scanning probe thermal microscopy, is shown schematically in Figure 2.16(a) alongside a photomicrograph of a probe in Figure 2.16(b). Other smaller probes that have been evaluated are based on micromachined probes.[18] In the type shown in Figure 2.16, the sensing element is an exposed length of Pt–Rh wire. The sample absorbs the mid-IR radiation at its characteristic absorption frequencies; the molecular vibrations then relax via a non-radiative process; the energy is then converted to heat in the form of a thermal wave; the temperature fluctuations (photothermal response) at the surface are recorded as a function of frequency; temperature rises in the probe induce changes of electrical resistance; these are amplified and the signal is fed into the external input of the FT-IR spectrometer; and a photothermal mid-IR spectrum is produced. Figure 2.16(c) also shows the simplicity of sampling whereby the probe is just brought into contact or near contact with the sample surface at the point onto which the mid-IR beam is focused. While this approach overcomes the problems associated with diffraction effects, and its lateral spatial resolution is in principle only limited by the probe element size, in a rapid scan FT-IR spectrometer set-up, account has

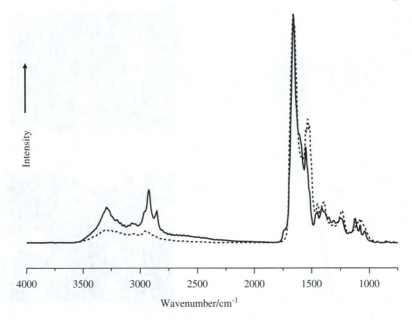

Figure 2.17   Photothermal IR microspectroscopy (PTMS) spectra of: control PC-3
cells (solid line), and benzol[a]pyrene-treated PC-3 cells (dashed line).[19]

to be taken of the balance between the thermal diffusion depth and the optical
absorption length.[19] Figure 2.17 shows example photothermal (non-synchro-
tron) IR microspectroscopy (PTMS) spectra that have been reported from
control and benzo[a]pyrene-treated prostate (PC-3) cells.[19]

2.5   FT-IR Mapping and Imaging Techniques

(See also Chapter 7.) As explained in section 2.3.3, data from a series of mid-IR
spectra recorded from consecutive, contiguous, masked or selected regions of a
sample may be processed to produce, for example, a species concentration map
of the sample area. This map may be based on such as relative absorbance band
intensity, an absorbance band ratio, a principal component (PC) derived from a
multivariate analysis routine, or a cluster from a classification analysis. The
map may be presented as a grey-scale image or a false-colour image based on
such as an intensity difference or species type.

   Mapping a property from mid-IR spectra over a sample area of any size
other than small is inherently a time-consuming process, since many spectra of
adequate signal-to-noise ratio must be recorded in sequence from each defined
element of the area. Because of the small focal area of the mid-IR beam within a
synchrotron-sourced mid-IR FT-IR microscope, mapping remains at this
moment in time by far the most commonly employed method for generating
species-differentiating maps (*e.g.*, cancerous *vs.* non-cancerous tissue). As this

chapter is being composed, however, developments are known to be underway exploring the possibilities of using limited size multi-element array detectors for producing more efficiently species-differentiated images. The problem lies in optimally utilising the brightness of the synchrotron-sourced mid-IR radiation across a multi-element array detector without annulling the advantage that use of the mid-IR radiation from a synchrotron source offers.[20]

2.6 Mid-infrared Spectroscopy: Data Analysis Techniques

(See also Chapter 6.) Data analyses techniques may be segregated into three or four types. The first category is based around the simple exploitation of Beer's Law, and involves comparison of relative band intensities or band areas associated with marker (diagnostic) peaks. The second is concerned with the so-called pre-processing of sets of spectra prior to further processing by multi-variate statistical quantitative or classification analysis. Since these multivariate analysis procedures are concerned essentially with analysing and comparing variances within a dataset (series of spectra), the majority of pre-processing techniques are associated with removing from a dataset those variances that are irrelevant in terms of the purpose of the interrogation of the dataset. Examples of these will be discussed below. The third and fourth categories may be considered as those for multivariate statistical analysis, either for quantitative comparisons or classification purposes, of pre-processed datasets. More details of the purpose and application of these procedures and pre-processing steps feature in later chapters within this book, particularly that by Lasch and Petrich (Chapter 6) and many of the Case Study chapters, 9–13.

As recorded, mid-IR spectra from a series of samples most likely suffer from several variances that make them less than optimum for comparative analysis purposes. These include absorption intrusions from atmospheric water vapour and carbon dioxide, differences in sample thickness (pathlength), and variable background absorption profiles. To optimise a dataset for multivariate analysis, it is usually recommended that the influences of these irrelevant variances be removed or minimised prior to multivariate analyses, otherwise they will be highlighted in and dominate the statistical analysis of variance within the dataset. The optimal means of eliminating absorption features from atmospheric water vapour and carbon dioxide is to ensure that the mid-IR spectrometer is efficiently purged with dry nitrogen or air; residual absorptions from these gases may be removed from spectra by using a spectral subtraction routine. Variable non-specific background absorptions, for example sloping baselines introduced by scattering, which introduce such as essentially relatively slowly varying background intensity changes, either essentially linear or shallow curves, may be compensated for by subtracting an appropriate linear, segmented multi-linear, or curved baseline. A common method utilised to remove baseline effects is to take the second-derivative of an absorbance spectrum; this produces spectra with a mostly flat baseline.[13] Variances related to differences in sample pathlength may be overcome by normalising a dataset of spectra; this may be achieved by a number of specific procedures. These include normalisation, sometimes with respect to the intensity of one of the most intense bands, usually either the Amide I band at *ca.* $1650\,cm^{-1}$ or Amide II band at *ca.* $1550\,cm^{-1}$, or other procedures, which include such as vector normalisation. Subtracting the mean spectrum (*mean centring*) of a dataset from each member of the dataset is another commonly employed procedure that reduces the sources of variance within a dataset and optimises it for further data processing.

For medical diagnostic applications, the prime subject matter of this book, the most commonly used multivariate analysis routines employed are those used for classification analysis. These enable the separation of the spectra within a dataset into classes/groups that are most similar (or dissimilar) to each other, thereby, for example, discriminating between those spectra associated with differing levels of cancer aggressiveness. Common among these procedures are principal component analysis (PCA) and hierarchical cluster analysis (HCA); others include neural networks (NN) and fuzzy logic (FZ). Figure 2.18 illustrates a simple example of separation (classification) by HCA of a series of input spectra recorded from a tissue sample into those recorded from regions of stroma and those recorded from regions of tumour.[10]

It is not the purpose of this chapter to provide a detailed explanation and considered comparative argument for the many pre-processing and multivariate data analysis procedures that are used, later chapters in this book and references cited therein provide useful sources to such information; the purpose of this section in this chapter was to provide a very brief insight into the uses of the procedures for those unfamiliar with the manner in which mid-IR spectra are used to assign a state of cancer aggressiveness to such as a biopsy or cell sample.

2.7 How Does Mid-infrared Spectroscopy Relate to and Differ from Near-infrared, Far-infrared and Raman Spectroscopy?

2.7.1 Fundamental Molecular Vibrations: Mid-infrared and Raman Bands

As stated in sections 2.2.2 and 2.2.3, mid-IR spectroscopy is used to generate a spectrum over the range $4000–400\,cm^{-1}$; this is the region in which the fundamental vibrations of organic molecules occur. Also, in section 2.2.3, it was stated that molecules undergo a series of molecular vibrations, which have characteristic mid-IR absorption bands, depending on whether there is a dipole moment change or not during the molecular vibration.

Infrared and Raman spectra (*vide infra*) occur as a consequence of transitions between quantised vibrational energy states.[2,4,21] In developing an

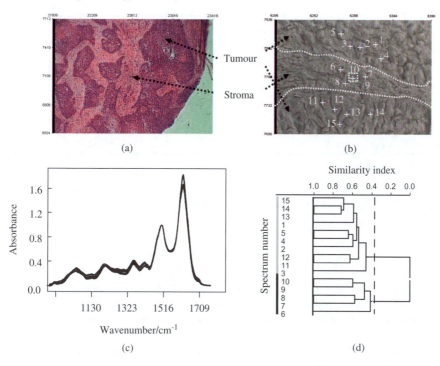

(a)   (b)

(c)   (d)

Figure 2.18 (a), (b) Visual images of an oral tissue section: (a) H&E stained, 4×
magnification; (b) unstained, 32× magnification, showing the centres,
marked with a +, of the 10 μm×10 μm areas from which FT-IR
microscopy spectra were recorded. The white dotted lines on (b) have
been superimposed to highlight the boundaries between the tumour and
stromal regions. The white dashed square around point 10 delineates the
aperture size. (c) Pre-processed SRS FT-IR absorbance spectra over the
range 1806 to 938 cm^{-1} recorded from the 15 10 μm×10 μm areas marked
on Figure 2.18(b), and used as input data for multivariate analysis using
Pirouette® software. (The abscissa scale reversal is a consequence of
loading the data into the Pirouette software for multivariate analysis.)
(d) Dendrogram from an HCA of the 15 spectra shown in Figure 2.18(c).
Reproduced from reference [10] with permission of John Wiley & Sons,
Ltd., Chichester; © 2008.

understanding of vibrational spectroscopy, the simplest model of a molecule to
consider is that of a diatomic molecule, which may be considered as two atoms
of masses m_1 and m_2 connected by a bond, represented by a spring. This 'spring'
has a property of a force constant f. The vibrational frequency of this spring
(bond), v, is given by:

$$v = \frac{1}{2\pi}\sqrt{\frac{f}{m}} \qquad (2.11)$$

Table 2.1   Calculated wavenumber of the fundamental stretching vibration for some diatomic groups.

Group	Reduced mass (amu)	Force constant ($N m^{-1}$)	Wavenumber (cm^{-1})
O–H	0.94	700	3600
N–H	0.93	600	3300
C–H	0.92	500	3000
C–C	6.00	425	1100
C=C	6.00	960	1650
C=O	6.86	1200	1725

amu = atomic mass unit

in which, m, known as the *reduced mass*, is given by:

$$\frac{1}{m} = \frac{1}{m_1} + \frac{1}{m_2} \tag{2.12}$$

Remember, wavenumber $\tilde{v} = \frac{v}{c} = \frac{1}{\lambda}$, see eqn (2.2). A force constant for a single bond involving a C atom and an H atom is in the region of $5\,N\,cm^{-1}$. The force constant for a double bond is about $10\,N\,cm^{-1}$; the force constant for a triple bond may be considered as about $15\,N\,cm^{-1}$. Using such values, one can determine fundamental vibrations for various diatomic groups to have the wavenumber values given in Table 2.1.

A free (non hydrogen-bonded) O–H group has a fundamental stretching mode vibration that gives rise to a sharp mid-IR absorption band in the vicinity of $3600\,cm^{-1}$, (hydrogen bonding results in a decrease in the absorption wavenumber position and broadening of the absorption band profile); $1725\,cm^{-1}$ is a position in a mid-IR spectrum that lies in about mid-range of the region in which absorption bands of the fundamental vibrational modes of carbonyl groups are observed, *etc.*

As a bond vibrates, consider for example in a longitudinal motion, that is compression and extension (stretching) between the two atoms, the motion could simply be described as *harmonic*, which would imply that the molecule could vibrate with any amplitude, and thus possess any amount of energy. However, a bond cannot be compressed below a certain value, because of the repulsive force between the atoms, since the closer they come together, the more strongly they repel each other; nor can it be extended beyond a certain value, otherwise bond break (ionisation) would occur; the actual representation of a diatomic bond vibration is better represented by that of an *anharmonic* oscillator, see Figure 2.19. Within this anharmonic potential (*Morse* potential) energy versus displacement curve, quantum mechanics stipulates that molecules can only exist in definite (*quantised*) states. In a classical mechanics picture (following *Hooke's Law*) for a harmonic oscillator these quantised states would be equidistant; in an anharmonic oscillator however, with increasing interatomic distance, the energy levels become more closely spaced. Infrared spectra occur as a consequence of transitions between these quantised vibrational

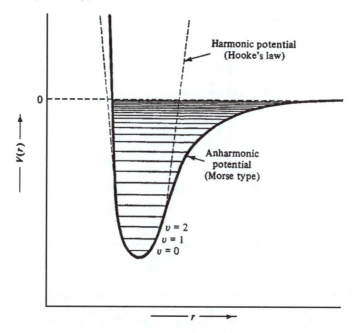

Figure 2.19 Potential energy, $V(r)$, of a diatomic molecule as a function of atomic displacement, r, during a vibration for a harmonic oscillator (dashed line) and an anharmonic oscillator (solid line). Reproduced from reference [4] by permission of John Wiley & Sons, Inc., Hoboken; © 2007.

energy states. A fundamental absorption is associated with the transition between the lowest energy level, E_0, $v=0$, and the next energy level, E_1, $v=1$. (The first overtone, which for X–H vibrations (see next section) are observed as absorptions in the near-IR region, occur as a consequence of a $v=0$ to $v=2$ transition.)

The transitions (see Figure 2.20) involved with Raman spectroscopy are discussed only briefly below, in section 2.7.3, but in more detail in Chapter 4 by Byrne, Sockalingum and Stone.

As mentioned in section 2.2.3, not all fundamental molecular vibrations are IR active. If we return to our simplest model of a diatomic molecule, in which there is only one molecular vibration, that is the stretching vibration (caused by stretching and compression) of the bond, and consider that each of the two atoms connected by the bond has a charge associated with it, then if these two atoms are identical, as for example in nitrogen (N_2) and oxygen (O_2), then during the molecular vibration there will be no net change of dipole moment during the motion. Thus, because of the *selection rules* that govern whether a molecular vibration is IR and/or Raman active, this particular molecular vibration does not give rise to an IR absorption band. However, these stretching vibrations of N_2 and O_2 are both Raman active. In contrast, for example, a heteronuclear diatomic organic molecule such as NO or CO,

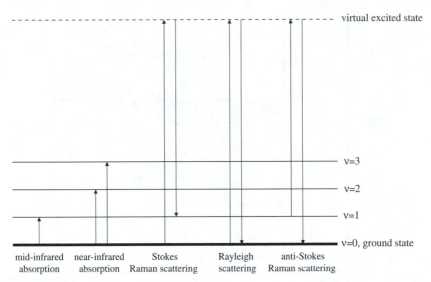

Figure 2.20   Energy level diagram depicting the energy level transitions associated with
mid-infrared and near-infrared absorption and Rayleigh and Raman
scattering processes. (Rayleigh and Raman scattering are a consequence of
the elastic and inelastic scattering of monochromatic radiation, respectively; see Chapter 4.) Reproduced from *Vibrational Spectroscopic Methods in Pharmaceutical Solid-state Characterization*, J.M. Chalmers and G.
Dent, pp. 95–138 in *Polymorphism in the Pharmaceutical Industry*, ed. R.
Hilfiker, Wiley-VCH Verlag GmbH & Co. KGaA, Weinheim (2006).

because there is a net change of dipole moment during the stretching vibration,
does give rise to IR absorption features.

If there are N atoms in a non-interacting molecule then it is considered to
have $3N$ *degrees of freedom*. Three of these are the translational motions in
mutually perpendicular directions; while another three represent rotational
motion of the molecule about the three mutually perpendicular axes. Thus a
non-linear molecule has $3N$–6 vibrational degrees of freedom. That is a non-
linear molecule has $3N$–6 possible normal modes of vibration. (For a linear
molecule, such as CO_2, this reduces to $3N$–5 possibilities.) In reality, the
number of distinct bands observed in a recorded mid-IR (or Raman) spectrum
from a condensed-phase sample (for example, a tissue biopsy or cell) is much
less; a normal mode of a species may be forbidden from giving rise to a mid-IR
absorption band because of the requirement of the selection rule (see above) or
forbidden; symmetry elements within a molecule may lead to more than one
mode in the molecule having the same vibrational frequency, leading to
degeneracy; similar modes may have absorption bands that overlap closely; and
the $3N$-6 statement says nothing about the relative intensities of bands within a
mid-IR (or Raman) spectrum – some will be very much stronger than others,
depending, in the case of mid-IR spectroscopy, on the strength of the dipole
moment change associated with the vibrational mode.

2.7.2 Near-infrared Spectroscopy

Near-infrared (NIR) spectroscopy[22] is associated with absorption bands appearing within the wavenumber region, 14300 to 4000 cm^{-1}. It is however much more common with this region to refer to the spectral range in terms of wavelength rather than wavenumber; the equivalent wavelength range is 700 nm to 2.5 μm (2500 nm). The absorption bands that are observed in the near-IR are those associated with overtone and combination modes of fundamental X–H stretching vibrations, where X=C, N and O. The near-IR region is often sub-divided into shorter ranges. These are the PbS (lead sulphide) region, which is 1100–2500 nm ($\sim$9100 to 4000 cm^{-1}) and the silicon region, which is 700 to 1100 nm (14300 to 9100 cm^{-1}).[16] The higher the overtone, the higher the frequency, and the shorter the wavelength: first and second overtones give rise to absorption bands in the PbS detector region, third and fourth overtones occur in the Si detector region. Binary and ternary combinations of C–H, N–H and O–H stretching vibrations with bending modes of a molecule also occur in the PbS detector region; higher combination modes occur in the Si detector region.

Compared with mid-IR spectroscopy, NIR spectroscopy is rarely used as a molecular characterisation tool; bands in the NIR region are much weaker than the fundamental bands observed in the mid-IR region, they also tend to be broader and much more overlapped. However, NIR spectra do offer some particular advantages for quantitative and classification analyses, which have seen it widely used in certain industries, notably the food and pharmaceutical industries. These advantages are related to the better sensitivity of detectors available, and hence comparatively higher SNR spectra, and less demanding sample preparation requirements, so that relatively thicker samples (including diffusely scattering samples) may be used; for example, whereas to record a transmission mid-IR spectrum from a tissue sample for qualitative analysis purposes may typically require a specimen of thickness 10 μm or less, in the first overtone region this would only need to be about 1 mm. Also, the NIR region is transparent to glass/quartz, and can be used with silica fibre-optics. A recent book contains details of developments of a NIR spectroscopic methodology for the classification of breast tissue aggressiveness.[23]

The author is not aware of any advantages of using NIR radiation emanating from a synchrotron over that available in a conventionally sourced NIR spectrometer, or any published work of synchrotron-sourced applications in this region. The major limitation in using this region, compared with the mid-IR region, is its lack of relatively easily derived molecular structure related diagnostic information.

2.7.3 Far-infrared/THz Spectroscopy

The far-IR region, as its name implies, lies to the long wavelength (low wavenumber) side of the mid-IR region; it may be considered as from about 400 cm^{-1} extending to about 10 cm^{-1}.[24] Recent developments in novel instrumentation for terahertz (THz) spectroscopy, particularly within the pharmaceutical industry[25–27] and security applications,[28,29] have created renewed interest in spectroscopy within this region. The THz region may be considered as over-lapping and extending to the longer wavelength region of the far-IR region. The new THz spectrometers cover the approximate range of about 1.3–$133.3\,cm^{-1}$, equivalent to $\sim 40\,GHz$ to $4\,THz$ and $7.7\,mm$ to $75\,\mu m$ wavelength.[25,30]

The far-IR-THz region is where low-frequency motions of molecules occur. These include those associated with entire molecule vibrations. Examples are skeletal bending modes and some torsion vibrations. It also includes nearest neighbour van der Waals bonding, inter-molecular interactions of hydrogen-bonded molecules, intra-molecular stretching modes involving heavy atoms, and phonon modes.[24,31] At this stage in the development of medical diagnostic applications of IR spectroscopy, in this spectral region, relative to the mid-IR region, there is little research or method development interest.

2.7.4 Raman Spectroscopy

In contrast to the far-IR-THz region, Raman spectroscopy is currently con-sidered a spectroscopic tool that does have a very high potential for medical diagnostic applications (see Chapter 4).

Raman spectroscopy is considered as the vibrational spectroscopy comple-ment to mid-IR spectroscopy. This is because, while both are vibrational spectroscopy techniques, they have different *selection rules* that govern whether a molecular vibration gives rise to a mid-IR absorption band or a Raman scattering band. As stated above in section 2.2.3, in order for a molecular vibration to produce a mid-IR absorption band it must undergo a change in dipole moment change during the molecular vibration; the selection rule for giving rise to a band in a Raman spectrum is that during the molecular vibration there must be change of *polarisability*. Since Raman spectroscopy is a scattering technique, the requirements for sample preparation are much less than those for a mid-IR spectroscopy examination.

In mid-IR spectroscopy, radiation from a broadband source passing through or incident upon a sample is selectively absorbed, according to the conditions outlined above. Raman spectroscopy is a consequence of selective inelastic scattering by molecular vibrations of a monochromatic radiation incident on a sample (see Figure 2.20); Figure 2.21 shows typical sampling arrangements for measuring a Raman spectrum (a Raman spectrum can also be observed using a transmission measurement geometry, although it is more common to use a back-scattering geometry with detection at $90°$ or $180°$).

Because Raman spectroscopy is an inelastic scattering technique, which couples the incident radiation with the molecular vibrations, any excitation laser wavelength can be used. Since Raman spectroscopy is usually undertaken using a visible or near-IR excitation laser, conventional optics can be used, and higher spatial resolution is achievable. The excitation laser wavelength may be chosen to match that of a visibly absorbing species (chromophore) and thereby generate a greatly enhanced Raman spectrum from the chromophore, such a spectrum is known as a resonance Raman spectrum; the excitation wavelength

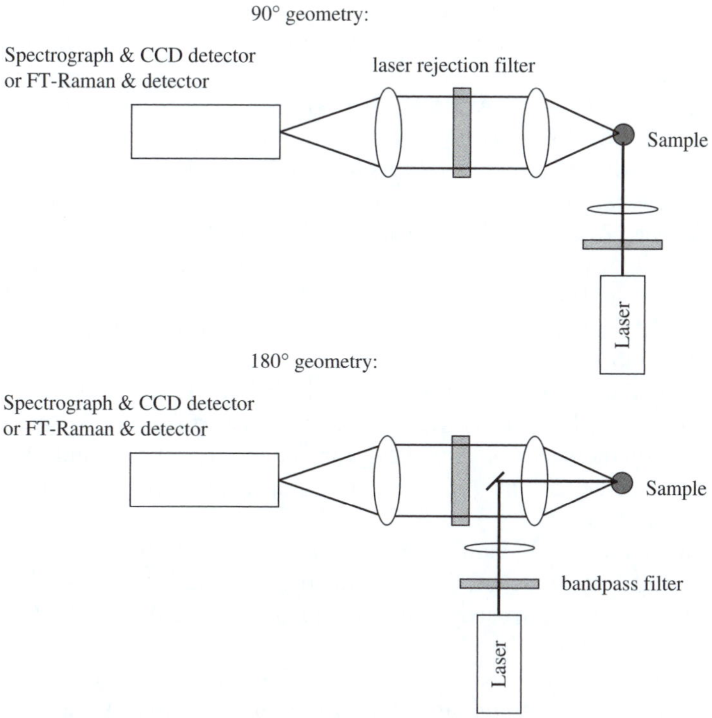

90° geometry:

180° geometry:

Figure 2.21 Commonly used Raman scattering and collection geometries. Reproduced from *Vibrational Spectroscopic Methods in Pharmaceutical Solid-state Characterization*, J. M. Chalmers and G. Dent, pp. 95–138 in *Polymorphism in the Pharmaceutical Industry*, ed. R. Hilfiker, Wiley-VCH Verlag GmbH & Co. KGaA, Weinheim (2006).

may also be selected so as to avoid or minimise interference from fluorescence. Raman is of particular interest to bio-spectroscopy as water is an extremely weak Raman scatterer, and it is thought that Raman may become the technique of choice for *in vivo* diagnostics. Raman is however an intrinsically weak effect and thus there are significant technological challenges facing its development for diagnostic purposes. The Chapter 4 gives a more thorough description of the origins and history of the Raman effect and its applications for diagnostic imaging in comparison to FT-IR.

References

1. J. M. Hollas, *Modern Spectroscopy*, 3rd ed., John Wiley & Sons Ltd., Chichester, 1996.
2. H. F. Shurvell, *Spectra-Structure Correlations in the Mid- and Far-IR*, in *Handbook of Vibrational Spectroscopy*, eds. J. M. Chalmers and P. R. Griffiths, **Vol. 3**, John Wiley & Sons Ltd., Chichester, 2002, pp. 1783–1816.

3. D. W. Mayo, F. A. Miller and R. W. Hannah, *Course Notes on the Interpretation of Infrared and Raman Spectra*, John Wiley & Sons, Inc., Hoboken, 2004.

4. P. R. Griffiths and J. A. de Haseth, *Fourier Transform Infrared Spectrometry*, 2nd ed., John Wiley & Sons, Inc., Hoboken, 2007.

5. A. J. Sommer, *Mid-infrared Transmission Microspectroscopy*, in *Handbook of Vibrational Spectroscopy*, eds. J. M. Chalmers and P. R. Griffiths, **Vol. 2**, John Wiley & Sons Ltd., Chichester, 2002, pp. 1369–1385.

6. G. P. Williams, *Synchrotron and Free Electron Laser Sources of Infrared Radiation*, in *Handbook of Vibrational Spectroscopy*, eds. J. M. Chalmers and P.R. Griffiths, **Vol. 1**, John Wiley & Sons Ltd., Chichester, 2002, pp. 341–348.

7. J. M. Chalmers, N. J. Everall, K. Hewitson, M. A. Chesters, M. Pearson, A. Grady and B. Ruzicka, "Fourier transform infrared microscopy: some advances in techniques for characterisation and structure-property elucidations of industrial material", *Analyst*, 1998, **123**, 579–586.

8. L. Miller and R. J. Smith, Synchrotrons versus globars, point-detectors versus focal plane arrays: Selecting the best source and detector for specific infrared microspectroscopy and imaging applications, *Vib. Spectrosc.*, 2005, **38**, 237–240.

9. M. C. Martin and W. R. McKinney, *BL 1.4.3: 10-micron spot size achieved for high spatial resolution FTIR spectromicroscopy*, Advanced Lights Source (ALS) publication, available on http://infrared.als.lbl.gov/pubs/spotsize.pdf (accessed on 16 March 2008).

10. S. E. Fisher, A. T. Harris, J. M. Chalmers and M. J. Tobin, *Head and Neck Cancer: a clinical overview and observations from synchrotron-sourced mid-infrared spectroscopy investigations*, in *Vibrational Spectroscopy for Medical Diagnosis*, eds. M. Diem, P.R Griffiths and J. M. Chalmers, John Wiley & Sons Ltd., Chichester, 2008, pp. 123–154.

11. M. J. Tobin, M. A. Chesters, J. M. Chalmers, F. J. M. Rutten, S. E. Fisher, I. M. Symonds, A. Hitchcock, R. Allibone and S. Dias-Gunasekara, *Infrared microscopy of epithelial cancer cells in whole tissues and in tissue culture, using synchrotron radiation*, in *Applications of Spectroscopy to Biomedical Problems*, Faraday Disc. 126 2004, pp. 27–39.

12. M. J. Romeo, R. K. Dukor and M. Diem, "*Introduction to Spectral Imaging and Applications to Diagnosis of Lymph Nodes*", in *Vibrational Spectroscopy for Medical Diagnosis*, eds. M. Diem, P. R. Griffiths and J. M. Chalmers, John Wiley & Sons Ltd, Chichester, 2008, pp. 1–25.

13. M. J. Romeo, S. Boydston-White, C. Matthäus, M. Miljkovíc, B. Bird, T. Chernenko, P. Lasch and M. Diem, *Infrared and Raman Microspectroscopic Studies of Individual Human Cells*, in *Vibrational Spectroscopy for Medical Diagnosis*, eds. M. Diem, P. R. Griffiths, J. M. Chalmers, John Wiley & Sons Ltd., Chichester, 2008, pp. 27–70.

14. N. J. Harrick, *Internal Reflection Spectroscopy*, Wiley Interscience, New York, 1967.

15. F. M. Mirabella and N. J. Harrick, *Infrared Reflection Spectroscopy: Review and Supplement*, Harrick Scientific Corporation, New York, 1985.
16. J. M. Chalmers and P. R. Griffiths, *Sampling Techniques and Fiber-Optic Probes*, in *Applications of Vibrational Spectroscopy in Pharmaceutical Research and Development*, eds. D. E. Pivonka, J. M. Chalmers and P. R. Griffiths, John Wiley & Sons Ltd., Chichester, 2007, pp. 19–49.
17. S. Kawata and Y. Inouye, *Near-field Optical Microscopy*, in *Handbook of Vibrational Spectroscopy*, eds. J. M. Chalmers and P. R. Griffiths, **Vol. 2**, John Wiley & Sons Ltd., Chichester, 2002, pp. 1460–1471.
18. L. Bozec, A. Hammiche, M. J. Tobin, J. M. Chalmers, N. J. Everall and H. M. Pollock, Near-field photothermal Fourier transform infrared spectroscopy using synchrotron radiation, *Meas. Sci. Technol.*, 2002, **13**, 1217–1222.
19. A. Hammiche, L. Bozec, M. J. German, J. M. Chalmers, N. J. Everall, G. Poulter, M. Reading, D. B. Grandy, F. L. Martin and H. M. Pollock, Mid-infrared microspectroscopy of difficult samples using near-field photothermal microspectroscopy, *Spectroscopy*, 2004, **19**(2), 20.
20. G. L. Carr, O. Chubar and P. Dumas, *Multichannel detection with a synchrotron light source: design and potential*, in *Spectrochemical Analysis Using Infrared Multichannel Detectors*, eds. R. Bhargava and I. W. Levin, Blackwell Publishing Ltd., Oxford, 2005, pp. 56–84.
21. *Infrared and Raman Spectroscopy Methods and Applications*, ed. B. Schrader, VCH Verlagsgesellschaft mbH, Weinheim 1995.
22. Authors various, in *Handbook of Vibrational Spectroscopy*, Vols. 1–3, eds. J. M. Chalmers and P. R. Griffiths, John Wiley & Sons Ltd., Chichester 2002.
23. S. H. Yan and S. Wallon, *Human Breast Cancer Identification by Near Infrared (NIR) Spectroscopy and Radiorespirometry*, IM Publications, LLP, Chichester, 2009.
24. P. R. Griffiths, *Far-infrared Spectroscopy*, in *Handbook of Vibrational Spectroscopy*, eds. J. M. Chalmers and P. R. Griffiths, **Vol. 1**, John Wiley & Sons Ltd., Chichester, 2002, pp. 229–239.
25. J. M. Chalmers and G. Dent, *Vibrational Spectroscopic Methods in Pharmaceutical Solid-state Characterization*, in *Polymorphism in the Pharmaceutical Industry*, ed. R. Hilfiker, Wiley-VCH Verlag GmbH & Co. KgaA, Weinheim, 2006, pp. 95–138.
26. C. J. Strachan, T. Rades, D. A. Newnham, K. C. Gordon, M. Pepper and P. F. Taday, Using terahertz pulsed spectroscopy to study crystallinity of pharmaceutical materials, *Chem. Phys. Lett.*, 2004, **390**, 20–24.
27. J. A. Zeitler, K. Kogermann, J. Rantanen, T. Rades, P. F. Taday, M. Pepper, J. Aaltonen and C. J. Strachan, Drug hydrate systems and dehydration processes studied by terahertz pulsed spectroscopy, *Int. J. Pharm.*, 2007, **334**, 78–84.
28. Y. C. Shen, T. Lo, P. F. Taday, B. E. Cole, B. W. Tribe and M. C. Kemp, Detection and identification of explosives using terahertz pulsed spectroscopic imaging, *Appl. Phys. Lett.*, 2005, **86**, 24116.

29. M. C. Kemp, P. F. Taday, B. E. Cole, J. A. Cluff, A. J. Fitzgerald and W. R. Tribe, Security applications of terahertz technology, *Proc. SPIE-Int. Soc. Opt. Eng.*, 5070, **44–52**, 2003.
30. P. F. Taday and D. A. Newnham, Technological advances in terahertz pulsed systems bring far-infrared spectroscopy into the spotlight, *Spectrosc. Eur.*, 2004, **16**(5), 22–25.
31. T. L. Threlfall and J. M. Chalmers, *Vibrational spectroscopy of solid-state forms – Introduction principles and Overview*, in *Applications of Vibrational Spectroscopy in Pharmaceutical Research and Development*, eds. D. E. Pivonka, J. M. Chalmers and P. R. Griffiths, John Wiley & Sons Ltd., Chichester, 2007, pp. 263–292.

CHAPTER 3

Infrared Synchrotron Radiation Beamlines: High Brilliance Tools for IR Spectromicroscopy

AUGUSTO MARCELLI[a] AND GIANFELICE CINQUE[b]

[a] Istituto Nazionale di Fisica Nucleare, Laboratori Nazionali di Frascati, Via Enrico Fermi, 40, I-00044 Frascati, Italy; [b] Diamond Light Source, Harwell Science and Innovation Campus, Chilton, Didcot, Oxon OX11 ODE, UK

3.1 Introduction

In the last decade we have witnessed impressive progress in the applications of infrared (IR) microspectroscopy, especially in the field of life sciences and more specifically for medical diagnosis. Different initiatives, meetings and conferences have been dedicated to the study of IR spectroscopy applied to investigate or to classify single cells or tissue sections. The key tools for such research are the IR microscope used for imaging and Fourier transform IR interferometer (FTIR) for spectral analysis.

In this context, the use of synchrotron radiation (SR) as a photon source has been suggested because of its brilliance advantage, namely in the range of two to three orders of magnitude through small apertures at the sample position, over conventional blackbody (Globar) sources that equip most commercial FTIR microscopes.

RSC Analytical Spectroscopy Monographs No. 11
Biomedical Applications of Synchrotron Infrared Microspectroscopy
Edited by David Moss
© Royal Society of Chemistry 2011
Published by the Royal Society of Chemistry, www.rsc.org

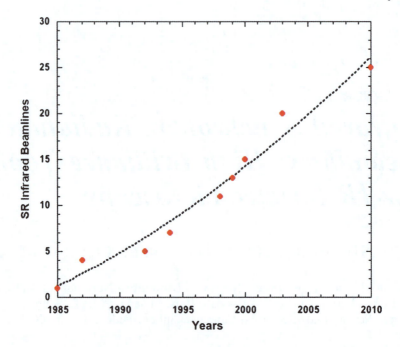

Figure 3.1   Geometrical increase in the total number of infrared (IR) beamlines
worldwide versus year of operation (excluding synchrotron radiation
facilities that have been shut down).

With respect to Globar sources, the high SR brilliance is thus expected to
enhance the spatial resolution while still maintaining a high signal to noise
ratio, or to allow work on a broader spectral range while using a diffraction-
limited IR microprobe. Indeed, the major contributions that SR may offer are
the highest brilliance[i] and spatial resolution for a broadband IR source, with
the further advantage of time resolution. In practice, IR synchrotron light
sources allow collection of high quality images with a lateral resolution below
the size of a human cell within minutes without any damage to the system under
analysis, an essential requirement for all biological studies.

In Europe several synchrotron IR beamlines are operational at the existing
SR facilities of ANKA (Karlsruhe), BESSY (Berlin), DAΦNE (Frascati),
DIAMOND (Oxford), ELETTRA (Trieste), ESRF (Grenoble), MAX (Lund),
PSI (Villigen) and SOLEIL (Paris). Many others are available in the USA/
Canada, at NSLS (Brookhaven), CSR (Madison), CLS (Saskatoon), ALS
(Berkeley), and in Asia/Oceania in Australia, China, Japan, Korea and Sin-
gapore. The growth observed in recent years (Figure 3.1) has been driven by
many concurrent demands but in particular by biomedical research.

Several recent experiments have demonstrated that, with a dedicated IR set-
up, there is a great advantage of using synchrotron radiation IR sources for

---

[i] In this chapter brilliance is used to mean the flux density per unit area and solid angle.

FTIR imaging at high sensitivity and high spatial resolution. Indeed, experiments lasting only a few minutes allow performance never achievable with any conventional broadband IR source in terms of S/N (signal to noise) on spot sizes of a few microns.

In this chapter we will try to illustrate what is meant by an IR synchrotron radiation beamline and to discuss the major advantages of these instruments for the research examples described in this chapter and in the other chapters of the book.

3.2 Infrared Synchrotron Radiation: Historical Background

About fifty SR sources are operational nowadays throughout the world, and this number is still increasing. Whether classified as second or third generation light sources, whether partially or fully SR dedicated, they represent undoubtedly the most powerful and brilliant sources in the X-ray range. Although the SR emission fully covers the microwave and IR regions, its use in this low energy domain has developed at a much slower rate. In the new millennium and after three decades of SR usage, the aim of extending the unique performance of the synchrotron source to the IR domain has been achieved by several dedicated beamlines in different countries. With their high brilliance, polarization and broadband range, SR sources allow experiments that are outside the capabilities of conventional sources from the near-IR up to the far-IR range. Observations and studies of the interaction of electromagnetic radiation with condensed matter, in the framework of the linear spectroscopy theory, contributed significantly to the understanding of, for example, solid state physics, surface science, organic and polymer chemistry, mineralogy and environmental sciences as well as biology and medicine. This is certainly true for the X-ray region of the radiation spectrum and, since the nineteenth century,[1] particularly true for a non-ionizing radiation such as IR. In fact, IR spectroscopy may probe the stretching and vibrations of molecules in the mid-IR range, also known as the IR "fingerprint" spectral region, but also phonons, excitons and polarons, *e.g.* the low energy excitations of solids, the forces between a surface and an adsorbate, and many other low-energy phenomena of fundamental importance in condensed matter, biophysics, materials science, *etc.*

The IR part of the electromagnetic spectrum is a very wide wavelength region that covers at least three orders of magnitude from 1 to 1000 μm (corresponding in wavenumber from 10000 to 10 cm^{-1}, respectively). In this region intense sources are available for the experimentalist, *e.g.* the Globar to cover the mid-IR energy region or the mercury lamp for the far-IR domain.

Although there were proposals and attempts to use SR also in the IR domain, until the mid-1980s these were the only broadband IR radiation sources available. However, spectral coverage and flux density are the main limitations of these sources, so that storage rings provide unique opportunities for experiments that require high brilliance in the mid-IR, as well as high intensity up to the far-IR domain, with the added bonuses of pulse structure for time resolved spectroscopy and/or a well-defined polarization state for ellipsometry.

In the last few decades, important technological developments in sources, spectrometers and detectors have enabled a significant improvement in spectroscopic methods for the IR region. As an example, in the 1960s a giant step forward in IR equipment was made with the introduction of Michelson interferometers coupled to the fast Fourier transform algorithm implemented on computers. With these new instruments it was possible to perform an on-line back Fourier transform, an indispensable mathematical tool to achieve an IR analysis in a broad IR range (see Chapter 2). Chemists discovered IR spectroscopy in the same period and used this rapid technique to investigate the most complex molecular structures. The physicist Gerhard Herzberg played a crucial role in the evolution of IR spectroscopy and was rewarded in 1971 with the Nobel Prize in Chemistry. Later, compact Fourier transform spectrometers working with conventional sources became available to an ever-broader scientific community. The emittivity of a Globar, which can be roughly approximated by that of a blackbody heated to 1500 or 2000 K, generally peaks at $\lambda \sim 5\,\mu m$ and decreases rapidly for $\lambda > 20\,\mu m$, the region of the far IR where mercury lamps start showing better performance. However, just a small fraction of the power emitted by a conventional source on a 4π solid angle can be effectively focused on the sample; therefore spectra of small samples collected with conventional sources are affected by a poor signal intensity as well as poor signal to noise ratio.

With the continuous developments, IR spectroscopy became the most widely used characterization technique in industry and in many technological processes, *e.g.* in food sciences. However, modern applications such as IR microspectroscopy require intense, brilliant and broadband sources, so that research has been oriented towards new IR sources. Actually, Fourier transform spectroscopy strongly benefits from a broadband non-thermal source such as SR that not only fulfils these requirements but also exhibits well-defined polarization properties and time structure.

Although suggested for the first time in 1966,[2] and then in the early 1970s in the USA, the use of infrared Synchrotron Radiation (IRSR) developed at a slow pace. The first attempts at extending the use of the synchrotron source to the IR region date back to the 1970s, when pioneering observations were made at Stoughton by Stevenson and coworkers,[3] and at Orsay by Lagarde and coworkers (Figure 3.2).[4]

However, only at the beginning of the 1980s was a port dedicated to the extraction of IRSR built by Yarwood on the SRS ring at Daresbury.[5] One of the young members of the original team at Daresbury (Figure 3.3), Takao Nanba, returned to Japan in 1985 and started the construction on the UVSOR ring, one of the first SR dedicated rings, of the first IRSR beamline that later opened to users.[6,7]

Almost contemporary measurements of the characteristics of SR in the IR region were performed at BESSY.[8,9] A couple of years later, in 1987, Gwyn

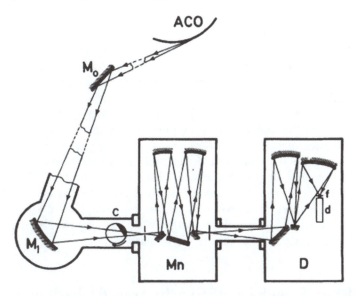

Figure 3.2 Schematic of the early ACO IR beamline set-up with mirrors (splitting M0 and focussing M1), rotating chopper finger (C), grating monochromator system (Mn) and detector module (D) including filter (f) and Golay cell (d). (Reprinted from ref. 4)

Figure 3.3 Photograph of the Yarwood infrared team at Daresbury. Courtesy of T. Namba (first from the left).

Williams opened the first IR beamline in the USA, at the NSLS ring at Brookhaven.[10] This initiative triggered a rapid development of the IRSR in the USA. The first European IRSR beamlines were realized in the early 1990s at MAX (Lund) by Bengt Nelander,[11] and at SUPERACO at Orsay by Pascale Roy and coworkers.[12] At Daresbury, a new IR group led by Michael Chester restarted IRSR research at SRS.[13]

In the early 1990s when a new project for a low energy, high current collider was approved in Frascati, the construction of a new IRSR beamline known as SINBAD (*Synchrotron INfrared Beamline At DAΦNE*) was proposed.[14,15] The new collider DAΦNE, which was designed to work with high current (>1 A) and at low energy (0.51 GeV), is indeed an ideal source of IR radiation. Infrared SR is now growing at a rapid rate and an increasing demand for new IR beamlines is emerging all over the world. Nowadays there are more than twenty dedicated IRSR beamlines in the world (see Figure 3.1) with a continuously increasing number of interdisciplinary users.

3.3   Basic Principles of Synchrotron Radiation

Research using SR started at the beginning of the 1960s with pioneering experiments using electron synchrotrons both at Frascati (1.1 GeV accelerator) in Europe and at the National Bureau of Standards in the USA (180 MeV accelerator) at low energies, *i.e.* in the soft X-rays and VUV (vacuum ultraviolet) domains.

A synchrotron is a particular type of cyclic particle accelerator in which both the magnetic field (to deflect particles) and the electric field (to accelerate particles) are synchronized with bunches of charged particles travelling in a closed orbit. By increasing these two electromagnetic parameters appropriately as the particles gain energy, their path can be held constant as they are accelerated. One of the main disadvantages of such accelerators as radiation sources was their instability, which resulted from the continuous injections. A step forward in machine accelerators occurred with the construction of e^+/e^- storage ring colliders.

The development of more stable accelerators encouraged first the construction of parasitic SR facilities and then fully dedicated experimental facilities. Dedicated storage ring accelerators are now characterized by the use of straight sections between bending magnets, with the overall shape of a round-cornered polygon. These fully dedicated SR sources opened at the end of the 1980s and hosted the first insertion devices, special multipole magnet sources called wigglers and undulators.[16] Neglecting the emission of insertion devices, the maximum energy contained in the spectrum of a cyclic accelerator is typically limited by the strength of the magnetic field(s), the minimum radius (maximum curvature) of the particle path and the energy of the accumulated particles.

The formula that gives the behaviour of the maximum radiated energy (frequency) as a function of the orbital energy is similar to Planck's law for a blackbody thermal source. However, the behaviour of blackbody thermal

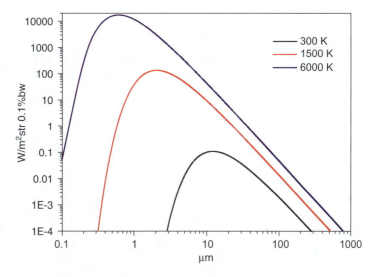

Figure 3.4 Blackbody power density emission according to Planck's law for room temperature, typical Globar of 1500 K and the temperature of the sun, ∼ 6000 K.

radiation is quite different from that of non-thermal radiation (Figure 3.4). In fact, as the temperature changes, the emission of a blackbody is described by a family of curves that do not have a common limit.[17]

Conversely, SR is a non-thermal source whose emission is polarized (in the orbital plane) and described by a family of curves with a characteristic critical energy, all having a common limit at low energy. This particular property has practical importance because at a fixed wavelength, the photons emitted in the IR and visible parts of the spectrum are independent of the energy of the electrons circulating in a storage ring for orbital energies above 500 MeV. Consequently, at IR wavelengths the energy fluctuations due to instabilities do not affect the photon emission, and an accelerator, as soon as the number of orbiting electrons remains constant, is an absolutely standard radiation source.

Synchrotron radiation is a continuous radiation source, and photons at all energies are available at the same time for time resolved spectroscopy and/or polarized experiments from the far-IR up to the X-ray region.

3.3.1 Synchrotron Radiation Properties

Among the various properties of the SR emission that hold true for the entire domain of emission, we may underline brilliance, collimation, polarization, stability and time structure. In addition, SR is a continuous source from the low to the high-energy domain and all wavelengths are available simultaneously for experiments in a wide IR range.

3.3.1.1   *Brilliance*

Brilliance is the photon flux density (photons/s/mA/mrad horizontal 0.1% bandwidth) taking into account the electron beam divergence, the opening radiation angle and the electron beam source size (flux/mrad vertical/mm^2). The gain in brilliance of an IRSR source with respect to a conventional blackbody source such as a Globar is at least two orders of magnitude in the mid-IR. A practical limitation to the theoretical brilliance achievable is given by the horizontal angular acceptance of any beamline front end, which in practice means that an SR source can be diffraction limited only in the vertical plane (see Figure 3.5).[10]

A Globar is essentially a blackbody source whose emission power is a function of the temperature of the emitter. Using the XOP package,[18] the flux of a blackbody, with 1 mm^2 source area and 1 mrad(H)×1 mrad(V) angular divergence, can be calculated. As the temperature increases, the maximum flux shifts to high energy. If we consider a thermal source working at 1800 K, the corresponding flux in the IR at 1000 cm^{-1} (10 μm) is ∼7×10^6 phot/sec/0.1%b.w., several orders smaller than that of IRSR. However, to perform spectromicroscopy IR, the photon brilliance is more important than the flux (Figure 3.6).

In a practical case, using a bending magnet source this quantity is a function of the wavelength λ and varies with the characteristics of the optical elements and with the horizontal angle of acceptance.[19] The latter is determined by the

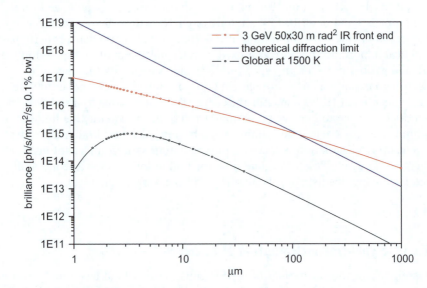

Figure 3.5   Calculated synchrotron radiation (SR) brilliance of a 3 GeV SR source (IR front end 50×30 mrad^2) in comparison to a conventional source and the ideal SR diffraction limited source (the latter for bending radiation only). (Source: DLS, Design Report of the IR beamline.)

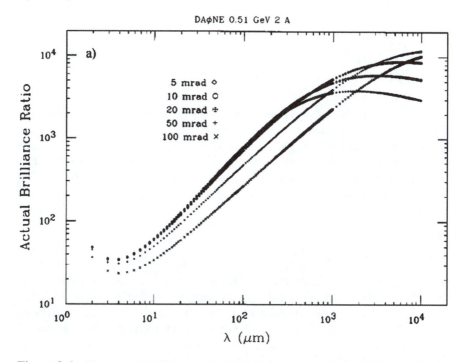

Figure 3.6 The actual brilliance ratio (ABR) function of the DAΦNE source, as calculated for different values of the horizontal collecting angle. (Reprinted from ref. 19.)

size of the exit port placed in front of the magnet that defines the length of the arc of electron trajectory that contributes to the emission. At long wavelengths the brilliance gain is of the order of 10^3 or even more. If we consider the losses of the beamline that transfers the radiation under vacuum to the sample, *e.g.* the reflectivity and the transmission of the installed mirrors and windows respectively, and the absorption due to the residual gas along the radiation path, as well as other geometric constraints that limit the optical transmittance of the beamline, an actual reduction of one order of magnitude of the gain compared with a conventional source, measured at the sample position, is realistic. However, one easily obtains, in the far-IR, a gain that ranges from 10 to 100 for samples smaller than 1 mm in diameter, and this is a great advantage for all experiments that involve IR microscopy when the actual brilliance ratio of SR versus Globar is calculated (see Figure 3.6).

3.3.1.2 Collimation

The brilliance of SR is not constant in the entire energy domain of emission because it is determined by the combined effect of the horizontal and vertical collection angle and of the natural divergence θ_{nat} of the SR. In the IR domain the latter is much larger than in the UV or X regions and depends on the photon wavelength λ through the relation:

$$\theta_{nat} = 1.66(\lambda/\rho)^{1/3} \tag{3.1}$$

where ρ is the radius of curvature of the electron trajectory.[20] Equation (3.1) shows that the divergence is already larger than 10 mrad at $\lambda = 1$ μm and that it increases significantly from the near-IR to the far-IR. Collection of this large emission requires large optical elements and a complex optical layout first to reject the high energy radiation and then to transport only the IR radiation from the exit port to the experimental area, usually located several metres from the magnet.

3.3.1.3 Polarization

Owing to its relativistic character, SR emission is linearly polarized in the orbital plane. However, since the early pioneering work in the visible region,[21] several investigations performed both in the VUV and X-ray regions demonstrated that an observer collecting the emission above and below the orbital plane see it as elliptically polarized. When observed at a particular angle over the orbit, the radiation is circularly polarized.

The polarization properties of IRSR are very promising for many applications, in particular in the far-IR region. By placing a slit on the exit port, in principle one can then select the desired degree of circular polarization and the flux of the emitted radiation. Dealing with polarization, we may define three polarization rates, of which two are more important: the linear polarization rate (P1) and the circular polarization rate (P$_3$), defined in terms of the corresponding photon fluxes (Figure 3.7). As an example for the case of SINBAD, experimental observations indicate that $> 80\%$ circularly polarized light can be obtained by a slit that selects 50% of the total flux available. Interesting applications have been already explored successfully, such as the extension to the IR of circular dichroism studies in magnetic materials,[22,23] and many other vibrational circular dichroism effects can be revealed by such intense and brilliant sources, also in materials of biological interest.

3.3.1.4 Stability

Another remarkable feature of IRSR in comparison with a thermal source, whose intensity varies at random, is that the power delivered is intrinsically stable in time, and, as outlined above, for an energy E of the electrons > 0.5 GeV is directly proportional to the current circulating in the beam. The circulating current can be monitored in real time and there is a trend to operate SR dedicated storage rings with a constant current (top up mode) so that IRSR provides the spectroscopists with an absolute radiation source.

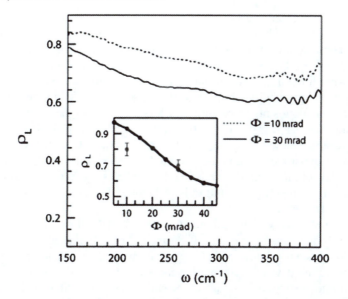

Figure 3.7 The linear polarization degree ρ_L of the IR synchrotron radiation at the sample site measured at DAΦNE in the far-IR region. The inset shows a comparison between ray-tracing simulation results and experimental data. [Reprinted from ref. 24.]

On the other hand, the IRSR spot or intensity at the sample may fluctuate if spatial instabilities of the electron beam occur. In fact, beam fluctuations are important in the IR region where the source size is not diffraction limited,[15] and this value depends on the parameters of the storage ring and in particular on the energy of the ring. Fluctuations may be amplified also by optical systems and may affect critical applications such as IR microscopy, with regard to the sample illumination position, or FTIR spectroscopy, with regard to the spectral noise (a time-modulated noise, after being back Fourier transformed, is "seen" as a spectral component or more generally as artefacts which cause a S/N degradation). This problem can be solved by an optical feedback that automatically corrects the electron orbit by acting on the electromagnetic fields along the ring, and modern storage rings exhibit unprecedented observed stabilities. It is also worth mentioning the active feedback system approach using piezo-controlled optics and a beam position monitor to cope with electron orbit instabilities at the lower end of the frequency range.[25]

3.3.1.5 Time Structure

Another significant advantage for experiments in the IR domain is the pulsed structure of the IRSR. Electrons in storage rings travel in *bunches,* which have lengths ranging from 1 to 10 mm. When *bunches* are diverted by a magnetic field, *e.g.* in a dipole magnet or a wiggler, they radiate, producing short pulses of light from tens of picoseconds up to a few nanoseconds depending on the ring and its mode of operation. These light pulses are suitable for investigating time-dependent phenomena in various systems up to the nanosecond time scale.[26,27] Even if IR lasers can produce shorter pulses at much higher power levels, only IRSR exhibits a continuum spectral output in the entire IR domain and can provide the shot-by-shot high stability required for spectroscopy. IRSR can then be used for pump-probe experiments using standard FTIR spectroscopic techniques. Actually, using Fourier transform spectroscopy and photovoltaic detectors in the mid-IR region, a time resolution better than 100 µs can be achieved.

Real time experiments on non-reproducible phenomena are then possible on the time scale of milliseconds, while in the so called step-scan mode much faster processes can be investigated, but only if the phenomena under study are cyclic and can be triggered.[28]

This opportunity will be extended to the far-IR when fast detectors are available. Indeed the typical cut-off frequency of a liquid-helium cooled bolometer is presently below 500 Hz, even if much faster detectors based on superconducting devices are being developed.

In the mid-IR region fast detectors with a response time of hundreds of picoseconds are already available and the IR emission can also be used with success for the diagnostic investigation of the accelerators.[29]

3.4   What is an SR Beamline?

The most common sources of IRSR are bending magnets. Nonetheless one beamline, at Orsay at SUPERACO, was designed to extract radiation from an insertion device and operated for several years. In fact, charges entering and exiting a magnet of an electron storage ring emit "edge radiation" at wavelengths that are long compared with the critical wavelength of the SR emitted by a bending magnet.[30] Far-field edge radiation is emitted along the straight section axis in a hollow cone with its maximum at an angle $\theta \approx 1/\gamma$ where γ is the relativistic mass factor (Figure 3.8). This radiation may also be used as a bright IR source. Several IR beamlines utilize the edge radiation nowadays, either exclusively (ANKA, ALS), together with bending radiation (Australian Synchrotron, Soleil, Diamond) or partially (Elettra).[31–34]

In the case of a dipole magnet, the considerable size of the source and the large intrinsic divergence of IRSR require an optical layout including large optics to transfer and focus the beam to the entrance pupil of an interferometer or spectrometer. The standard extraction system of IRSR from the bending magnet of a storage ring is based on a flat mirror, placed in front of the exit port, that deflects the radiation cone on a focusing mirror and rejects the large associated X-ray flux and heat load (Figure 3.9). In this way a large solid angle can be collected, *e.g.* 90×90 mrad at Brookhaven (U4IR port). If the electron beam energy E is high (> 1 GeV), the thermal load due to the X-ray flux on the first optical elements has to be considered. Indeed the power density associated with the "hard" X-ray radiation in the central region of the extraction mirror is consistent.

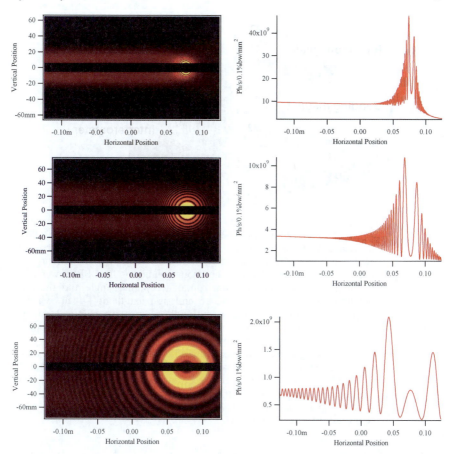

Figure 3.8 IR emission from Diamond IR bending magnet. From top, simulations by SRW[35] at 2 µm, 10 µm and 100 µm wavelengths. Left: 2-D intensity plots. Right: linear horizontal intensity traces at the vertical position of maximum bending radiation intensity, namely 9 mm, 20 mm and 40 mm, respectively. The blind region at the centre is due to the first mirror slot of ~ 10 mm vertical size.[35] (Source: DLS, Design Report of the IR beamline.)

The first mirror has to be cooled and may also be partially shielded by an absorber when the beam energy is very high or the distance from the source is short. The second mirror, typically a toroid or an ellipsoid, focuses the radiation on a window that isolates the first section of the beamline, in ultra-high vacuum because of the open connection to the storage ring, from the second section. The early beamlines were equipped with windows made with expensive natural diamonds, type IIa, typically below 2 cm^2 in area. Diamond is chosen for its excellent hardness and chemical stability, and for its "flat" transmittance in the whole IR domain (except for an absorption band at ~ 2000 cm^{-1}, *i.e.* ~ 5 µm). Nowadays, the availability of large synthetic CVD diamond films makes the use of diamond much simpler. Moreover, synthetic diamond can be

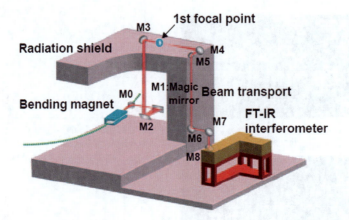

Figure 3.9 Schematic view of the front end of BL43IR. The first plane mirror M0 was located in the crotch chamber 2.66 m from the centre of the source with acceptance angles of 36.5 mrad (H) and 12.6 mrad (V). The mirror has a narrow slit of 2 mm width in the orbital plane to allow SR power to go through the slit. The large horizontal acceptance angle and the large bending radius of 39.3 m give an emission length of 1.44 m, which is impossible to focus using conventional spherical or elliptical mirrors. A magic mirror M1 located 4.4 m from the source reflects the photon beam and focuses above the ceiling of the radiation shield. A long beam pipe between the plane mirrors is used as an acoustic delay line and a wedged diamond window separates and protects the UHV section of the storage ring from the low vacuum section of the beam transport line. The SR beam is transported to the FTIR interferometer located 20 m from the source by parabolic mirrors (M4, M8) and plane mirrors (M5, M6, M7).[36] (Courtesy of Taro Moriwaki)

grown with a wedge angle of the order of 1° or more, in order to suppress multiple reflections.[37] After the diamond window (DW) the IR light can be transmitted to the sample by the optical system in a lower vacuum regime for reducing the IR absorption of air, and in particular of water vapour and carbon dioxide which are strong IR absorber molecules (Figure 3.10).

As already mentioned, bending magnets are not the only sources of IR radiation. Interesting IR sources are also the edge of a magnet, where the field experienced rapidly drops to zero, and a wiggler or an undulator. Unlike uniform dipoles, due to coherence effects, different emission patterns can be associated with each source.[12,31] For $\lambda \gg d/\gamma^2$, where d is the length required to deflect an electron through an angle of the order of $1/\gamma$, the edge radiation may be even brighter than that emitted from the uniform magnetic field region.[32–34] Similar effects are found at both edges of an undulator.[38] The first beamline designed by Bosch to exploit the edge effect is operational at the Aladdin storage ring of Stoughton (USA), as well as at ANKA, in Karlsruhe.[39] For all beamlines where the exit port views only a central portion of the magnet, the edge effect is negligible and the intensity can be reliably calculated by the use of ray tracing simulation of the electron source. As an example, the ratio

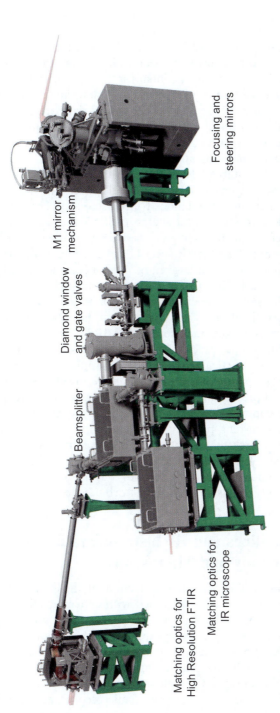

Figure 3.10 Schematic of the Australian Infrared beamline showing (from right) synchrotron beam entering front end optics (M1, M2, M3, M3a mirrors), diamond exit window, beamsplitter optics vessel and matching optics boxes for the two end station instruments. (Courtesy of Mark Tobin.)

$I_{SR}(\omega)/f_{BB}(\omega)$ of the SR intensity with respect to a blackbody (BB) has been calculated *vs. d,* and compared with the corresponding calculations of the actual brilliance ratio.[19] Data were normalized to a beam current of 1 A under the assumption that the intensity is linear with the current. Therein, BB is a mercury lamp for $\lambda = 100\,\mu m$ and a Globar for $\lambda = 2\,\mu m$. The data show that at $100\,\mu m$ $I_{SR}(\omega)/I_{BB}(\omega) \sim 2$ for a full aperture ($\sim 4\,mm$) and approaches a maximum value of 20 for $d = 1\,mm$. The experimental data are even higher than theory; however the agreement is quite satisfactory. Deviations observed in the far-IR for the smallest apertures are due to diffraction effects, which are hardly reproduced by ray tracing simulations.

In spite of the slow start of IR research with SR and the competition of available intense and compact IR conventional sources, the interest in IRSR research and applications is continuously increasing all over the world.

What are the main reasons for the success of IRSR in recent years? Actually, as emphasized before, in the IR region the energy of the electron beam (for $E > 0.5\,GeV$) does not affect the SR spectral distribution, while the intensity is substantially proportional to the large horizontal opening angle and the current circulating in the storage ring. As a consequence almost all rings are equivalent in terms of IR emission, while current and stability are more significant and both have increased significantly in recent years to have become the qualifying parameters. In fact, the experiments run using IRSR beamlines have become the most demanding ones, such as high-pressure studies, high resolution microspectroscopy, in particular for biological applications, reflectance and absorption spectroscopy in surface science, time resolved spectroscopy, *etc.*

To summarize, at present an ideal IR source to be used for IR spectroscopy and microspectroscopy is a low emittance, stable low energy storage ring with the highest possible current, possibly working in top-up mode, to perform the most demanding research that is not possible with conventional sources.

The top-up mode of operation is certainly beneficial for IR spectroscopy in terms of constant photon flux at the samples, but it is questionable for FTIR spectroscopy stability unless coupled to fast orbit feedback for e-orbit.

3.5 SR Beamlines and IR Instrumentation for Spectroscopy and Microscopy

The unique features of IRSR allow collection of higher quality data in a number of experiments usually performed with conventional sources. Moreover, IRSR has expanded the limits of IR spectroscopy and has opened new fields of application, where the severe experimental conditions prevent the use of standard sources. Among the areas that have benefitted from the availability of SRIR facilities, we have firstly to mention microscopy. Since Marcello Malpighi, who was considered the pioneer of histology in Bologna and Pisa during the seventeenth century using optical microscopes, advances in this field have always involved the search for new methods to obtain high contrast images, and three Nobel prizes have been awarded to discoveries related to the microscope. Although in the mid-IR range the gain in brilliance of an SR is clear and beneficial for IR microspectroscopy, at longer wavelengths, *i.e.* in the far-IR or even THz region, IRSR sources are the most powerful broadband sources for IR spectromicroscopy.

3.5.1 IR Spectromicroscopy

(See also Chapter 2, Section 2.3.2, and Chapter 7.) Infrared microspectroscopy is a unique technique that combines microscopy and spectroscopy for the purposes of microanalysis. Spatial resolution, within a microscopic field of view, is the goal of the modern IR micro-spectroscopy for applications in condensed matter physics, materials science, biophysics and now in biomedicine (Figure 3.11). Infrared microscopy is a non-destructive analytical and imaging technique, which achieves contrast via the intra-molecular vibrational modes. The method is the same as that used in X-ray microscopy, where contrast is achieved by recording spectra before and after the absorption edges of an element contained in the specimen. The gain in brilliance of IRSR with respect to blackbodies is remarkable at shorter wavelengths, where an IR microscope can work without experiencing major intensity problems. The use of IR microscopes coupled to SR sources has rapidly given outstanding results because the high brilliance of this source allows one to reduce the apertures of the instrument to define geometrical areas of a few microns in the mid-IR region, *e.g.* at the diffraction limit, with good S/N.[40]

The real goal of FTIR microscopy, in particular when using an SR source, is to achieve the best spatial resolution at the sample location, *i.e.* the micrometric scale, for example to allow true single cell analyses. Infrared Mercury Cadmium Telluride detectors (*i.e.* MCT single element detectors) optimized in the mid-IR range are characterized by a high responsivity and a typical cut off at $\sim 500\,\mathrm{cm}^{-1}$ or higher, depending on the doping that optimizes either the spectral range or the sensitivity. A confocal microscope may generate 2D spectral maps with a single element detector using an automated sample stage capable of accurate positioning.[42,43] Although they have a low throughput, the use of small apertures achieves images with a high spatial resolution. The drawback of the use of small pinholes or small rectangular apertures is the degradation of the S/N ratio and the need to increase the collection time of each spectrum and of the associated image. In fact, time and spatial resolution are inversely correlated parameters and thus have to be balanced in order to optimize acquisition with a single element detector. The parameter to monitor is the signal to noise ratio:

$$S/N = SP/NP \tag{3.2}$$

where SP and NP are the source power and the noise power, respectively, both measured in watts.[44] Actually, NP can be determined using the formula.

$$NP = (A_d/t)^{1/2}\,1/D^* \tag{3.3}$$

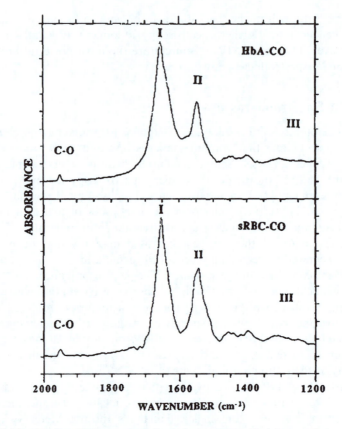

Figure 3.11 Globar FTIR spectra of a solution of purified human haemoglobin A
and the first spectrum of a single erythrocyte in aqueous media exposed
to carbon monoxide.[41] Upper panel: The spectrum of purified HbA
carbonyl (8 mM) at $2\,cm^{-1}$ resolution. Lower panel: The spectrum of a
single adult human red cell at $8\,cm^{-1}$ resolution. The stretch band for
bound CO and the amide I, II and III bands appear near 1950, 1650,
1550 and $1300\,cm^{-1}$, respectively. (Reprinted from ref. 41.)

where A_d is the detector area and D^* is the specific detectivity. If we consider the
source power constant, the S/N ratio increases as the noise power decreases, *i.e.* it
increases on reducing the detector area and/or increasing its detectivity. The
typical size of a single element detector for IR microscopy ranges generally
between $250\,\mu m \times 250\,\mu m$ and $50\,\mu m \times 50\,\mu m$ to better match the standard apertures of IR microscopes and objectives with magnification typically between $15\times$
and $36\times$. As a consequence, working with a conventional source at high spatial
resolution, *e.g.* with an aperture $\leq 20\,\mu m \times 20\,\mu m$ thus smaller than the detector
area, the S/N strongly decreases while the detector noise remains constant owing
to a significant reduction in the photon throughput but an intrinsic constant
detector area. In fact, neglecting possible non-linear contributions, if we consider
a detector of $100\,\mu m \times 100\,\mu m$ area with a microscope aperture of $10\,\mu m \times 10\,\mu m$,

the IR flux illuminating the detector is reduced by two orders of magnitude while its NP is constant. In this case, looking at Eqn (3.2), working with a better source is not useful. If we define the source power as that reaching the detector in an FTIR instrument, we may write:[44]

$$SP(\lambda) = B(\lambda)\, \xi \theta \Delta v \qquad (3.4)$$

where $B(\lambda)$ is the brilliance of the light source, practically defined as the source energy per unit of the six-dimensional phase space volume it occupies (flux/mm^2/mrad^2), θ is the instrumentation experimental throughput or the solid angle acceptance (or *etendue*), Δv is the resolution bandwidth, and ξ is the optical efficiency. Equation (3.4) is the integral of the brilliance over the acceptance coordinates of a spectrometer. If we consider that SR brilliance is two or three orders of magnitude greater than that of a standard IR source, this brilliance advantage is globally lost while working at high spatial resolution using single element detector of standard dimensions, and the best actual SR spatial resolution performances are limited to $3\,\mu m \times 3\,\mu m$.[45] Because the size of a non-thermal SR source is naturally small, the radiation is emitted in a narrow angular range ($\sim$mrad), yielding a high throughput at small aperture sizes. As a consequence, although they have not been designed for these sources, SR-based FTIR microscopy systems perform best, owing to the higher brilliance of an SR source and the lack of thermal noise. In particular, the achievable spatial resolution, previously limited by the brilliance, now can became diffraction-limited.[40,46]

Using a confocal microscope, an IR spectrum is collected using two sets of apertures: one placed before and one after the sample to define a small area with negligible contributions from the surroundings. Energy from outside the sample area may still reach the detector and contribute to the signal. Considering the optical system and the wavelength of the measuring radiation, the aperture(s) play an important role in defining the effective spatial resolution of an instrument. In fact, a single point IR microscope, working with a dual aperture confocal layout, achieves a better resolution than an imaging system without apertures (see the following section). However, it is clear that a trade-off exists between spatial resolution wavelength, signal to noise ratio and collection time, and the acquisition time is also a relevant parameter in image collection because it can be associated with the dose released to the sample and its dynamics.

In order to evaluate roughly the experimental capabilities, the theoretical optimum spatial resolution of a confocal IR microscope with Cassegrain optics is about λ/NA.[47] Measurements performed with standard resolution targets or with a knife-edge confirmed that the spatial resolution of all imaging systems is governed by both the wavelength and the numerical aperture (NA) of the microscope. Combining a confocal microscope with an SR source, the spatial resolution may be increased as close as possible to the ideal diffraction limit ($1.22\,\lambda/D$ for an "ideal lens" of diameter D). However, the diffraction limit is only one of the parameters to consider when determining the experimental spatial resolution. In addition, mirror objectives, shape and size of apertures, and detector geometry and size contribute to the effective spatial resolution as determined by source brilliance and sample characteristics.

Moreover, the usual concept of lateral spatial resolution is associated with a resolution criterion, *i.e.* the ability to record the separation of two closely spaced objects inside a sample. As an example, using the Rayleigh criterion, two points can be separated when the central maximum of the first Airy disk is placed at the distance of just one radius. However, under ideal conditions this criterion is valid only when the two points have the same intensity, a condition that corresponds to a minimum contrast of 26.4%.[48] Again, one should consider that the spatial resolution is only a parameter whose relationship with the contrast is fundamental for imaging, notably the correlation with the contrast in the sample, because it is degraded by the contrast transfer function (CTF) of the imaging system, a function that depends on the spatial frequency. Image contrast and spatial resolution can be replaced by the concept of visibility, *i.e.* the ability to recognize two closely spaced features inside a sample as being separate components. The latter is directly related to the S/N ratio and can be used to determine and interpret patterns of intensity that fulfil statistical requirements.[49] The concept is particularly useful when optical aberrations affect the image and/ or when the resolution becomes a critical parameter, as in the investigation of cells and tissues where the level of morphological heterogeneity is critical (Figure 3.12).[50]

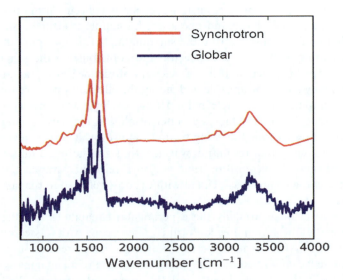

Figure 3.12   IR absorbance spectra (after vector normalization and in arbitrary units) obtained via SR (in red) or conventional source (in blue) from a single cell area of $5\,\mu m \times 5\,\mu m$. Experiment performed at Diamond beamline B22 in reflection mode using the $74 \times$ objective, resolution $4\,cm^{-1}$ and 128 scans.

3.6 Synchrotron Radiation and Imaging IR

(See also Chapter 7.) Fourier-transform infrared (FTIR) synchrotron radiation (SR) microspectroscopy is a powerful molecular probe of biological samples at cellular resolution. In complex systems, such as biological systems, the IR spectrum is the sum of the contributions gathered from all biomolecules. The global molecular information contained in a sample and provided by the FTIR analysis, in particular in the mid-IR region, is a critical advantage for characterizing biological sample contents.[51] Furthermore, no internal standard, contrast agent or even staining is required to obtain high quality spectra from a biological sample. At present a synchrotron source has the capability of providing IR radiation through a pinhole of a few microns two to three orders of magnitude higher than a conventional Globar such as those available in commercial FTIR instrumentation.[24,52] This superior S/N ratio allows imaging with a spatial resolution down to the diffraction limit, or analysis of thicker samples while maintaining a high spatial resolution.[44]

As a consequence, among the analytical techniques able to provide molecular imaging of a biological system, the availability of FTIR microscopy combined with SR sources is completely changing the existing scenario.[53] These FTIR microscopes are now available in almost all SR facilities around the world and a wide range of applications has been proposed, from the analysis of single cells[54] to complex tissues,[55] and from the characterization of the physiological status of a sample[56] to sophisticated disease pattern recognition.[57] Nevertheless the combination of high brilliance and high spatial resolution is still limited by the lack of optimized instrumentation for microscopy.

New IR focal plane array (FPA) detectors implemented at ultra-bright SR facilities will extend performance and overcome the existing limitations, possibly allowing the achievement within the next few years of the ultimate sensitivity necessary for high contrast molecular imaging.

In a microscope equipped with an FPA, the pixel size is generally smaller than the diffraction limit, which means in practice that oversampling takes place and in principle could be exploited to perform point spread function (PSF) deconvolution of IR images. Actually, in IR imaging applications the SR spot has to cover a significant part of the area imaged by the FPA, apparently reducing the brilliance advantage of a mid-IR IRSR source. The first experiments that matched an FPA IR detector (64×64) with SR were successfully performed at ANKA in 2005.[58] At present, available FPA arrays are suitable to match brilliant IR source with a moderate expansion of the beam, maintaining the high S/N ratio characteristic of the SR emission. The installation of FPA detectors for IR spectral imaging is in progress in many IRSR facilities.[59] The trend seems to be for using the high magnification objective >50× for better matching with the SR effective size at the FPA for imaging.

Infrared imaging was originally developed for remote sensing for military purposes, later applied for astronomical applications and only in the early 1990s for pioneering IR spectroscopical applications.[60] Spectral images collected with FPAs, or by scanning with linear arrays, are a three-dimensional block of data that spans one wavelength and two spatial dimensions. The spectroscopic resolution is defined by the number of wavelength readout elements, *e.g.* bands or channels, while the spatial resolution at a fixed wavelength is determined by the size of the pixel and the parameters of the optical system.[61]

In fact, confocal single element mapping achieves higher spatial resolution than apertureless imaging systems such as FPA, in particular when dealing with extremely brilliant SR sources. However, not all experiments need the ultimate resolution and, as addressed by recent experimental data, SR emission is characterized by a sufficient photon flux density to illuminate the array detector in order to operate with a high S/N.[58]

The FPA detectors enable different imaging modalities. In particular, multiple spectra covering the area of interest may be collected with a resolution of a few microns (depending on the pixel size) and tiled together. Tests performed with conventional instrumentation have clearly shown that combining the sensitivity and the speed of read-out of the latest generation of FPAs can scale down the time requirement from hours to minutes. Except on the most modern storage rings where an almost continuous refill occurs, in standard accelerators injections are performed over 20 to 60 minutes to hours while the current is continuously decreasing. Therefore, acquisition of IR images of cells and tissues with FPA may be affected by illumination changes, and almost everywhere the contribution of SR beam instabilities still needs to be identified and minimized. Instabilities are one of the main issues when working with FPA detectors at IR beamlines because they contain thousands of pixels: 64×64 or 128×128 pixels are typically used for these devices, and it is important to work with signal uniformity over the entire field of view. Using a conventional source the noise is primarily related to spatial non-homogeneity, while SR fluctuations may introduce a non-negligible contribution of temporal noise that, when combined with the spatial noise and other additional contributions coming from optics and electronics, could affect the S/N and introduce wavelength-dependent variations.

The massive and fast data collection of array detectors enables a drastic reduction in the image acquisition time. For example an $n \times n$ pixel FPA detector may provide up to an n^2 time saving compared with a single element detector, *e.g.* if $n = 64$ a gain of more than three orders of magnitude can be obtained with read out around 1 kHz and noise characteristics similar to single element detector acquisition.

3.7 Biomedical Applications at IRSR Beamlines

(See also Chapter 1.) Important applications in biological research have emerged in recent years. Fourier transform IR microspectroscopy using SR may collect data with high resolution on a time scale of 10–100 s and up to areas of a few microns, opening a new scenario: high contrast IR imaging of entire cells and tissues. In this field, IR spectroscopy has important advantages compared with other techniques: the use of non-ionizing radiation (*vs.* X-ray absorption spectroscopy) and the almost negligible damage to samples under analysis (*vs.* Raman spectromicroscopy). Infrared spectroscopy is a unique technique for the purposes of microanalysis because of its high spatial resolution within a microscopic field of view and the physico-chemical identification of the molecular components via their vibration fingerprints.

Thanks to this specific sensitivity, IR microscopy achieves contrast via the intra-molecular vibrational modes and it is particularly advantageous for identifying organic matter constituents such as H–, C–, O– or N–bonds. While in light microscopy the image contrast is achieved with stains or fluorescent materials, with IR the use of chemical reagents or stains is not necessary. Image contrast is simply produced from intrinsic IR absorption bands, as is the case in X-ray microscopy, where contrast is achieved by recording spectra before and after the absorption edges of an element contained in the specimen. The first spectrum of a single red blood cell with and without carbon monoxide was measured in 1988 using a microscope connected to a conventional FTIR spectrometer.[41] The use of IR microscopes coupled to SR sources started a few years later and the high brilliance of this source made it possible to reach spatial resolutions at the diffraction limit with high S/N.

The IR spectrum of a biological system exhibits characteristic absorption bands (characteristic molecular vibrations) that appear at different positions with different relative intensities (see Chapter 2, section 2.2.3). Each spectral difference reflects a potential structural difference in the system. Therefore, vibrational spectroscopy allows the definition of sets of IR marker absorption bands for the various conformations. In Figure 3.12 we showed an IR spectrum of a biological specimen. This kind of spectrum, covering a wide spectral range, can be recorded with high spatial resolution using an IR microscope by averaging data from a reasonable number of scans. The arrows indicate, in the near-IR and mid-IR ranges, the main IR bands that can be used to recognize and distinguish different samples and materials.[62] For instance, the amide I profile ($1600-1700 \, \text{cm}^{-1}$) can help to identify either the protein configuration or changes occurring in tissue.[63] Even if the IR spectroscopy of whole cells needs major improvements in order to achieve full reliability, it represents a new challenging research field that may be used to detect pathologies or diseases within human cells.[64]

Among the analytical techniques able to provide molecular imaging of biological samples, a growing interest in FTIR microscopy has emerged during the last decade. This is mainly due to the global molecular information of the sample provided by the FTIR analysis, which is a critical advantage for characterizing biological sample contents.[51] This technique is based upon the absorption of IR light by vibrational transitions in covalent bonds. In complex systems such as biological ones, the IR spectrum is the sum of the contributions gathered from the proteins, lipids, nucleic acids and metabolic contents. Another important feature of this technique is that no internal standard, contrast agent, or even staining, is required to obtain high quality data from a biological sample. Thus, a wide range of applications has been proposed.[54,57]

However, limits still exist when applying FTIR microscopy to biological tissues, such as the lack of sensitivity of detectors for analysing thick tissue sections (signal saturation above $A \sim 1.7$ for most MCT detectors) and, at the opposite end of the spectrum, a limited spatial resolution for work on single cells (in practice $\sim \lambda$, *i.e.* $\sim 3\,\mu m$ for lipids and $\sim 10\,\mu m$ for nucleic acids). In the first case, hard tissues cannot be analysed owing to their high density of molecular contents at available sample thicknesses, *e.g.* for bones or teeth, sections are usually thicker than $5\,\mu m$, even when using diamond saws.[65] For the same reason, it is not possible to work on soft tissue sections with a thickness > 30–$50\,\mu m$.[55] In the second case, single cell analysis requires working at a resolution of a very few microns, which is currently not accessible with conventional laboratory instruments.

Interesting experiments performed on single cells using IRSR microscopy are still few in number.[66–70] In fact, single cell analysis at high spatial resolution is possible using an IRSR microscope,[45] but accurate evaluations of the advantages of SR over a conventional set-up are not yet available. The quality of FTIR spectra obtained using SR or Globar sources is roughly comparable up to a $20\,\mu m$ aperture, but the S/N ratio is very poor below this spatial resolution for experiments performed with a Globar.[71] While SR undoubtedly has an advantage over a thermal source for single cell analysis, with large apertures, *e.g.* $> 50\,\mu m$, the SR still does not show a great advantage over confocal microscopy (Figure 3.13).[72]

Nevertheless, one may wonder about the significance of the spatial resolution parameter for single cell analysis. Indeed, the sub-cellular components of a eukaryotic cell have sizes comparable to or smaller than the spatial resolution achievable by FTIR microscopy. For example the nucleus is 1–$3\,\mu m$, the mitochondria $\sim 1\,\mu m$, the ribosomes, lisosomes and vesicles $< 1\,\mu m$, *etc.* The size of the FTIR probe in the mid-IR is significantly larger than those for VUV, X-ray and NMR, which all work below a spatial resolution of $1\,\mu m$. However, with a spatial resolution of about $1\,\mu m$, FTIR microscopy provides molecular information on sub-cellular components of cells unavailable by other techniques.

The analysis of tissues by IRSR microscopy achieves the highest performance in hard tissues such as bone,[73–79] or chondrocytes,[75] which benefit from the brilliance of SR by allowing transmission mode analyses of thin (4–$6\,\mu m$) and un-decalcified tissue sections. Comparable results could be obtained using a conventional laboratory system, but with tissue thicknesses between 2 and $4\,\mu m$.[65,80] Thus, SR allows analysis of bone tissue sections $\sim 50\%$ thicker than Globar sources in transmission mode.

Early bone development in the mouse skeleton can also be investigated to test the hypothesis that specific compositional properties determine the stiffness of this hard tissue. As an example, Figure 3.14 shows a comparison of X-ray computed tomography (CT) images and IR phosphate/protein images of the tibias of mice selected at different ages.[81] X-ray and IR may be combined to characterize chemical properties and microstructure of bone. Both play an important role in defining the micromechanical properties of the skeleton

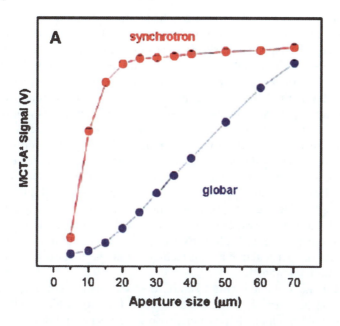

Figure 3.13 Infrared signal measured through various aperture sizes using a synchrotron source versus a Globar source. Data collected with a confocal IR microscope and a single-point detector. (Reprinted from ref. 72.)

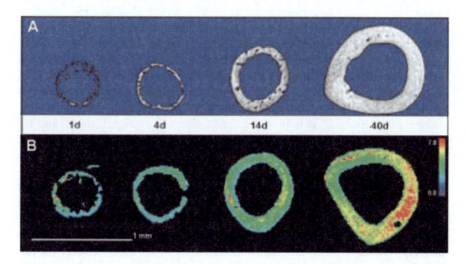

Figure 3.14 Comparison of X-ray computed tomography images (top panel) and IR phosphate/protein images of sections of mouse tibia from animals at 1, 4, 14 and 40 days of age. Images collected with both techniques are plotted on the same intensity scale for direct comparison. (Reprinted from ref. 81.)

during growth, so that a better mechanistic understanding of the underlying processes may enable the diagnosis, prevention, and treatment of bone diseases.

However, an exhaustive comparison is still lacking, and in general a limited number of SR-based FTIR microscopy studies are available on hard tissues.

The complexity of the materials and phenomena investigated demands new experimental methods based on concurrent and possibly simultaneous use of different techniques. The potential offered by X-ray methods is well recognized in materials science and, recently, it has been proposed to combine X-ray atomic and structural analysis with IR molecular analysis to produce a unique new tool for materials science research.[82] Pioneering time resolved experiments combining X-rays and IR radiation were performed more than a decade ago at SRS, the Daresbury storage ring, which used time-resolved FTIR with small angle X-ray scattering to investigate an organic polymer structure.[83] Concurrent IR and X-ray experiments may provide unique and accurate information about processes, leading to a deeper understanding of materials and phenomena, in particular imaging techniques at the highest available spatial resolution are combined.

An IR image obtained with a standard source instrument is the result of a compromise between spatial resolution and sensitivity, which limits its application to small biological specimens with a very small amount of molecular content, such as single cell. In principle, FT-Raman imaging may offer higher spatial resolution and sufficient sensitivity for the analysis of cells. Apart from the size of the detector and the magnification of the objectives used, the laser beam excitation also spreads over the sample above the micrometric illuminated spot size. In addition, excluding the fluorescence issues common to organic samples and visible (blue–green) laser excitation, the Raman signal in the IR may be quite low because of its scaling with $1/\lambda^4$. This limits the sensitivity of analysis to significant amounts of organic materials, or requires laser intensities that could affect the sample integrity.

IRSR sources provide a brilliance at least two orders of magnitude higher than a conventional source, resulting in a higher photon flux at small apertures, *i.e.* below $10 \, \mu m \times 10 \, \mu m^2$.[84] The advantage of an SR source for IR microscopy of cells has been demonstrated using a single element detector.[45] However, obtaining an IR image at high spatial resolution takes several hours, and the spatial resolution achievable with a single detector remains limited to a few microns when IR microscopes are equipped with large single element detectors. The introduction of FPA detectors with individual elements having a dimension of $40 \, \mu m \times 40 \, \mu m$ may resolve this problem, although the shape of an IRSR source is different from that of a Globar source. The IRSR beam illumination at the sample can be effective from ~ 10–$30 \, \mu m$ in diameter but is typically asymmetric. In addition, an energy distribution that decreases several times in intensity is intrinsic when moving from the centre to the edge, because the far-IR has a larger emission angle at the SR source and is more diffraction limited at the sample. The use of an FPA may reduce the acquisition time of an image significantly, accompanied by a large gain in S/N, thanks to the smaller size of the individual detectors of the array. Nevertheless, they require a spread of the

IRSR spot to achieve a quasi-homogeneous distribution of the light on the detector area.

Figure 3.15 shows the performance of FTIR imaging of individual cell growth on a window transparent to both visible and IR light. A Bruker Hyperion 3000 microscope equipped with a 64 × 64 FPA of detectors was used, coupled to a Bruker Equinox 55 spectrometer with a very high current SR ring (DAΦNE).[107] The optical set-up and alignment were the same, except for the first mirrors at the entrance of the spectrometer which select alternately one of the two IR sources. Although the IRSR source flux per pixel on the FPA detector may be greater than that of a Globar, still an higher noise can be present on the final IR image owing to the huge difference in the signal level between pixels fully illuminated (5–10% with SR) and those not illuminated. In fact, a Globar source with a 15 × objective is more efficient in the determination

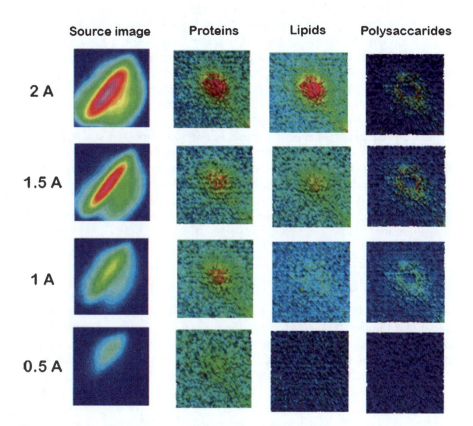

Figure 3.15   FTIR imaging of a single cell grown on an Si–O substrate. Background and sample images were collected at the same nominal current values. Spectral image acquisitions were performed with 32 × 32 pixels of the FPA detector. From left to right, images were obtained by integrating the spectra in the protein (1715–1600 cm^{-1}), lipid (3020–2880 cm^{-1}) and polysaccharide (1152–951 cm^{-1}) regions, respectively.

of the distribution of amide I (1715–1600 cm^{-1}, the spectral range of integration), fatty acid chains (3020–2880 cm^{-1}) and osidic residues (1152–951 cm^{-1}) absorptions on the final IR image. Using a 36×objective, the final IR image corresponds to a pixel size of about ~1.1 μm×1.1 μm, so that the IRSR source illuminates many more pixels than the 15×objective. However, the photon count for both sources is one order of magnitude lower, and this difference affects the quality of spectra extracted from the IR images at concurrent cellular locations.

To optimize the SR source IR imaging of a small sample such as a single cell it is possible to set the FPA parameters to collect data only using ~30×30 pixels via a 36×objective, which is sufficient to cover an entire cell with a final IR image dimension of ~40 μm×40 μm (see Figure 3.15). The SR source at DAΦNE[24] is characterized by an optimized focus spot of about 25 μm × 15 μm (with a stored current of ~1.5–2 A), which could be used slightly unfocused to distribute the intense photon flux onto the largest number of pixels (see Figure 3.15). The defocusing procedure of the IRSR source allows distribution of the photon flux among pixels, to perform a better comparison among spectra of contiguous pixels and to determine with greater accuracy the absorption distribution of organic cell contents.

The images in Figure 3.15 show a series of IR images obtained in about two minutes using the SR source at DAΦNE at different current values. The experiment demonstrated that a high current SR source may provide information on cellular organic contents that is not accessible with a conventional source. Actually, as underlined in section 3.3, because SR is a non-thermal source, its spectral distribution at low energy is the same for all storage rings with electron beam energy E greater than 0.5 GeV, and small intensity absorptions are directly correlated with the value of the current. The data show that only a stable beam at high current supports the analysis of polysaccharide contributions.

3.8   Status and Perspectives of IRSR Facilities

Around the world, more than twenty dedicated IR beamlines are now operational (see Figure 3.1), and all cover the mid-IR and nearby spectral regions (Figure 3.16). In spite of the pioneering experiments performed in France, Britain and Germany in the 1970s, many facilities are available in the USA and Asia. In Europe the development of IRSR has proceeded at a slower rate, and in order to bridge the gap with the USA and Japan, in 1995 the European Union funded a network coordinated by the Department of Physics of Sapienza University. The network was aimed at providing a forum for the European groups involved in the realization and the exploitation of IRSR sources. This initiative has helped greatly to realize the IR beamline at Frascati and to better exploit those of Orsay and Daresbury, two early SR facilities that are now decommissioned. Moreover, the network has organized the first international meetings in this field (University of Rome III in 1995,

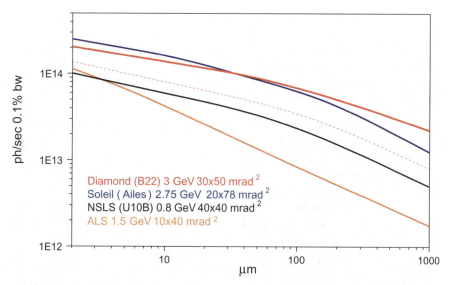

Figure 3.16 Spectral flux estimated for some IR beamlines now operational in Europe and the USA. Related front end sizes are given and the data refer to a nominal current of 500 mA. (Modified from data provided by P. Roy.)

LNF-INFN Frascati in 1996 and LURE-Orsay in 1997) and has published the first book entirely devoted to IRSR.[85] In recent years several other meetings and workshops devoted to IRSR and its applications have been organized in the USA and Europe, where old and new SR facilities offer scientific opportunities for research in the IR domain.

The imaging techniques available in many IRSR facilities are now implemented in FPA or linear array detectors (see Chapter 7).

At present, at diffraction-limited wavelengths the spatial resolution of images collected by FPAs illuminated by SR sources is limited mainly by the physical size of the pixel. In the future, in these apertureless systems, with the high S/N available with SR, it will be possible to introduce a multiple aperture mask at an intermediate image plane to reach the ultimate spatial resolution of the confocal layout. Alternatively, using high spatial oversampling methods an enhancement in resolution could be achieved by point spread function deconvolution. Moreover, a significant effort to improve the existing packages and to develop efficient new procedures to manage and evaluate the huge amount of three-dimensional data collected by FPAs is required.[50]

Teams working at synchrotron IR beamlines all around the world have installed and tested or are going to install array detectors on their IR microscope end-stations. The scenario is changing continuously, and many facilities with IR microspectroscopy beamlines are planning the installation of mid-IR arrays in the near future.

With regard to IR sources, in addition to the development of IR edge radiation[34] in the recent years, interest is steadily increasing in the science of radiation in the THz region (see http://www.thznetwork.org).[86] The availability of high power, short-pulse radiation in this low energy range is very important for studying collective excitations in solids, protein conformational dynamics, superconductor band gaps and electronic and magnetic scattering. Generating radiation of significant intensity in this frequency range, between microwaves and IR, is not straightforward and this region is often referred to as the "terahertz gap".

With respect to the coherence properties of IRSR, the energetic electron bunches in storage rings produce the emission of incoherent SR; however they should also produce a roughly comparable power output of coherent radio-frequency radiation.[87,88] However, preliminary experiments did not show evidence of coherent enhancement of the SR emission in the long-wavelength IR domain for bunch lengths of about 30 cm.[89] Additional investigations were necessary, using modern synchrotron sources with shorter bunch length, to observe coherence effects in the IR region. In 1989, however, the first observation of THz coherent SR (CSR) in the wavelength of 4.5–25 cm^{-1} was reported.[90] Carr and co-workers also reported the observation of coherent SR from the NLSL VUV ring and partially characterized this emission. Multi-particle coherent emission occurred in the very far IR from bunched electrons, and peaked near a wavelength of 7 mm, much shorter than the nominal electron bunch length, indicating the presence of a density modulation within the bunch.[91] In the last few years, there has been significant progress in the understanding of CSR in electron storage rings. A model accounting for the CSR terahertz random bursts observed at high single bunch current in many rings has been proposed,[92,93] and more recently, at BESSY Il, coherent emission generated by a controlled, steady-state process was observed in the 1–0.3 mm wavelength range,[94] and experimentally verified (Figure 3.17).[95,96] Stable CSR was produced for the first time in a storage ring, and a theoretical model accounting for the observations was developed.[97] A new technique based on the energy modulation of a fraction of the electron beam by a femtosecond laser pulse, which can generate intense sub-picosecone pulses of THz CSR, has also been developed and characterized.[98–101]

Ring-based sources completely optimized for the generation of coherent terahertz SR, and which exploit the full complement of the CSR production mechanisms mentioned above, have been proposed in the USA[103] and in Italy.[104]

The main experimental studies in the far-IR or THz regions performed nowadays involve spectroscopy and the imaging of materials, in particular those of biomedical interest. Given the intrinsic long wavelength of this radiation (1 THz is equivalent to 33 cm^{-1} or 300 μm), diffraction plays a major role and the optical detection method intrinsically limits its use to sub-milli-metre types of spectroscopy. On the other hand, when new detection systems are applied, such as with near field or non-optical methods, the spatial resolution limitation can be overcome, provided that a brilliant IR source is used, in the case of SR especially with coherent emission mode. For example, Figure 3.18 demonstrates the feasibility of the use of intense THz radiation for

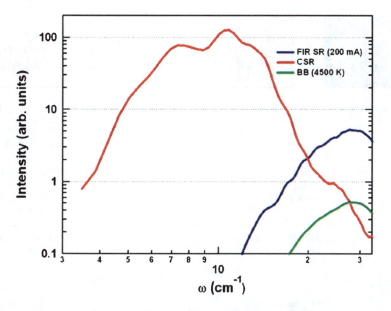

Figure 3.17 Measured far-IR intensity for the BESSY coherent synchrotron radiation (CSR) source, compared with mercury arc and Globar conventional
thermal sources. In the picture the turn-on of the CSR source below
$2\,cm^{-1}$ is a real effect of the CSR emission process, while the drop off at
the low frequency end is due to a combination of diffraction losses in the
optical path of the beamline and to contributions of optical components
in the interferometer.[102] (Reprinted from ref. 102.)

the diagnosis of tooth decay; X-ray radiography and far-IR in different modes
are shown for comparison. In contrast to investigation using confocal imaging,
both the near-field technique and the sub-THz radiation can be used to image
bulky tooth samples. Simulated buried caries lesions were produced by drilling
cavities 1 mm in diameter in the proximal region of the tooth, and the cavity
was filled with hydroxyapatite. The near-field image was obtained by utilizing a
200 μm wire cone, while the confocal imaging was performed with the same
optical set-up without the near-field cone. In confocal imaging geometry the
tooth cannot be resolved spatially, and the image is strongly blurred, as one
would expect from diffraction due to the long wavelengths involved.

Among the new attempts to overcome the limitations of SR and IR imaging is the IRENI (InfraRed ENvironmental Imaging) beamline approach
(Figure 3.19) at SRC.[105] Developed together with the University of Wisconsin–
Milwaukee the goal is to bring the acquisition times for IR maps down to one
minute by using an FPA illuminated by several IRSR beams coming from a
multiple beamline optics system (see Chapter 7). The beamline extracts
320 hor. × 25 vert. $mrad^2$ from a dedicated bending magnet. This fan of light is
separated into 12 beams and rearranged into a 3 × 4 beam bundle with the help
of a total of 48 mirrors. The bundle is then sent into a commercial IR

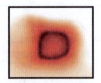

- shadow image
- x-ray

- far-field @ 1 mm (0.3 THz)
- con-focal geometry
- bursting mode

- near-field @ 1 mm
- 200 μm aperture
- bursting mode

- near-field @ 1 mm
- 200 μm coax. cone
- 80 μm resolution
- low alpha mode

Figure 3.18　Feasibility tests of THz imaging for the diagnosis of tooth decay. From left to right: an X-ray shadow image of the investigated tooth, a broadband THz image taken in confocal geometry, and two images obtained with the near-field probe. The second near-field image was obtained by utilizing the 200 μm wire cone, giving a spatial resolution of about 130 μm. In confocal geometry the image is blurred by diffraction owing to the long wavelengths involved. (Courtesy of U. Schade.)

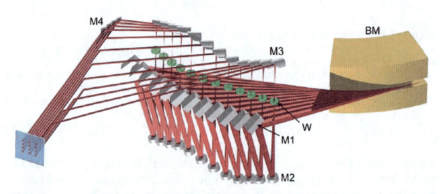

Figure 3.19　Schematic (not to scale) of the new IRENI beamline at the SRC (Madison). For clarity, only 4 out of the 12 M4 mirrors are shown in this diagram of the beamline. The first optical components are 12 identical toroidal mirrors (M1) located 2 m from the source, working in unity t o collect the available horizontal fan of radiation. A set of 12 identical paraboloidal mirrors (M3), with 250 mm focal lengths, "collimate" the beams and deflect them at 90° in the horizontal plane. The colli- mated beams are then combined and rearranged to form a 4×3 matrix by a set of 12 plane mirrors (M4) and directed by a single flat mirror to the interferometer and to the microscope. (Reprinted from ref. 105.)

microscope equipped with a 128×128 pixels FPA. The first IR maps have been obtained for biomarine samples by the system in "defocused mode" in order to achieve a more homogeneous sample illumination, while the optical system that leads to the detector is in focus.

3.9 Conclusions

Decades of effort have outlined the technological challenges involved in coupling IR instruments to SR sources. Nevertheless, it is noteworthy that the coupling of IR microscopes to IRSR beamlines is still at the early stage of development rather than in a routine application phase.

Synchrotron radiation microscopy may now be used to investigate highly inhomogeneous and small samples with a very limited amount of organic matter, such as a single cell. However, owing to the inhomogeneous distribution of the SR–IR photon flux, the natural shape of an IRSR source does not match the existing array detectors. Nevertheless, the results achieved clearly demonstrate the great advantage of the IRSR source for FTIR imaging at high sensitivity and high spatial resolution within a few minutes, when experiments can be performed at the highest current available on stable storage rings.

Moreover, several laboratories are currently commissioning array detectors of small size for IR microscopes to improve detector performance at small apertures. An alternative to pure confocal geometry, the use of FPA detectors optimized in the mid-IR range represents the best way to collect IR images rapidly over large tissue areas. However, the alignment and optimization of these components in commercial systems remains a challenge owing to optical limitations and, when present, to SR instabilities. A huge effort has been made already, many ideas have been implemented and others are under investigation at third generation storage ring facilities in order to improve stability, and as a consequence imaging performance, in terms of both spatial and time resolution.

An optimal coupling between an IRSR source and FTIR imaging instrumentation equipped with small FPA detectors may push the established limits of FTIR imaging towards a "real time" imaging of single-cell phenomena.

New opportunities in the IR domain are represented also by concurrent spectroscopic experiments investigated on a short timescale. As an example, simultaneous IR and X-ray experiments may be applied to understanding dynamical processes and phase transitions.[82,106]

Recently a simple chemical–physical system has been considered to perform a "proof of concepts" experiment in order to verify the possibility of performing time resolved analysis in the far-IR domain with a rapid scan method, using the emission of the SR characterized by bunches from tenths of picoseconds to a few nanoseconds long.[106] The optical layout combined two different sources, a conventional thermal source in the near- to mid-IR range and SR in the far-IR region, to allow the recording of time-resolved IR absorption spectra in a wide interval, ranging from the NIR up to the THz region. The experiment also showed that rapid scan acquisition can be performed in the time domain of seconds and that the THz region is suitable for the investigation of evaporation processes or phase transitions.

In recent years, thanks to the availability of IR beamlines at SR facilities, new applications and opportunities have emerged. In particular, IRSR spectroscopy is going to have a new and important role in the life sciences.

Advancements are continuous, and in the future, thanks to the properties of SR, the existing limits of FTIR imaging should be overcome. There is thus a "brilliant" future for IRSR spectromicroscopy and imaging. The SR has advantages over other systems and, thanks to the continuous improvement in the performance of these light sources, important results are expected in biological and biomedical applications.

Acknowledgements

We want to acknowledge all the people who kindly provided material, especially about the early phase of SR IR and the most recent progresses in the field we tried to summarize. A special thanks to helpful friends and colleagues from ex-LURE and Daresbury facilities, as well as from SOLEIL, ANKA, BESSY, MAX, Elettra, ESRF, the Hefei and Australian Synchrotrons, CLS, ALS, NSLS and SRC. One of the Authors (GC) wants also to thank Paul Dumas and Oleg Chubar at Soleil for the SRW code and the help with simulations.

Finally, our gratitude to coworkers and collaborators at Diamond and DAΦNE IR beamlines for the help in some of the experimental results shown.

References

1. E. F. Nichols, *Phys. Rev.*, 1893, **1**, 1.
2. F. C. Brown, P. L. Hartman, P. G. Kruger, B. Lax, R. A. Smith and G. Vineyard, *Synchrotron Radiation as a Source for the Spectroscopy of Solids*, NRC Solid State Panel Subcommittee Rep., March, 1966.
3. J. R. Stevenson, H. Ellis and R. Bartlett, *Appl. Optics*, 1973, **12**, 2884.
4. P. Meyer. P. Lagarde, *J. Phys.*, 1976, **37**, 1387.
5. J. Yarwood, T. Shuttleworth, J. B. Hasted and T. Nanba, *Nature*, 1984, **317**, 743.
6. T. Nanba, Y. Urashima, M. Ikezawa, M. Watanabe, E. Nakamura, K. Fukui and H. Inokuchi, *Int. J. Infr. Mill. Wav.*, 1986, **7**, 1769.
7. T. Nanba, *Rev. Sci. Instrum.*, 1989, **60**, 1680.
8. E. Schweizer, J. Nagel, W. Braun, E. Lippert and A. M. Bradshaw, *Nucl. Instrum. & Meth.*, 1985, **A239**, 630.
9. E. Schweizer, J. Nagel, W. Braun, E. Lippert and A. M. Bradshaw, *Nucl. Instrum. & Meth.*, 1986, **A246**, 163.
10. G. P. Williams, *Nucl. Instr. & Meth.*, 1990, **A291**, 8.
11. B. Nelander, *Vibr. Spectrosc.*, 1990, **9**, 29.
12. P. Roy, Y. L. Mathis, A. Gerschel, J. P. Marx, J. Michaut, B. Lagarde and P. Calvani, *Nucl. Instrum. & Meth.*, 1993, **A325**, 568.
13. D. A. Slater, P. Hollins, M. A. Chesters, J. Pritchard, D. H. Martin, M. Surman, D. A. Shaw and I. Munro, *Rev. Sci. Instrum.*, 1992, **63**, 1547.
14. A. Marcelli and P. Calvani, 1993, *LNF-INFN Report 93/027*.
15. A. Nucara, P. Calvani, A. Marcelli and M. Sanchez del Rio, *Rev. Sci. Instrum.*, 1995, **66**, 1934.

16. A. Marcelli, *Insertion devices*, in Proceedings of the International School of Physics E. Fermi Course CXXVIII on *Biomedical Applications of Synchrotron Radiation*, ed. E. Burattini, SIF, Varenna, 1996, p. 21.
17. V. Montelanici, Internal Note 1972, LNF-72/056.
18. M. Sanchez del Río and R. J. Dejus, in *Synchrotron radiation instrumentation*, AIP Conference Proceedings Volume 705, eds. T. Warwick, J. Stöhr, H. A. Padmore and J. Arthur, 2004, p. 784.
19. A. Marcelli, E. Burattini, A. Nucara, P. Calvani, G. Cinque, C. Mencuccini, S. Lupi, F. Monti and M. Sanchez del Rio, *Nuovo Cimento D*, 1998, **20**, 463.
20. W. Duncan and G. P. Williams, *Appl. Opt.*, 1983, **22**, 2914.
21. M. Ado Yu and P. A. Cherenkov, *Sov. Phys. Dokl.*, 1957, **1**, 517.
22. S. Kimura, *UVSOR Activity Report*, 1997, p. 60.
23. S. Kimura, *Jpn. J. Appl. Phys.*, 1999, **38**(Supplement 1), p. 392.
24. M. Cestelli Guidi, M. Piccinini, A. Marcelli, A. Nucara, P. Calvani and E. Burattini, *J. Opt. Soc. Amer. A*, 2005, **22**, 2810.
25. W. R. McKinney, M. C. Martin, M. Chin, G. Portman, M. E. Melczer and J. A. Watson, *ALS Compendium*, 2000, http://infrared.als.lbl.gov/.
26. G. L. Carr, R. P. S. M. Lobo, J. La Veigne, D. H. Reitze and D. B. Tanner, *Phys. Rev. Lett.*, 2000, **85**, 3001.
27. R. P. S. M. Lobo, J. La Veigne, D. H. Reitze, D. B. Tanner and G. L. Carr, *Rev. Sci. Instrum.*, 2002, **73**, 1.
28. G. D. Smith and R. A. Palmer, in: *Handbook of Vibrational Spectroscopy*, eds J. M. Chalmers and P.R. Griffiths, Wiley, Chichester, 2002, **Vol. 1**, p. 626.
29. A. Bocci, A. Marcelli, E. Pace, A. Drago, M. Piccinini, M. Cestelli Guidi, A. De Sio, D. Sali, P. Morini and J. Piotrowski, *Nuclear Instr. and Meth. A*, 2007, **580**, 190.
30. R. Cöisson, *J. Physique Lettres*, 1984, **45**, L89.
31. Y.-L. Mathis, P. Roy, B. Tremblay, A. Nucara, S. Lupi, P. Calvani and A. Gerschel, *Phys. Rev. Lett.*, 1998, **80**, 1220.
32. R. A. Bosch, *Nucl. Instr. & Meth. A*, 1997, **386**, 525.
33. R. A. Bosch, *Nuovo Cimento D*, 1998, **20**, 483.
34. R. A. Bosch, *Nucl. Instr. & Meth. A*, 2000, **454**, 497.
35. O. Chubar and P. Elleaume, *Proceedings EPAC98*, 1998, 1177.
36. H. Kimura, T. Moriwaki, N. Takahashi, H. Aoyagi, T. Matsushita, Y. Ishizawa, M. Masaki, S. Ohishi, H. Okuma, T. Nanba, M. Sakurai, S. Kimura, H. Okamura, N. Nakagawa, T. Takahashi, K. Fukui, K. Shinoda, Y. Kondo, T. Sata, M. Okuno, M. Matsunami, R. Koyanagi and Y. Yoshimatsu, *Nucl. Instr. Methods Phys. Res. A*, 2001, **467–468**, 441.
37. P. Dore, A. Nucara, D. Cannavò, G. De Marzi, P. Calvani, A. Marcelli, R. Sussmann, A. J. Whitehead, C. N. Dodge, A. J. Krehan and H. H. Peters, *Appl. Opt.*, 1998, **37**, 5731.
38. M. Castellano, *Nucl. Instr. & Meth. A*, 1997, **391**, 375.
39. A.-S. Müller, I. Birkel, S. Casalbuoni, B. Gasharova, E. Huttel, Y.-L. Mathis, D. A. Moss, N. Smale, P. Wesolowski, E. Bründermann, T. Bückle and M. Klein, Proceedings EPAC08, Genoa, 2008, p. 1056.

40. G. L. Carr, *Rev. Sci. Instrum.*, 2001, **72**, 1613.
41. A. Dong, R. G. Messerschmidt and J. A. Refnner, *Biochem. Biophys. Res. Commun.*, 1988, **156**, 752.
42. M. Minsky, U.S. Patent #3013467, 1957, *Microscopy Apparatus.*
43. *Handbook of biological confocal microscopy*, ed. J. B. Pawley, Plenum Press, New York, 1995.
44. T. D. Smith, *Nucl. Instr. & Meth. A*, 2002, **483**, 565.
45. N. Jamin, P. Dumas, J. Moncuit, W.-H. Fridman, J.-L. Teillaud, G. L. Carr and G. P. Williams, *Proc. Natl. Acad. Sci. USA*, 1998, **95**, 4837.
46. P. Dumas, N. Jamin, J. L. Teillaud, L. M. Miller and B. Beccard, *Faraday Discuss.*, 2004, **126**, 289, ; discussion 303.
47. K. Nishikida, *Thermo Electron Corp.*, 2004, App. Note 50717.
48. *Handbook of biological confocal microscopy*, ed. J. B. Pawley, Plenum Press, New York, 1995, chapt. 2.
49. A. Rose, Television pickup tubes and the problem of noise, *Adv. Electron.*, 1948, **1**, 131.
50. P. Lasch and D. Naumann, *Biochim. Biophys. Acta*, 2006, **1758**, 814.
51. C. Petibois and G. Déléris, *Trends Biotechnol.*, 2006, **24**, 455.
52. P. Dumas, G. D. Sockalingum and J. Sulè-Suso, *Trends Biotechnol.*, 2007, **25**, 40.
53. I. W. Levin and R. Bhargava, *Ann. Rev. Phys. Chem.*, 2005, **56**, 429.
54. D. Ami, A. Natalello, A. Zullini and S. M. Doglia, *FEBS Lett.*, 2004, **576**, 297.
55. C. Petibois, K. Gionnet, M. Goncalves, A. Perromat, M. Moenner and G. Déléris, *Analyst*, 2006, **131**, 640.
56. D. C. Malins, K. E. Hellstrom, K. M. Anderson, P. M. Johnson and M. A. Vinson, *Proc. Natl. Acad. Sci. USA*, 2002, **99**, 5937.
57. D. C. Fernandez, R. Bhargava, S. M. Hewitt and I. W. Levin, *Nat. Biotechnol.*, 2005, **23**, 469.
58. D. Moss, B. Gasharova and Y.-L. Mathis, *Infrared Phys. Technol.*, 2006, **49**, 53.
59. C. Petibois, M. Piccinini, M. A. Cestelli Guidi, G. Déléris and A. Marcelli, *Nature Phot.*, 2009, **3**, 177.
60. P. J. Treado, I. W. Levin and E. N. Lewis, *Appl. Spectrosc.*, 1992, **46**, 1211.
61. L. M. Miller and R. J. Smith, *Vib. Spectrosc.*, 2005, **38**, 237.
62. D. L. Wetzel and S. M. LeVine, in *Infrared and Raman Spectroscopy of Biological Materials*, ed. H.-U. Gremlich and Bing Yan, Marcel Dekker, New York, 2001, p. 101.
63. L. M. Miller, P. Dumas, N. Jamin, J.-L. Teillaud, J. Miklossy and L. Forro, *Rev. Sci. Instrum.*, 2002, **73**, 1357.
64. M. Diem, S. Boydston-White and Luis Chiriboga, *Appl. Spectr.*, 1999, **53**, 148A.
65. R. Mendelsohn, E. P. Paschalis, P. J. Sherman and A. L. Boskey, *Appl. Spectrosc.*, 2000, **54**, 1183.

66. N. Jamin, P. Dumas, J. Moncuit, W.-H. Fridman, J.-L. Teillaud, G. L. Carr and G. P. Williams, *Proc. Natl. Acad. Sci. USA*, 1998, **95**, 4837.

67. H.-Y. N. Holman, M. C. Martin, E. A. Blakely, K. Bjornstad and W. R. Mckinney, *Biopolymers*, 2000, **57**, 329.

68. E. Gazi, J. Dwyer, N. P. Lockyer, J. Miyan, P. Gardner, C. Hart, M. Brown and N. W. Clarke, *Biopolymers*, 2004, **77**, 18.

69. P. Heraud, B. R. Wood, M. J. Tobin, J. Beardall and D. McNaughton, *FEMS Microbiol. Lett.*, 2005, **249**, 219.

70. A. J. Bentley, T. Nakamura, A. Hammiche, H. M. Pollock, F. L. Martin, S. Kinoshita and N. J. Fullwood, *Mol. Vision*, 2007, **13**, 237.

71. D. A. Moss, M. Keese and R. Pepperkok, *Vib. Spectrosc.*, 2005, **38**, 185.

72. L. M. Miller and P. Dumas, *Biochim. Biophys. Acta*, 2006, **1758**, 846.

73. L. M. Miller, V. Vairavamurthy, M. R. Chance, R. Mendelsohn, E. P. Paschalis, F. Betts and A. L. Boskey, *Biochim. Biophys. Acta*, 2001, **1527**, 11.

74. R. Y. Huang, L. M. Miller, C. S. Carlson and M. R. Chance, *Bone*, 2003, **33**, 514.

75. L. M. Miller, J. T. Novatt, D. Hamerman and C. S. Carlson, *Bone*, 2004, **35**, 498.

76. M. J. Tobin, M. A. Chesters, J. M. Chalmers, F. J. Rutten, S. E. Fisher, I. M. Symonds, A. Hitchcock, R. Allibone and S. Dias-Gunasekara, *Faraday Discuss.*, 2004, **126**, 27, ; discussion 77.

77. M. Petra, J. Anastassopoulou, T. Theologis and T. Theophanides, *J. Mol. Struct.*, 2005, **733**, 101.

78. M. E. Ruppel, D. B. Burr and L. M. Miller, *Bone*, 2006, **39**, 318.

79. A. J. Burghardt, Y. Wang, H. Elalieh, X. Thibault, D. Bikle, F. Peyrin and S. Majumdar, *Bone*, 2007, **40**, 160.

80. A. L. Boskey, S. Gadaleta, C. Gundberg, S. B. Dot, P. Ducy and G. Karsenty, *Bone*, 1998, **23**, 187.

81. L. M. Miller, W. Little, A. Schirmer, F. Sheik, B. Busa and S. Judex, *J. Bone Miner. Res.*, 2007, **22**, 1037.

82. A. Marcelli, D. Hampai, W. Xu, L. Malfatti and P. Innocenzi, *Acta Phys. Pol. A*, 2009, **115**, 489.

83. W. Bras, G. E. Derbyshire, D. Bogg, J. Cooke, M. J. Elwell, B. U. Komanschek, S. Naylor and A. J. Ryan, *Science*, 1995, **267**, 996.

84. E. Levenson, P. Lerch and M. C. Martin, *Infrared Phys. Technol.*, 2006, **49**, 45.

85. *Infrared Synchrotron Radiation*, ed. by P. Calvani and P. Roy, Compositori, Bologna, 1998.

86. M. S. Sherwin, C. A. Schmuttenmaer and P.H. Bucksbaum, in *Opportunities in THz Science*, DOE-NSF-NIH Workshop 2004.

87. G. P. Williams, *Rep. Prog. Phys.*, 2006, **69**, 301.

88. F. Curtis Michel, *Phys. Rev. Lett.*, 1982, **48**, 580.

89. G. P. Williams, C. J. Hirschmugl, E. M. Kneedler, P. Z. Takacs, M. Shleifer, Y. J. Chabal and F. M. Hoffmann, *Phys. Rev. Lett.*, 1989, **62**, 261.

90. T. Nakazato, M. Oyamada, N. Niimura, S. Urasawa, O. Konno, A. Kagaya, R. Kato, T. Kamiyama, Y. Torizuka, T. Nanba, Y. Kondo, Y. Shibata, K. Ishi, T. Ohsaka and M. Ikezawa, *Phys. Rev. Lett.*, 1989, **63**, 1245.

91. G. L. Carr, S. L. Kramer, J. B. Murphy, R. P. S. M. Lobo and D. B. Tanner, *Nucl. Instrum. Methods A*, 2001, **463**, 387.

92. G. Stupakov and S. Heifets, *Phys. Rev., ST Accel. Beams*, 2002, **5**, 52002.

93. M. Venturini and R. Warnock, *Phys. Rev. Lett.*, 2002, **89**, 224802.

94. M. Abo-Bakr, J. Feikes, K. Holldack, G. Wüstefeld and H.-W. Hübers, *Phys. Rev. Lett.*, 2002, **88**, 254801.

95. J. M. Byrd, W. P. Leemans, A. Loftsdottir, B. Marcelis, M. C. Martin, W. R. McKinney, F. Sannibale, T. Scarvie and C. Steier, *Phys. Rev. Lett.*, 2002, **89**, 224801.

96. M. Abo-Bakr, J. Feikes, K. Holldack, P. Kuske, W. B. Peatman, U. Schade, G. Wüstefeld and H.-W. Hübers, *Phys. Rev. Lett.*, 2003, **90**, 094801.

97. F. Sannibale, J. M. Byrd, A. Loftsdóttir, M. Venturini, M. Abo-Bakr, J. Feikes, K. Holldack, P. Kuske, G. Wüstefeld, H.-W. Hübers and R. Warnock, *Phys. Rev. Lett.*, 2004, **93**, 094801.

98. A. A. Zholents and M. S. Zolotorev, *Phys. Rev. Lett.*, 1996, **76**, 912.

99. K. Holldack, S. Khan, R. Mitzner and T. Quast, *Phys. Rev. Lett.*, 2006, **96**, 054801.

100. J. M. Byrd, Z. Hao, M. C. Martin, D. S. Robin, F. Sannibale, R. W. Schoenlein, A. A. Zholents and M. S. Zolotorev, *Phys. Rev. Lett.*, 2006, **96**, 164801.

101. S. Bielawski, C. Evain, T. Hara, M. Hosaka, M. Katoh, S. Kimura, A. Mochihashi, M. Shimada, C. Szwaj, T. Takahashi and Y. Takashima, *Nature Phys.*, 2008, **4**, 390.

102. U. Schade, K. Holldack, M. C. Martin and D. Fried, *Proc. SPIE-Int. Soc. Opt. Eng.*, 2005, **5725**, 46.

103. J. M. Byrd, M. C. Martin, W. R. McKinney, D. V. Munson, H. Nishimura, D. S. Robin, F. Sannibale, R. D. Schlueter, W. Thur, J.-Y. Jung and W. Wan, *Infrared Phys. Technol.*, 2004, **45**(5-6), 325.

104. F. Sannibale, A. Marcelli and P. Innocenzi, *J. Synch. Rad.*, 2008, **15**, 655.

105. M. J. Nasse, R. Reininger, T. Kubala, S. Janowskib and C. Hirschmugl, *Nuclear Instr. Meth. Phys. Res. A*, 2007, **582**, 107.

106. P. Innocenzi, L. Malfatti, A. Marcelli, M. Piccinini, D. Sali and U. Schade, *J. Phys. Chem. A*, 2009, **113**, 9418.

107. M. Moenner and A. Marcelli, *Anal. Bioanal. Chem.*, 2010, **397**, 2123.

CHAPTER 4

Raman Microscopy: Complement or Competitor?

HUGH J. BYRNE,[a] GANESH D. SOCKALINGUM[b] AND NICK STONE[c]

[a] Focas Institute, Dublin Institute of Technology, Kevin Street, Dublin 8, Ireland; [b] Unite MeDIAN, CNRS UMR6237-MEDyC, UFR Pharmacie, Université de Reims, 51 rue Cognacq-Jay, 51096 Reims Cedex, France; [c] Biophotonics Research Group, Gloucestershire Royal Hospital, Great Western Road, Gloucester GL1 3NN, UK

4.1   Introduction

The objective of this chapter is to provide an introduction to Raman spectroscopic microscopy and its potential for biochemical analysis and clinical diagnostic applications, such that it can be compared and contrasted to the techniques of synchrotron and bench-top mid-FTIR spectroscopy discussed elsewhere in this book. Raman spectroscopy is a complementary technique to mid-IR absorption spectroscopy with established capabilities for materials and process analysis. As a bioanalytical and diagnostic technique, similar to FTIR spectroscopy, its potential has been demonstrated although there are many differing technical considerations to be addressed. Raman has potentially significant advantages as well as drawbacks compared to FTIR techniques. Here we endeavour to outline these benefits and pitfalls and project the complementary and competitive usage of Raman techniques.

RSC Analytical Spectroscopy Monographs No. 11
Biomedical Applications of Synchrotron Infrared Microspectroscopy
Edited by David Moss
© Royal Society of Chemistry 2011
Published by the Royal Society of Chemistry, www.rsc.org

4.2   Raman Spectroscopy – a Brief History

The Raman effect was proposed and demonstrated by Sir C. V. Raman in 1928,[1] and independently by G. Landsberg and L. Mandelstam.[2] Inspired by the accepted inelastic scattering of X-rays, Raman proposed a "new type of secondary radiation" or "modified" scattering which resulted from the effect of the fluctuations from the normal state of atoms and molecules associated with vibrations. He demonstrated that in addition to elastic (Rayleigh or Mie) scattering in which radiation scattered by a material has the same energy (frequency/wavelength), light can be inelastically scattered through a gain or loss of photon energy to the molecular vibrations of the material. The spectrum of the inelastically scattered radiation represented a fingerprint of the molecular vibrations within a material. The observation of the Raman effect gave rise to the field of Raman spectroscopy, a versatile alternative to infrared (IR) spectroscopy and now a common analytic laboratory tool. C. V. Raman was awarded the Nobel Prize in physics in 1930 for his work, and in 1998 the Raman effect was designated an ACS National Historic Chemical Landmark in recognition of its importance in materials and process analysis.

   The Raman effect is extremely weak, and the evolution from its discovery to a laboratory technique is principally one of technological development. In their original work, Raman and Krishnan used sunlight and narrow band optical filters. Mercury arc discharge lamps subsequently became the source of choice, the scattered radiation being recorded on photographic plates. The use of spectroscopic detection followed, but Raman spectroscopy remained largely a curiosity until the advent of the laser in the 1960s, providing monochromatic sources of significant brightness and intensity and variable wavelength such that the intrinsic limitation of the low efficiency of the scattering process could be overcome. Apart from being a weak process, Raman spectroscopy in the ultraviolet (UV)–visible regions suffered greatly from sample fluorescence, scattering and photodegradation, which made the technique less attractive for coloured samples. Nevertheless, Raman spectroscopy became a very popular research tool, for example in the analysis of phonons, electrons, and electron–phonon interaction in high T_c superconductors.[3] In the mid 1980s Raman went through a renaissance with FT-Raman set-ups that operated with near-IR lasers such as Nd^{3+}:YAG emitting at 1064 nm, and detection was done via In:Ga:As detectors. This system benefited from the same advantages as Fourier transform IR spectroscopy, viz. high throughput and multiplex advantages and high precision in the frequency scale. Some FT-instruments were built to accommodate both IR and Raman systems using the same interferometer. In the case of FT-Raman, the scattering sample acts as a polychromatic source. By exciting at high wavelength, both sample fluorescence and degradation could be circumvented but at the expense of a lower scattering process. The low sensitivity of the FT-Raman systems was a drawback for biological samples. Dispersive Raman came back into play with the revolution in charged coupled detector (CCD) arrays in the 1980s and 1990s, which added to the benefits of high laser source intensities. In addition to this, the development of narrow band laser line rejection filters meant that the huge losses in signal from traditional triple monochromator systems could be overcome with the combination of a filter set and a single spectroscopic grating. Furthermore, the significant reductions in acquisition time with multichannel signal detection enabled significant improvements in signal to noise ratio.[4] The combination of technology developments led to a new range of Raman spectroscopic microscopes in the 1990s, establishing Raman spectroscopy as a relatively inexpensive bench-top laboratory tool to rival conventional IR spectroscopy.

Raman is a scattering technique and can be induced in any wavelength region of the optical electromagnetic spectrum. Whereas IR absorption spectroscopy measures transitions in the low energy IR region of the spectrum, Raman spectroscopy can be carried out using UV, visible or near-IR sources, avoiding the need for non-conventional sample mounting in, for example, potassium bromide (KBr) disks or on calcium fluoride (CaF_2) windows as required for FTIR, although for thin samples contributions to the signal from the substrate can be significant. Its adaptability to common silica fibre probes could lead to *in vivo* diagnostic tools, although this is beyond the scope of this chapter.

Raman spectroscopy is viewed as a complementary technique to IR spectroscopy but has significant advantages for many applications, specifically biological. As will be outlined below, Raman is relatively insensitive to water, whose absorption bands often swamp IR spectra, and therefore has potentially significant advantages for *in vivo* diagnostics. The application of Raman spectroscopy to biomolecules and even tissues was first demonstrated as early as the 1960s,[5–7] and by the mid 1970s biomedical applications were explored.[8] Whole cell and tissue studies have been carried out on a range of pathologies,[9–13] and *in vivo* studies have demonstrated the potential use in diagnostic applications.[14–16] Very recent developments have included developments of Raman technologies to probe tissue biochemistry at a depth of many millimetres, leading to the prospect of *in vivo* diagnostics in harder to reach areas of the body.[17] Further developments in probe and other technologies as well as signal processing techniques will undoubtedly see the fulfilment of this potential.

For a detailed description of the basic principles of Raman spectroscopy, the associated instrumentation and potential for spectroscopic imaging, the reader is referred to some of the many excellent texts in the literature.[18–22] This chapter provides an introduction to Raman spectroscopy and how it is measured. It outlines some experimental considerations specific to biospectroscopy and explores applications from molecular through cellular to tissue imaging for biochemical analysis and disease diagnostics. The complementarities and potential advantages over IR spectroscopy [Fourier Transform (FTIR) and Synchrotron Fourier Transform (S-FTIR)] are described. Finally, the future potential of the development of Raman spectroscopy for biochemical analysis and *in vivo* disease diagnostics are projected.

4.3 What is Raman Spectroscopy?

Rayleigh or Mie scattering (elastic scattering) occurs when light scattered from a material is of the same frequency (or energy) as the incident light. Raman scattering (inelastic scattering) is a result of light that is scattered off a molecule or solid such that its frequency (or energy) differs from that of the incident light as a result of the interaction. In Raman scattering the energy increase (anti-Stokes) or decrease (Stokes) from the excitation is related to the vibrational energy spacing in the ground electronic state of the molecule, and therefore the shifts in energy of the scattered radiation from the incident frequency are a direct measure of the vibrational energies of the molecule. In Stokes Raman scattering, the molecule starts out in a lower vibrational energy state and after the scattering process ends up in a higher vibrational energy state. Thus the interaction of the incident light with the molecule creates a vibration in the material. In anti-Stokes scattering, the molecule begins in a higher vibrational energy state and after the scattering process ends up in a lower vibrational energy state. Thus a vibration in the material is annihilated as a result of the interaction. The frequency (or energy) differences between the Raman lines and the incident line are characteristic of the scattering molecules and are independent of the frequency of excitation. The process is often depicted as in Figure 4.1 with the aid of a virtual or polarized electronic state. It should be noted however that no electronic transition or "absorption" process is required. The Raman effect arises from the coupling of the induced polarization of scattering molecules (which is caused by the electric vector of the electromagnetic radiation) with the molecular vibrational modes.

Figure 4.2 shows a typical Raman spectrum for crystalline silicon. The parameter of interest is the frequency shift (directly proportional to the energy captured by or donated to the molecule of interest) from the laser illumination, and therefore the incident laser frequency is set to zero, the Stokes line being represented as a positive shift. The Stokes (positive) and anti-Stokes (negative)

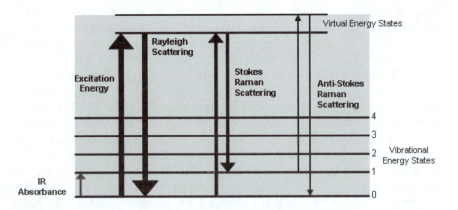

Figure 4.1   Schematic illustration of the transition states during Rayleigh and Raman scattering in a material, in comparison to IR absorption.

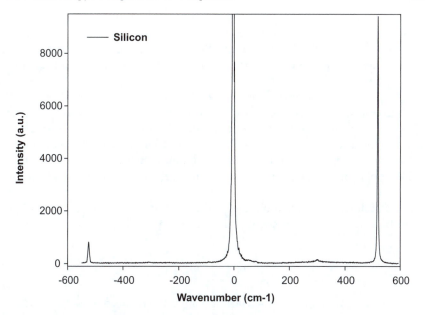

Figure 4.2 Raman spectrum of a silicon crystal showing, from left to right, anti-Stokes, Rayleigh, and Stokes lines. The frequency scale is expressed as the Raman shift with respect to the excitation wavelength; this is why the Rayleigh scattering is at $0 \, \text{cm}^{-1}$.

Raman lines corresponding to the optical phonons can be seen symmetrically shifted from the incident laser line. For ease of comparison to IR spectroscopy, frequency shifts are expressed in wavenumbers (cm^{-1}). The Stokes shift is most commonly measured at room temperatures, as from simple thermodynamics using Boltzmann's equation, there are very few vibrations in most materials at room temperature which can contribute to anti-Stokes scattering.

At room temperature, the number of molecules in an excited vibrational state will be low. This can be shown by using Boltzmann's equation:

$$\frac{N_V}{N_0} = \exp\left(-\frac{E_V}{kT}\right) \qquad (4.1)$$

where N_V/N_0 is the fraction of molecules in the vibrational state; E_V is the energy of the vibrational state; k is Boltzmann's constant and T is the absolute temperature. For example the C=C stretch oscillation ($1612 \, \text{cm}^{-1}$ shift) of a benzene ring requires 1.99×10^{-20} Joules of energy to excite the oscillation from the ground state. Using the above equation, the fraction of benzene molecules in the excited vibrational state at $20 \, ^\circ\text{C}$ is 0.0078. Hence it is obvious that, at room temperature, incident photons are much less likely to encounter molecules in an excited state. Therefore the likelihood of Stokes radiation, whereby the molecule captures a portion of the incident photon's energy, is that much greater than the alternative anti-Stokes, since the anti-Stokes radiation can only occur if the molecule is in an excited vibrational or rotational state. The relative signal strengths of the Stokes to anti-Stokes scattering will change with the temperature of the probed material, and indeed can be used as a measure of temperature.

The Raman effect can be induced by light of all frequencies. However, the cross-section for an inelastic scattering process is proportional to λ_{in}^{-4}, where λ_{in} is the wavelength of the incident photon. For example, photons of 300 nm wavelength have a cross-section of scattering sixteen times greater than photons at 600 nm, assuming that there are no resonance effects, which may occur for incident photons having energy near an electronic absorption line of the molecule.

In a simplified diatomic molecule, in the harmonic oscillator approximation, as in the case for IR spectroscopy, the frequency of vibration is given by:

$$\omega_k = (k/m_r)^{1/2} \tag{4.2}$$

where ω_k is the frequency of the vibration, m_r is the reduced mass, calculated by $m_1 m_2 / m_1 + m_2$, where m_1 and m_2 are the masses of the bonded atoms respectively, and k is the force constant of the vibration, related to the bond energy. In a complex molecule, the vibration of each bond can couple to the incident photons generating a vibrational spectrum on both the Stokes and anti-Stokes sides. As in IR spectroscopy, the frequency positioning of a Raman band is characteristic of a molecular bond or group vibration and the combination of bands represents a characteristic fingerprint of that molecule. It follows that any changes to the fingerprint can be used to monitor or characterize physical or chemical changes at a molecular level.

Not all vibrational modes are "Raman active", however, and the strength of the scattering or the scattering cross-section is governed by selection rules. Whereas electric dipole transitions of IR (and UV–visible) absorption require a change of the dipole moment of the material as a result of the transition, Raman scattering requires a change in the polarizability of the bond as a result of the transition. Thus, while Raman spectroscopy is based on a very different photophysical process to the more frequently used FTIR spectroscopic technique, the two vibrational spectroscopic techniques are, in fact, very complementary. Thus, one has access to molecular level information via two different physical processes. In a molecule with a centre of symmetry, a change in dipole is accomplished by loss of the centre of symmetry, while a change in polarizability is compatible with preservation of the centre of symmetry. In a centrosymmetric molecule, asymmetric stretching and bending will be IR active and Raman inactive, while symmetric stretching and bending will be Raman active and IR inactive. In this case, IR and Raman spectroscopy are mutually exclusive. For molecules without a centre of symmetry, each vibrational mode may be IR active, Raman active, both, or neither. Symmetric stretches and bends, however, tend to be Raman active. Vibrations that are strong in an IR spectrum, those involving strong dipole moments, are usually weak in a Raman spectrum. Likewise, those polarizable but non-polar vibrations that give very strong Raman bands usually result in weak IR signals.

As a crude rule of thumb, those modes that are not Raman active tend to be IR active. By extension, symmetric modes tend to be stronger in Raman than in IR spectroscopy and *vice versa*. For example, hydroxyl or amine stretching vibrations, and the vibrations of carbonyl groups, are usually very strong in an FTIR spectrum, and usually weak in a Raman spectrum. However, the stretching vibrations of carbon double or triple bonds and symmetric vibrations of aromatic groups are very strong in the Raman spectrum. In terms of biochemical analysis, Raman has the particular advantage of minimal interference from the highly polar water vibrations so is a good choice for biological samples with a view to live conditions and *in vivo* measurements.

Figure 4.3 shows, for example, the Raman spectrum of the amino acid phenylalanine with illustrative band assignments. Particularly strong in the spectrum is the stretch of the highly polarizable aromatic ring at $1004 \, \text{cm}^{-1}$, also called ring breathing mode. This feature is seen prominently in all Raman spectra of cells and tissue. Figure 4.4 shows the Raman spectra of the amino acids arginine and lysine, and the dipeptide formed between them. Notable is the emergence of the band at $\sim 1650 \, \text{cm}^{-1}$, the so called Amide I band, common to all peptides and proteins.

Since its discovery in 1928, Raman spectroscopy has evolved in terms of the fundamental understanding of the process, instrumentation and applications. More advanced techniques such as Resonant Raman Spectroscopy (RRS)[22–24]

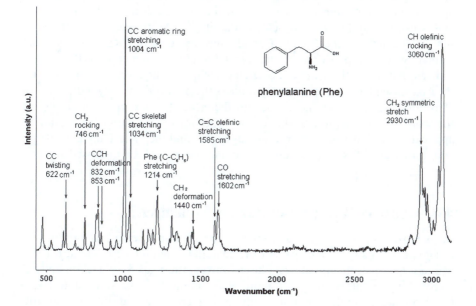

Figure 4.3 Raman spectrum of phenylalanine powder.

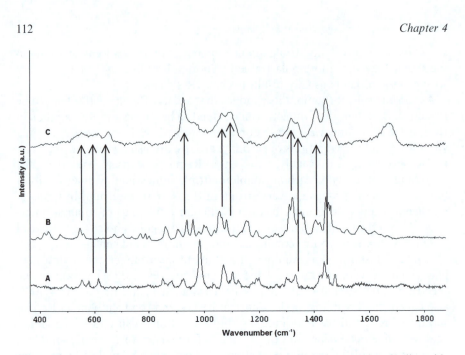

Figure 4.4   Raman spectrum of a A) arginine powder, B) lysine powder, C) dipeptide
formed between arginine and lysine.

have found many applications in photophysics and photochemistry. As substrates and media for Surface Enhanced Raman Spectroscopy (SERS)[25,26] are
becoming more reliable and reproducible, the technique is finding increased
applications for biological, chemical and bioanalytical characterization with
high sensitivity and hence low detection limits.[27,28] More advanced techniques
such as Coherent Anti-Stokes Raman Spectroscopy (CARS),[29] Stimulated
Raman Spectroscopy (SRS),[30] and Hyper Raman Spectroscopy (HRS)[31] have
evolved. Although these are extremely powerful techniques in their own right,
their increased technical complexity renders them, at present, beyond the realm
of routine diagnostic applications, and therefore they are considered beyond
the scope of this chapter. In the following sections the basic instrumentation,
applications to biospectroscopy and diagnostics and the underlying advantages
and drawbacks of Raman spectroscopic microscopy will be discussed.

4.4   How is Raman Scattering Measured?

In its most simple form, Raman spectroscopy is implemented using a monochromatic light source, a dispersion element and a light detector (Figure 4.5a).
Modern day instruments utilize a laser source, either gas (*e.g.*, Helium–Neon,
Argon Ion) or increasingly the more easily miniaturizable solid state lasers
(semiconductor diode, titanium sapphire). Depending on the wavelength,
powers of 10's to 100's of mW are typically employed. Particularly in the case

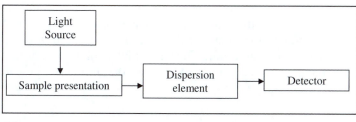

(a)

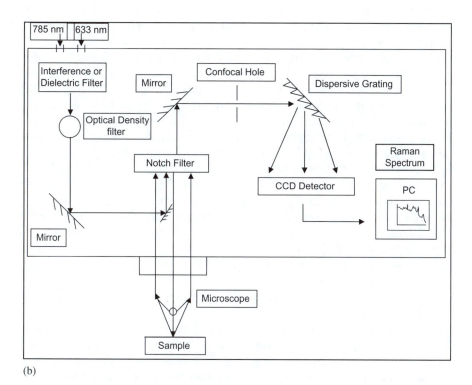

(b)

Figure 4.5   a) Typical set-up for Raman spectroscopy. b) Schematic optical layout of a Raman microspectrometer.

of gas lasers, it is imperative that the background plasma emission in the region of the laser line is minimized such that the weak Raman scattering can be observed and so a dielectric interference filter is used to clean up the excitation line.

The laser is directed onto the sample via a focusing lens, which in modern day systems is usually a microscope objective, often that of a commercially manufactured instrument (*e.g.*, Olympus, Leica, Nikon). Such an instrument allows a selection of objectives from the turret with varying magnification and numerical aperture and, for example, long working distances. With an objective of ×100, small spotsizes and micron level spatial and axial resolutions are achievable. The spot size is diffraction limited in a similar way to FTIR microspectroscopy, although the significantly shorter wavelength used in Raman spectroscopy leads to significantly higher spatial resolution than that available with IR radiation. The lateral resolution in the diffraction limit is given by:

$$D_x = D_y = 0.61\lambda/NA \tag{4.3}$$

where λ is the wavelength of the light and NA is the numerical aperture of the objective employed. Increased spatial resolution is therefore achievable with shorter wavelengths (UV) and high numerical aperture objectives.

In commercial microspectrometers, the Raman signal is generally collected in a backscattering geometry (Figure 4.5b); the microscope objective which delivers and focuses the laser also acts as the collection lens and collimates the reflected, Rayleigh and Raman scattered light. The collection efficiency is dependent on the numerical aperture of the objective. High numerical aperture is associated with high magnification objectives and therefore small spotsizes and high spatial resolution. Typically, spatial resolution can range from 0.5 μm to 1–2 μm when going from visible to near-IR lasers. Scattered light is collected from the focal depth of the objective and thus the choice of objective also governs the sampling depth in transparent materials. A high magnification objective gives a surface sensitivity (in transparent materials) of ∼ 1 μm, while a longer focal length ×10 objective can be used to sample the depth of a transparent liquid or solid (Figure 4.6).

The resolution in the z-direction is given by:

$$D_z = \lambda n/(NA)^2 \tag{4.4}$$

where n is the refractive index of the medium between the lens and the sample. Raman microscopes commonly operate in a confocal mode. Confocality has the advantage of providing improved z-resolution and better discrimination of the Raman signal from diffusely scattered radiation in inhomogeneous materials such as tissue. Before entrance to the spectrometer, the collimated radiation is imaged onto a variable aperture. The imaged spotsize is typically ∼ 100–200 μm, and radiation not emanating from the focal region of the microscope objective, or which is diffusely scattered by the sample, is blocked by the confocal hole, providing better depth resolution and discrimination of the Raman signal from other radiation.

Once collected, it is important to remove the strong contributions from the reflected or Rayleigh scattered laser light. This is commonly achieved using a holographic Notch filter or a dielectric stack (Figure 4.7) which has a spectrally narrow reflection band centred on the laser wavelength. The element acts as an almost 100% reflector of the laser radiation, which directs it into the microscope for illumination. The collimated backscattered radiation is incident on the element on its return, whereupon the reflected and Rayleigh scattered

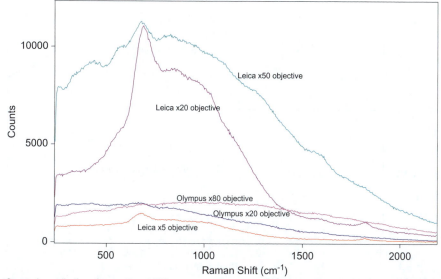

Spectral contributions from optics at 830 nm

Figure 4.6 Collection optics for Raman spectroscopy in the backscattering geometry. A comparison of spectral contributions from different illumination/collection optics. The spectra were excited from a clean chromium surface at 830 nm excitation (32 mW) and with an integration time of 30 s.

radiation is reflected, while the frequency shifted Raman bands are transmitted into the spectrometer. The spectral width of the element can be tailored to differing specifications but typically the Raman signal can be recorded to within 50–$100\,\mathrm{cm}^{-1}$ shift from the laser line. This routine specification is a significant improvement over the spectral responses of commercial FTIR systems in the far-IR region which typically have a lower limit of $\sim 400\,\mathrm{cm}^{-1}$.

The transmitted radiation is spectrally dispersed using a diffraction grating. The grating can be optimized for the operating wavelengths but typical gratings cover the entire optical range. Commonly, instruments are fitted with two or more interchangeable gratings to allow for low or high resolution measurements, covering the spectral range of interest in a single image or multiple images which can then be "stitched" using the instrument software. Operating at low resolution allows more rapid spectral recording and improves the signal to noise ratio by increasing the signal per wavenumber interval. Higher dispersion gratings can be employed where the fine structure of spectral features is to be resolved (*e.g.*, the Amide I band of proteins which gives information on secondary structure and conformation).

Spectrometer lengths are typically 300 mm for medium resolution or 800 mm for high resolution systems. The dispersion per pixel is thus typically $1\,\mathrm{cm}^{-1}$ for a 300 mm length spectrometer with an $1800\,\mathrm{mm}^{-1}$ grating operating at 633 nm and can be as low as $0.25\,\mathrm{cm}^{-1}$ for a high resolution system. The resolution of

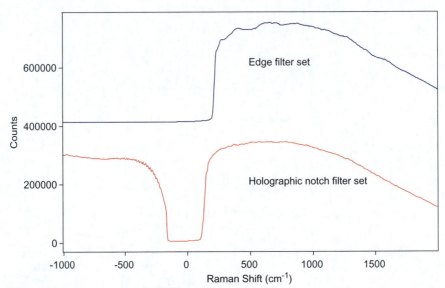

W-filament lamp spectrum measured with each filter set.

(a)

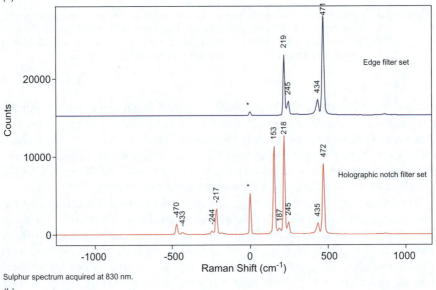

Sulphur spectrum acquired at 830 nm.

(b)

Figure 4.7 a) Comparison of white light spectra measured with a spectrometer fitted either with an edge filter set or a holographic notch filter set. Spectra were acquired for 60 s. The y-axis has arbritary units of intensity, whereas the x-axis represents the spectral energy in cm^{-1} relative to 830 nm (0 cm^{-1}). b) Comparison of sulfur spectra measured at 830 nm, with a Raman spectrometer fitted either with edge filter or holographic notch filter sets. c) Spectral contributions from some typical optical substrates at 830 nm, with 32 mW laser power at the sample, ×80 objective, t = 10 s.

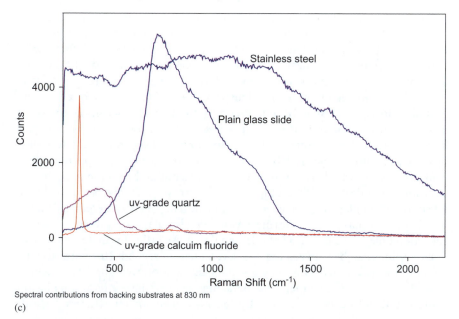

Spectral contributions from backing substrates at 830 nm

(c)

Figure 4.7 Continued.

the spectrometer is of course additionally determined by the entrance optics, and commonly instruments are fitted with a variable entrance slit. For comparison, commercial FTIR instruments commonly operate at $2-16\,\text{cm}^{-1}$ resolution and high resolution systems can achieve resolution for gas spectroscopy. The spectral resolution depends mainly on the displacement of the moving mirror of the interferometer.

The Raman signal is commonly collected using a charge coupled detector (CCD) device. Many different CCD options are available on commercial systems, including Deep Depletion CCDs which are required for the near-IR region, Back Thinned CCDs which can increase sensitivity, and electronic amplified CCDs which can increase the signal but also the noise. Recent improvements in CCD sensitivities mean that sufficient signal to noise can be achieved using electronic Peltier cooling, avoiding the inconvenience of cryogenic coolants. The Peltier effect is a thermoelectric effect whereby heat is displaced from a conducting material to another in the presence of an electrical current.

A further feature of modern Raman spectroscopic microscopes is that the laser is polarized, allowing determination of, for example, depolarization ratios, and molecular orientation in crystals or liquid crystals. Care must be taken to account for the polarization response of the vertically ruled diffraction grating however. To date, there have been few or no polarization dependent studies of biological materials although recent studies have demonstrated that polarization dependent Raman can detect structural changes in the extracellular matrix associated with basal cell carcinoma.[32]

The above describes the commonly utilized instrumentation required for dispersive Raman spectroscopic microscopy. In the past, Fourier transform Raman spectroscopy provided an alternative for coloured and fluorescent samples but the use of near-IR lasers at 1.064 μm together with In:Ga:As detectors reduced the sensitivity. Recent developments in laser rejection filters and CCD technologies have rendered dispersive techniques the preferred option.

Similar to conventional IR microscopy (mapping in opposition to recent imaging array detectors), Raman spectroscopic microscopy is usually performed as a point measurement, the sample area and depth being determined by the choice of the objective. Because it is an intrinsically weak phenomenon, relatively high power densities are required and simultaneous illumination over large areas and detection by multiple detector arrays, as can be performed with FTIR and focal plane arrays, is not easily achieved. To date, therefore, Raman imaging *per se* has been performed as a stepwise mapping process. Average collection times to achieve acceptable signal to noise ratios for materials such as tissue sections can be between ten and several tens of seconds (from low to high excitation wavelengths). Mapping a significant area of even 10×10 μm, with a 1 μm diameter spot, can therefore be a time consuming exercise. Significant effort has therefore been devoted by the instrument manufacturers towards improving sampling rates and mapping capabilities. Line mapping and continuous scanning modes have been introduced in many commercial systems although the weakness of the signals derived from biological samples remains a limiting factor.[33]

4.5   System Calibration

In a dispersive Raman set-up, the spectral dispersion is achieved by the action of a diffraction grating which distributes the light onto the multidetector CCD array. The system software keeps a record of which pixel of the CCD corresponds to which wavenumber of the spectrum. The process of assigning spectral positions to pixel number is one of calibration. The calibration can however change from day to day as the dispersion depends on the optical pathlength (distance times refractive index) between the grating and the detector. Small changes in temperature and/or humidity can thus affect the spectral calibration on a day to day basis and it is important to ensure that a rigorous calibration procedure is adhered to if direct inter-comparison of results is required. It is also important to note that the system sensitivity (intensity response) can vary depending on the laser line employed, the angle of the laser line rejection filter, the grating, the objective, the detector, and many other system parameters. Whereas FTIR spectra are taken as a ratio of the sample spectrum to a reference, no such facility is available in conventional Raman spectroscopy. Contributions due to sample substrate can also influence the results significantly (see Figure 4.7c for the contribution of different optical substrates).

Thus, for intra- and inter-laboratory comparison, it is important to calibrate the intensity axis also. Recommendations from manufacturers vary, and therefore the DASIM Raman Working Group has devised a calibration protocol based on a consensus of best practice. This protocol is outlined in Addendum A and shown schematically in Figure 4.8. In general, it is important to record the spectrum of the dark response and substrate to be used for the measurements in advance of a measurement set. Figures 4.9a and 4.9b show, respectively, the flowchart and an example of the Raman pre-processing procedures.

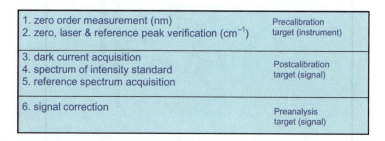

Figure 4.8 Raman calibration procedure.

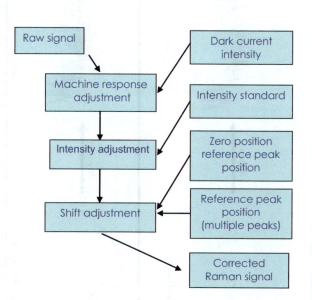

Figure 4.9 (a) Flow-chart showing Raman pre-analysis procedure. (b) Step 1 showing the subtraction of dark current from the raw Raman spectrum and the white light signal and the ration of the former to the latter to give a first corrected spectrum. (Courtesy of C. Gobinet.) (c) Step 2: starting from the corrected spectrum in step 1, the spectrum is smoothed, then the substrate and background are subtracted to give the final spectrum corrected for both instrument response and substrate contribution. (Courtesy of C. Gobinet.)

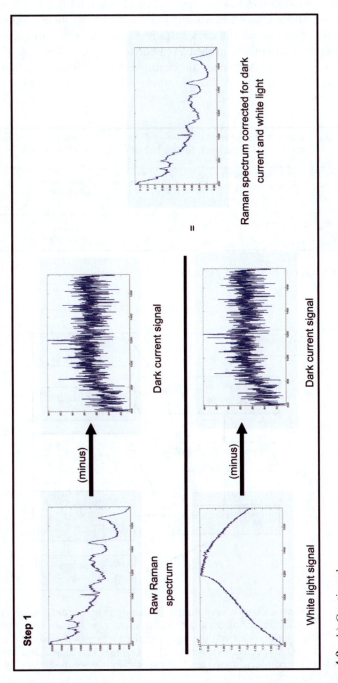

Figure 4.9 b) Continued.

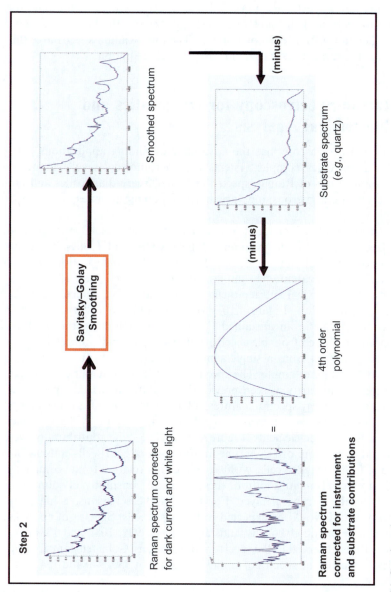

Figure 4.9 (c) Continued.

Figure 4.9b and 4.9c show a two-step procedure. Step 1 shows the subtraction of the dark current from both the raw Raman spectrum and the dark current spectrum. The former is then divided by the latter to give in the first instance a Raman spectrum corrected for dark noise and white light. Step 2 considers the latter spectrum and includes a Savitsky–Golay smoothing, followed by subtraction of the substrate (here quartz) and the background contributions (here a fourth order polynomial). The final Raman spectrum is thus corrected for instrument response and substrate contribution.

4.6 Raman Spectroscopy for Diagnostics and Biochemical Analysis

Raman (micro)spectroscopy has the advantage of finding applications going from isolated molecules, complex systems like macromolecules, cells, tissues, to humans. Applications of Raman spectroscopy to disease diagnostics and biological analysis are numerous and varied. The following outlines one study as an example.[34]

FFPP (Formalin-fixed paraffin processed) cervical tissue sections were characterized by the Registrar, National Maternity Hospital, Holles St, Dublin; the samples consisted of 20 normal and 20 invasive carcinoma sections from 40 individuals. Of the 20 carcinoma samples, 10 samples were identified as having various grades of cervical intraepithelial neoplasia (CIN), which were also marked for examination. Figure 4.10 shows the different cell types seen in normal cervical tissue in an unstained FFPP tissue section together with the Raman spectra recorded from basal cells, epithelial cells, and connective tissue. Spectra were recorded from a single sample and each spectrum represents a different spot within the sample. The spectra of the three different cell types do have a degree of similarity as seen previously for different tissue types. However, there are also many spectral features differentiating the different cell types. The spectra of basal cells show strong bands at 724, 779 and 1578 cm^{-1} which are characteristic of nucleic acids (Figure 4.10A). The same contributions were observed in the spectrum of DNA. The morphology of basal cells consists of a single line of tightly packed cells, with large nuclei in relation to the compacted surrounding cytoplasm. In addition, these cells are constantly dividing, providing cells to the parabasal layer. For both of these reasons, a high concentration of DNA would be expected in the basal cells. Spectra of epithelial cells have characteristic glycogen bands at 482, 849, 938, 1082 and 1336 cm^{-1} (Figure 4.10B) as observed in the spectrum of glycogen. Collagen contributions can be clearly seen in the spectra of connective tissue at 850, 940 and 1245 cm^{-1} (Figure 4.10C).

Figure 4.11a compares the Raman spectra collected from normal epithelial cells and invasive carcinoma from a selection of different patients. Glycogen contributions are clearly visible in the spectra from the normal epithelial tissue. The most obvious bands arise at 482, 849 and 938 cm^{-1} and are due to glycogen skeletal deformation, CCH aromatic deformation and CCH

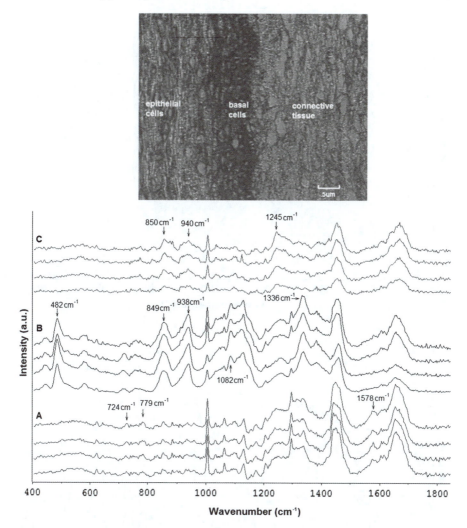

Figure 4.10 Top: Photomicrograph of unstained cervical tissue section, with different cell types identified. Bottom: Micro-Raman spectra recorded from basal cells (A), epithelial cells (B), and connective tissue (C) in cervical tissue sections. The main spectral features associated with each cell type are highlighted.

deformation respectively. However, there are also other glycogen contributions not as apparent, including a CC stretching band at 1082 and CH_3CH_2 wagging at $1336 \, cm^{-1}$. These glycogen bands (482, 849 and $938 \, cm^{-1}$) are absent in the spectra from invasive carcinoma, as well as a reduction in the intensity of the CC stretching mode ($1082 \, cm^{-1}$). Glycogen is known to be linked with cellular maturation and disappears with the loss of differentiation during neoplasia.[35] This agrees with the findings in other Raman and FTIR studies of

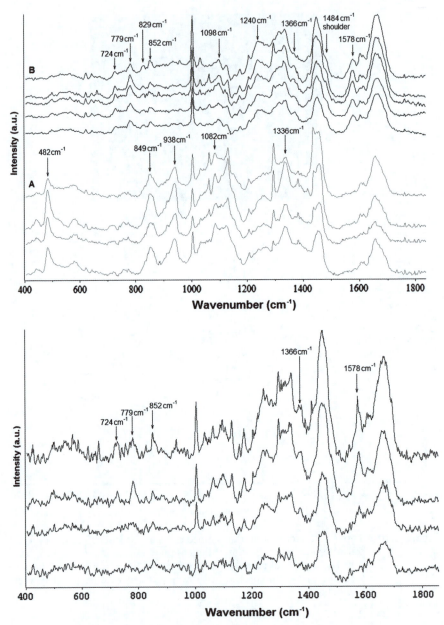

Figure 4.11 Raman spectra of normal cervical epithelial cells and invasive carcinoma cells (top), and (b) Raman spectra of cervical intraepithelial neoplasia (CIN) tissue (bottom). Assignments of the main Raman vibrational modes are detailed in Table 4.1.

Table 4.1 Peak position and assignments of main Raman vibrational modes.

Peak position (cm^{-1})	Assignment
622	C–C twisting
724	CH_2 deformation
746	CH_2 rocking
754	Symmetric ring breathing
779	Ring vibration
832	CCH deformation aliphatic
853	CCH deformation aromatic
873	C–C stretch
922	C–C stretching
1004	C–C aromatic ring breathing
1034	C–C stretching
1065	C–N stretch
1096	C–C chain stretching
1098	C–C stretch
1102	C–C stretch
1124	C–C skeletal stretch *trans*
1214	C–C stretch backbone carbon phenyl ring
1236	CN stretch, NH bending Amide III band
1240	CN stretch, NH bending Amide III band
1314	CH deformation
1337	CH_2 deformation
1335	CH_2 deformation
1366	CH_2 bending
1440	CH_2 scissoring
1484	CH_2 deformation
1548	NH deformation; CN stretch Amide II band
1578	C=C olefinic stretch
1585	C=C stretching
1602	C=O stretching
1660–1665	C=O stretch Amide I α-helix
2930	CH_2 stretching ($2930\,cm^{-1}$)
2932	CH_3 symmetric stretch ($2932\,cm^{-1}$)

epithelial tissues. The spectra of invasive carcinoma also show characteristic nucleic acid bands. These include prominent bands at 724, 779 and $1578\,cm^{-1}$, but also at 829, 852, 1098 and $1240\,cm^{-1}$. Distinct bands were also seen at $1366\,cm^{-1}$, a shoulder at $1484\,cm^{-1}$ and a band at $1578\,cm^{-1}$. An increase in the intensity of the Amide I band ($1655\,cm^{-1}$) was also observed in the spectra of carcinoma samples compared to the normal tissue samples. The increased nucleic acid and protein bands are a result of the increased proliferation of these tumour cells.

To investigate whether pre-malignant changes could be highlighted using Raman spectroscopy, 10 areas of neoplasia (CIN) from 10 different patients were marked by a pathologist and a selection of the resulting Raman spectra are shown in Figure 4.11b. A number of the spectral features identified in the invasive carcinoma samples were also observed in the CIN samples, such as the nucleic acid bands at 724, 779, 852, 1366 and $1578\,cm^{-1}$.

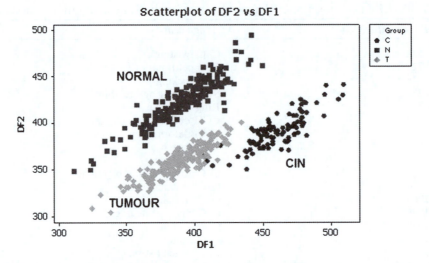

Figure 4.12   Linear discriminant analysis of the principal components of the first derivative, normalized, 10 point averaged spectra, over the entire spectral range. **C** = CIN, **N** = normal and **T** = invasive carcinoma.

This indicates that early biochemical changes can be identified using Raman spectroscopy.

Principal components analysis was used to reduce the number of parameters needed to represent the variance in the spectral data set. The principal components were then used to generate a linear discriminant model. All three tissue classes were successfully discriminated as shown in Figure 4.12. The classification model was tested using a leave one out cross-validation in which all but one spectrum were used to build the model. This model was then used to predict the remaining spectrum. This was repeated for all 498 spectra. Of 498 tissue spectra, 492 were correctly classified as normal, invasive carcinoma or CIN. The cross-validation misclassified six spectra, two of which were normal samples assigned as invasive carcinoma. The other four were either invasive carcinoma or CIN misclassified as either CIN or invasive carcinoma respectively. Importantly, no abnormal samples were classified as normal. Based on the cross-validation results, sensitivity and specificity values were calculated as 99.5% and 100% respectively for normal tissue, 99% and 99.2% respectively for CIN and 98.5% and 99% respectively for invasive carcinoma.

The results show the ability of Raman spectroscopy to classify cervical cancer and pre-cancer with high sensitivity and specificity. These classifications are based on biochemical changes known to accompany cervical cancer such as loss of differentiation and increased proliferation. This study shows the capability of Raman microspectroscopy to investigate not only the tissue but also the cells within the tissue, as it is known that a tumour can contain a heterogeneous population of cells.

4.7 Raman Microscopy and Imaging at Cellular and Subcellular Levels

The possibility of probing events at the single cell level is of great importance in disease diagnostics, in particular for cancer. Single cell analysis is an important issue both on a fundamental level, for understanding biological processes such as cell differentiation and proliferation, cell division and cell death, and on a clinical level for rapidly assessing cell phenotype or how a patient will respond to a given drug treatment. Very often, in real life samples, only a few cells are available for diagnostic purposes. Given the importance of developing non-invasive, cell-specific detection and monitoring methods, researchers are encouraged to develop low-cost, widely accessible, real-time detection and sensing technologies for living systems. Thus there is a real need for techniques capable of probing single cells. However, there are not many existing methods that can give access to high biomolecular information with cellular and sub-cellular resolution. Raman and IR microspectroscopies have such potential, as they can give spatially resolved biochemical information without the use of extrinsic labels and without being invasive or destructive to the studied system. Both IR and Raman techniques are truly label-free since the inherent vibrational signatures of the biochemical components of a cell are being observed. A significant advantage of Raman spectroscopic microscopy over FTIR microscopy is that of lateral spatial resolution. The micron- or submicron-level spatial resolution obtained with lasers and adapted optics helps to interrogate subcellular compartments. Furthermore, Raman techniques can be readily applied to single live[36] and fixed[37] cells.

Raman spectroscopy has proven its potential for the analysis of cell constituents and processes. However, sample preparation methods compatible with clinical practice must be implemented for collection of accurate spectral information. Micro-Raman spectroscopy as a non-invasive and non-destructive tool can therefore probe single living cancer cells while preserving cell integrity and functions, such as adhesion and proliferating capacities.[36] Figure 4.13 shows an example of micro-Raman spectra recorded from the nucleus and cytoplasm of a single live cancer cell using a 785 nm laser excitation and a ×100 water immersion objective. It highlights the differences in the biochemical and molecular composition between the cytoplasmic and nuclear cell compartments. Such spectral data are then compared in order to identify spectral signatures of the main macromolecules such as nucleic acids, lipids and proteins. Table 4.2 shows the main Raman bands observed when exciting the cells with a 785 nm laser. Based on these signatures and using multivariate statistical approaches, Raman maps of a single living cancer cell can thus be produced. Thus, Raman spectral imaging at the single living cell level represents a potential avenue for probing various cellular processes and monitoring for example cell–drug interactions. It can be developed into a rapid, high throughput, and automated diagnostic tool for screening cells from patients.

On a fundamental research basis, and as a complement to FTIR spectroscopy, Raman spectroscopy can be used to understand the processes underlying

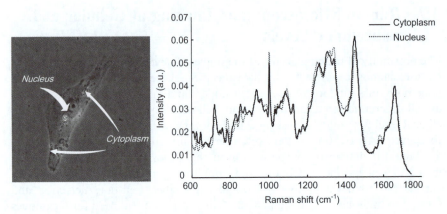

Figure 4.13 Photomicrography showing a single live cancer cell growing on a quartz window and the cellular compartments such as the nucleus and cytoplasm. Micro-Raman spectra corresponding to these compartments measured with an ×100 water immersion objective, 50 mW of a 785 nm laser, and a collection time of 20 s. (Courtesy of F. Draux.)

Table 4.2 Some of the major peaks that can be observed in the Raman spectrum (excitation at 785 nm) of single cells.

Wavenumbers	Assignments
787	DNA/RNA: ring breathing (C)
809	RNA: O-P-O stretching
1003	Prot: ring breathing Phe
1092	DNA/RNA: O-P-O stretching
1264	Prot: Amide III
1451	Prot: (C–H) bending
1486	DNA/RNA: ring mode (G,A)
1553	Prot: C=C stretching (Trp)
1575	DNA/RNA: ring mode (G,A)
1660	Prot: Amide I

cancer cell migration (metastasis) in model systems mimicking the extracellular matrix or the cancer cell's microenvironment.

Efforts to measure single cells in aqueous media by synchrotron IR microscopy in a flow system have been attempted but remain challenging.[38] However, in a study to monitor the response of cancer cells to an antitumour drug, it was shown that results obtained using synchrotron FTIR microscopy of fixed single cells gave comparable results to bench-top FTIR measurements of cell populations.[39] More recently, ATR-FTIR (attenuated total reflection-FTIR) imaging was used to monitor cancer cell compartments in natural aqueous media.[40] The latter mode may also reach higher spatial resolution down to 2–3 μm when using a high refractive index element like germanium. Combined with a synchrotron source and new imaging detectors, these modalities will open new avenues for applications in cytology as the performances are expected to be comparable to

Raman microspectroscopy. It must be noted that with modern micro-Raman systems, live or fixed cells can now easily be probed with green lasers operating at low powers without damaging them and with little or no parasitic fluorescence background.

An increase in the number of applications for single cell analysis is therefore foreseen and this will undoubtedly foster the development of Raman and IR as innovative approaches for spectral cytology. They can be automated into techniques for earlier detection of diseases with enhanced resolution, sensitivity and specificity.

4.8 Comparison to FTIR – Pros and Cons

4.8.1 Physical Principles

Fundamentally, Raman scattering is of different physical origin from IR absorption spectroscopy and many of the pros and cons in terms of applications stem from this fundamental difference and how it impacts on the regions of the spectrum probed, the sample response and the technological implications and limitations. A fundamental difference lies in the fact that Raman is a scattering process originating from a change in the molecule's polarizability whereas IR absorption is an electric dipole transition resulting in the absorption of a photon. The two processes are governed by different selection rules and thus while the two techniques can be considered complementary, the molecular fingerprint of a material obtained from Raman spectroscopy is different from that obtained from IR absorption spectroscopy. Figure 4.14 compares the mean Raman and FTIR spectra of oesophageal lymph nodes, and Table 4.3 lists the peak assignments.[41]

While both are rich in the so-called fingerprint region, it is clear that the Raman spectrum intrinsically contains significantly more information. There are many molecules such as amino acid residues, S–S disulfide bridges, C–S linkages from proteins, and nucleic acid signals that are more highlighted in Raman spectra. This may also result from the significantly higher resolution ($\sim 1\,cm^{-1}$) of typical research grade Raman spectrometers compared to FTIR instruments (~ 2–$8\,cm^{-1}$). The FTIR spectrum exhibits broad features due to overlapping bands. However, it is known that increasing the spectral resolution in IR does not significantly improve the band width. This higher biochemical information content, associated with a higher resolution, affords an ease of differentiation between for example normal and cancerous tissue for diagnostic applications (Figure 4.15). The wealth of information is often highlighted through the use of first or second order derivative spectra.

The key advantage of Raman for *in vivo* diagnostics is the low contribution of water to the Raman signal. Most human tissues contain around 70–80% water. Furthermore, the illumination and scattered light for Raman is usually of ideal wavelengths for transmission through optical fibres. This is certainly not the case for mid-IR light as overcoming the water signal with IR fibre probes is a real challenge.

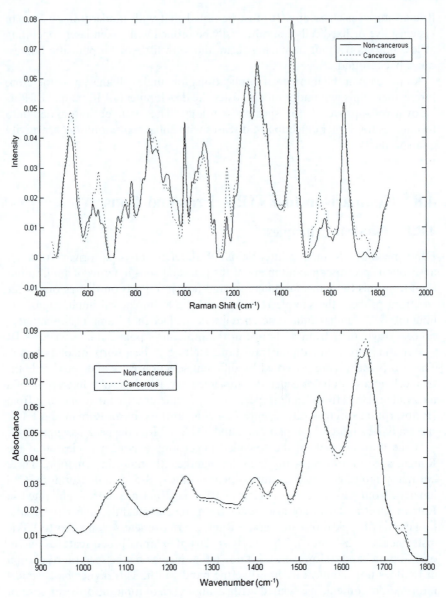

Figure 4.14   Mean Raman (left) and FTIR (right) spectra obtained from oesophageal lymph nodes. Both figures cover the "fingerprint" spectral range, which is the region of both spectra where most spectral features are found.[41]

4.8.2   Spatial Resolution

The spatial resolution of either measurement is determined by the diffraction limit and therefore the wavelength of the light used in the spectroscopic technique, as well as the imaging optics. For visible light (wavelength $\approx 0.5\,\mu m$),

Table 4.3 Key peaks and their assignments observed in Raman and FTIR spectra of lymph nodes.[41]

Biomolecule	Raman peaks (cm⁻¹)	Biomolecule	FTIR peaks (cm⁻¹)
Protein	**1659, 1451, 1319, 1246,** 1207, 1174, 1128, 1103, 1058, **1002,** 959, **936**	Protein	**1662, 1646, 1550, 1532, 1516,** 1471, 1453, 1403, 1387, 1238,1171, 979
Nucleic acids	1666, **1574,** 1483, 1459, 1414, 1377, 1336, 1304, 1291, **1253,** 1215, 1194, **1099,** **1066,** 1011,957, **913**	Nucleic acids	**1712,** 1662, 1643, 1602, 1576, 1527, 1493, 1406, 1370, 1327, 1238, 1212, **1088,** **1050, 1011, 962,** 917
Fatty acids	1636, 1464, **1441,** 1423, 1375, **1298,** 1175, **1129,** 1099, **1064,** 1029, 1011, 977	Fatty acids	**1705, 1689, 1464,** 1431, **1408, 1311, 1295,** 1271-1187, 1098, 940
Triglyceride	**1747, 1653, 1439, 1300, 1267,** 1117, **1079,** 1038	Triglyceride	**1747, 1466, 1380,** 1243, 1164, 1118, 1096
Carbohydrates	1461, **1382, 1337,** 1261, 1207, **1125,** 1084, **1049, 939**	Carbohydrates	1201, **1153, 1080,** 1055, **1020, 994**

this implies that spot sizes as low as 1 μm diameter are easily attainable whereas in the mid-IR (wavelength ≈ 5–10 μm) apertured spot sizes of 25 μm are typical with bench-top instruments. At such spatial resolution, subcellular detail is impossible to determine. The DASIM project has helped to advance synchrotron FTIR microscopy at cell and tissue levels. The high brilliance of the source enables apertures to be as low as $10 \times 10 \, \mu m^2$ or even $6 \times 6 \, \mu m^2$, making whole cell and intracellular measurements feasible (see Chapters 3 and 7). During the course of this project, much progress has also been made in terms of standardization protocols and understanding the contribution of scattering phenomena such as Mie scattering (refer to Chapter 8). In Raman spectroscopy, the diffraction limit applies to the incident monochromatic light source. In the case of FTIR, however, the incident light is not monochromatic, and operating at the diffraction limit can imply reduction of spectral range, as shown in Chapter 2. In terms of spatial resolution, synchrotron IR sources perform significantly better than conventional bench-top instruments due to the increased beam collimation. Attenuated total reflection imaging techniques have reported spatial resolutions of as low as 2 μm, although this is a significantly more specialized and complex technique.

In terms of axial resolution, for transparent samples, Raman microscopy is governed by the focal depth of the objective, whereas IR microscopy is governed by the thickness of the sample (in simple transmission mode). For a 100× objective, this is typically ∼ 1–2 μm and the axial resolution can be improved by taking advantage of confocal imaging conditions. An important consideration, however, is that whereas in the visible region the majority of cellular

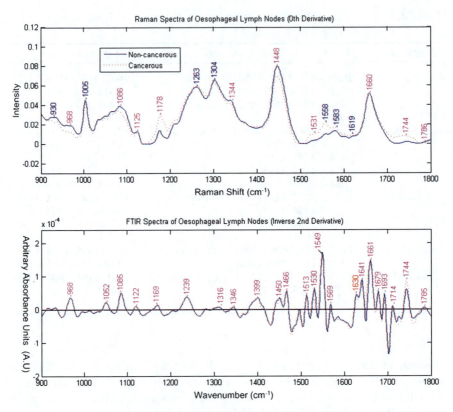

Figure 4.15 Plot of Raman (0th derivative) *vs.* FTIR (inverse 2nd derivative) mean spectra for non-cancerous and cancerous lymph nodes (numbers in purple indicate peaks shared in both Raman and FTIR while numbers in blue are only selective for Raman and those in red only selective for FTIR).[41]

components are transparent, many are strongly absorbing in the mid-IR. Thus the nucleus of a cell is optically very dense whereas the cytoplasm is sparse. This has led to many anomalous results indicating that the nucleus is deficient in, for example, nucleic acids compared to the surrounding cytoplasm.[42] The variations in optical density across cells and tissue have led to many confusing results, and effects such as the "dispersion artefact" have entered the vernacular of FTIR spectroscopy. This phenomenon is a direct result of the fact that FTIR is an absorbance technique applied to samples of significant inhomogeneity of optical density and is not manifested in Raman spectroscopy. This stresses the importance of pre-processing FTIR spectra of cells and tissues to avoid such pitfalls and misinterpreting the spectral information, which is not only composed of the sample's biochemical information but also information of physical origins. The physical origin of some of these effects in both transmission and transflection mode have recently been elucidated,[43,44] and a reliable method to remove these artefacts has been proposed (see also Chapter 8).[45]

4.8.3 Fluorescence and Scattering

While operating at visible wavelengths in Raman has the advantage of spatial resolution and optical transparency, there are also disadvantages. In general, Raman scattering is inherently a weak process and it suffers from the problem of fluorescence background. If the sample of interest is resonant with the illuminating wavelength, even a low efficiency luminescent emission can be sufficient to swamp the weak Raman signal on the Stokes side. The spectrum of thyroid tissue section in Figure 4.16 was recorded using 514.5 nm irradiation and the background registers in the spectrometer region ~ 525–630 nm (400 cm^{-1} to 3500 cm^{-1} Raman/Stokes shifted from 514.5 nm) and beyond. This background is commonly assumed to be fluorescence and this assumption has entered the biomedical spectroscopy literature.[46–49] Fluorescence from biomaterials has also been reported at irradiation wavelengths of 785 and even 830 nm.[48,49] It is important to remember, however, that for a material to fluorescence, it must absorb at the irradiating wavelength. Fluorescence spectroscopy has been explored as a potential probe of malignancies in for example skin, the principle chromophores being nicotinamide adenine dinucleotide (NADH), collagen, elastin, tryptophan, flavins and porphyrins.[50] However, excitation wavelengths are typically less than 400 nm and while some chromophores such as haemoglobin absorb at visible wavelengths, there is significantly less fluorescence higher than ~ 600 nm. It must be argued therefore that the large and

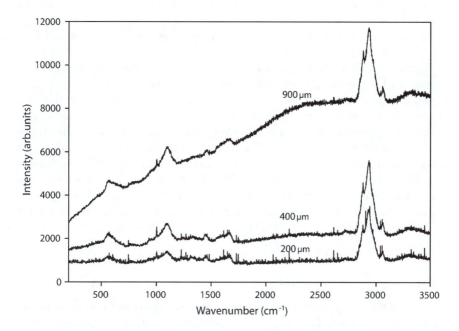

Figure 4.16 Unprocessed Raman spectra of thyroid tissue obtained at an excitation wavelength of 514.5 nm with varying confocal hole.

problematic background to the Raman spectra commonly observed at visible wavelengths may have contributions from a different physical origin.

The background to the Raman spectrum at visible wavelengths can have its origin in stray light from Mie scattering, such that sample morphology can play an important role and the spectral signal to background ratio can be improved by employing a true confocal configuration. Mie scattering occurs upon the interaction of radiation with particles of similar or larger dimensions compared to the incident radiation. Tissue sections have cellular and subcellular features of the order or 1–$10\,\mu m$. This is in the realm of Mie scattering of visible radiation and gives the sections a white diffuse appearance. It is not surprising therefore that these inhomogeneous samples give rise to a broad background to the Raman spectra. Mie scattering is an elastic process and, since the scattered radiation is of the same frequency as the laser, it should be removed by the holographic notch filter and prevented from entering the spectrometer. The transmission spectrum of the notch filter is however strongly angular dependent. Diffusely scattered radiation is not well collimated by the microscope objective and is thus incident on the notch filter at non-optimized angles. Although it is not well understood, it is assumed that this radiation is transmitted by the notch filter and enters the spectrometer as stray light where, although monochromatic, it appears dispersed across the CCD.

The problems of Mie scattering can be reduced by utilizing the confocality of the microscope configuration. The Mie scattered radiation is not collimated and thus is not transmitted by the confocal hole. Improvements of signal to background of a factor of ~ 10 have been demonstrated in this way (see Figure 4.16).

The measurements reported do not imply that resonant excitation of biological or other materials cannot give rise to fluorescent emission. While the background to the Raman spectra can be viewed as an inconvenience which can be removed either instrumentally or by background removal post spectral recording, fluorescent emission by definition implies a resonant excitation of a chromophore which, when present, through their emission or resonantly enhanced Raman signals could in themselves be valuable as diagnostic or analytical markers. However, as fluorescent efficiencies are in most cases sub unity, any excited state can give rise to local heating and/or photo-oxidative chemistry. With operating powers of $10\,mW$, this implies a power density of $\sim 10^{4}\,W\,m^{-2}$, and while these powers are required given the low efficiency of the Raman effect they can potentially cause significant sample degradation, especially at shorter wavelengths, where the photon energy is high.

4.8.4 Photodegradation

Reports of photodegradation in Raman spectroscopy are indeed numerous. In many cases, however, these are due to multiphoton resonances at high intensities in Optical tweezers or CARS experiments. In a study of photodegradation in Raman spectroscopy of living cells and chromosomes Puppels reported that while significant degradation was observable at 514.5 nm, no degradation was observable at 660 nm [51].[51] It should be noted, however, that even at 514.5 nm, the photon energy (2.4 eV) is not sufficient to cleave a covalent bond. Such a spectral dependence of the photodegradation process is reminiscent of that observed for conjugated organic polymers. In this case the mechanism is one of photo-oxidation whereby a photoexcited species transfers its energy to an oxygen molecule which now in its highly reactive singlet state attacks the donating species causing bond cleavage. In addition to photo-oxidative mechanisms, high power density and absorption can combine to produce thermal damage.[52-54]

It is often necessary to compromise collection efficiency in order to reduce the spotsize and therefore power density while maintaining illumination power to avoid sample degradation. In general in Raman spectroscopy, many intrinsic problems can be avoided by use of near-IR radiation, however. In this way the likelihood of photoexcitation of constituent molecules giving rise to photo-emission or photochemistry is reduced. Rayleigh and Mie scattering are similarly reduced. Commonly available near-IR sources supplied with commercial Raman spectrometers are 785 nm and 830 nm laser diodes. On the Stokes shifted side of the Raman spectrum, the "CH" region of the spectrum at 3000–3500 cm^{-1} is already shifted beyond the sensitivity range of silicon based CCD systems and therefore 785 nm illumination is accepted as the preferred choice for most biological applications.

4.8.5 Signal to Noise

Ultimately, the Raman effect is intrinsically weak and a limit of the technique is the signal to noise achievable. Raman spectroscopy for the measurement of biochemical changes in tissue depends as much upon the signal to noise (S/N) ratio as the magnitude of the measured Raman signal alone. The S/N ratio is a useful measure describing the quality of the spectrum; its inverse is the relative precision of the measurement, or the relative standard deviation from the true signal. This section aims to quantify the contributions to the Raman spectrum of each common component of noise or source of erroneous signal.

Shot noise is the dominant source of noise in dispersive Raman measurements. It is caused by the random probabalistic nature of light and matter. If the intensity of light is measured with a perfect noise-free detector, the standard deviation of the number of detected photons will be equal to the square root of detected photons.[55] Using an optimized Renishaw System 1000 Raman spectrometer to measure Raman scattering in oesophageal biopsy samples with 830 nm light, the strong C–H stretch band intensity at 1455 cm^{-1} was approximately 4500 counts in 30 s. Therefore the uncertainty of the measurement, due to shot noise, is 67 counts or ±1.5%. If the time of acquisition were reduced to 10 s, the measurement would yield 1500 ± 39 counts or ±2.6% uncertainty; a further reduction to an integration time of 1 s would give 150 ± 12 counts or ±8% uncertainty. This example shows the effect of reducing acquisition time leading to a reduction in the S/N ratio and the certainty of the measurement.

It must be noted that the shot noise will not only originate from the Raman scattering signal but also background signal originating from stray light and/or fluorescence. Therefore even if these signals can be subtracted, the shot noise contribution will remain superimposed on the measurement, sometimes completely obscuring the Raman spectrum. For example if the background induced in a sample produces 1000 counts per second at a particular wavenumber of a Raman scattering band that produces 100 counts per second, then the combined signal at the band position will be 1100 counts $\pm$ 33 counts from shot noise. Following subtraction of the background signal of 1000 counts the Raman band has intensity of 100 photons $\pm$ 33 photons or $\pm$ 33% uncertainty. At 830 nm, tissue scattering/fluorescence background has been shown to contribute around 100 counts per second (in this example), whereas the Raman signal can be between 5 and 50 counts per second.

The variation in pixel sensitivity and thermal noise across the CCD detector will superimpose a fixed pattern noise on the Raman spectrum. The effects of these can be reduced by binning several illuminated pixels in the intensity direction; or they can be corrected by collecting a spectrum of light that changes slowly with wavelength and dividing the Raman spectrum by the source spectrum. (This procedure has been outlined elsewhere.[56]) CCDs have an inherently low dark noise that will depend upon the pixel location, as there will be a temperature gradient from the centre to the edge of the chip, yielding fixed pattern noise plus random noise. CCD detectors are therefore commonly Peltier cooled to $-70\,°C$ to minimize thermal noise contributions.

The contribution to the signal in this example from the CCD [Renishaw RenCam (1998)] readout noise was approximately seven electrons per readout. Read noise is a random noise and therefore increases with the square root of the number of readouts per channel. In the case where an active region of 576×20 pixels is used, when reading out 20 pixels separately then the read noise would be around 32 electrons per wavenumber channel (this equates to approximately 6 counts with the Renishaw system in high gain mode). However, if the 20 pixels are binned and digitized together the read noise would be about seven electrons per channel.

Cosmic rays passing through the photosensitive region can produce thousands of photoelectrons. This effect results in a very strong sharp signal in the Raman spectrum. Quantification of spike noise is difficult due to the random nature of its occurrence. However, it is usually quite obvious to the observer when a spectrum of biological tissue contains a contribution from spike noise. These spikes can be erased from the spectrum or the whole spectrum can be discarded. They can also be circumvented by averaging several scans.

The effect of ambient lighting, another source of fixed pattern noise, on tissue Raman spectra should also be considered.

Source noise is caused by fluctuations in the irradiance of the incident light, which inherently causes fluctuations in the Raman scattering. Simultaneous measurement of all spectral components across the CCD array reduces the effect of this noise in an individual spectrum. However, comparison of one spectrum to the next for quantification of biochemical changes is complicated by source noise. The fluctuation in intensity and wavelength of the laser source should also be quantified. This is why the use of stable diode lasers is preferred.

An often neglected source of fixed pattern noise is that caused by instrument alignment and calibration errors. Unwanted information about the performance of the Raman instrument is added to the Raman spectrum. Calibration drift errors should therefore also be considered.

In summary, a typical Raman spectrum of fresh tissue, measured at 830 nm for 30 s with a laser power of 32 mW at the sample, will include a C–H stretch peak at 1445 cm^{-1}. If the intensity at this peak is for example 4500 counts then a contribution of 67 counts will be due to shot noise. The total signal of 4500 counts is made up of approximately 3000 counts of fluorescence/stray light signal and 1500 counts of inelastic scattering signal. The contributions from fixed pattern noise sources can be minimized by multiplying the spectra by correction files. Readout noise contributes about 6 counts and the dark current or thermal noise in the CCD contributes approximately 4 counts in the 30 s integration time. Hence measurement repeatability due to quantifiable noise contributions is approximately $\pm 5\%$ for the 1445 cm^{-1} peak measured in 30 s. This can be converted to a quantifiable S/N ratio of 20. The theoretical noise contributions agree well with the measured S/N ratio at 830 nm of 18.5.

4.9 Conclusions

Early detection of disease is critical to successful treatment and reduction of its impacts, *i.e.*, reduced morbidity and mortality. There have been significant advances in Raman technologies that could be exploited for the detection and tracking of molecules, signals or dynamic cellular events in living systems. The challenge is to advance these technologies further to enable the early detection of disease and to monitor disease progression and therapeutic efficacy. Its main advantages lie in the fact that it is chemical-free, offers high spatial resolution and is minimally invasive. It is readily amenable as a novel sensor for diagnostics in whole organisms as well as for miniaturized systems for point-of-care diagnostics. Raman therefore holds promise for bench-top and clinical applications.

Raman analysis holds inherent advantages over FTIR and synchrotron FTIR methodologies. These include higher spatial and spectral resolution and, for biological systems, the weak response of water. Operating at optical or near-IR wavelengths lends further advantages for fibre based *in vivo* applications. Although, in terms of spatial resolution, synchrotron-FTIR is comparable, cost and accessibility is an obvious issue. However, the combination of modalities, *i.e.*, synchrotron-FTIR microscopy with new imaging devices like focal plane arrays (see Chapter 7) or with an ATR imaging should in the future largely improve the potential of IR microscopy in both cell and tissue research.

One of the biggest challenges to moving Raman spectroscopy, as a diagnostic technology, from the laboratory to the health care system is the high computational burden of transforming measurements into some meaningful information that health care providers can use. This also applies to FTIR cell and tissue imaging. Therefore, a very important aspect will be the development of computational techniques and analytical tools for signal extraction/processing and computational modelling of living systems as a predictive tool for therapy, and dealing with large amounts of real-time diagnostic data coming from living systems. The main objective will be to find ways of providing meaningful diagnostic and monitoring information that can be captured efficiently, reliably and in real time (intelligent diagnostic). Such advances will make health care services more efficient, improve patient care and safety, reduce health care costs and/or create opportunities for remote care.

Among the future Raman based techniques, CARS, first reported by Duncan *et al.*,[57] can provide molecular specific contrast,[58,59] with 3D signal localization, due to the fact that multiphoton interactions are required to induce CARS signals in the volume of interest from highly focused, pulsed laser beams.[60] Due to its coherent signal generation it has an advantage over spontaneous Raman scattering microscopy, in that the signals are observed at higher energies relative to the excitation wavelengths. Therefore any fluorescence background from the sample will not interfere with the signal collected.[60] The most likely applications of this technology will be as a research tool for understanding carcinogenesis processes in *ex vivo* tissue specimens by optically dissecting the sample. It is already capable of rapidly providing high contrast molecular images at the cellular level and should be able to investigate intracellular pharmacokinetics by giving the distribution map of a specific molecule such as a drug.

Surface Enhanced Raman Spectroscopy can provide molecule specific enhancement of Raman signals,[61] by bringing the target molecule into close proximity with a roughened (nanometre scale) noble metal surface. Huge enhancement factors of the order of greater than 10^9 are possible and single molecule detection has been reported.[62] However, it has proved difficult with SERS to achieve reliable and reproducible results, a key requirement for clinical use. Improvements have been made with recent developments of novel substrates such as encapsulated nanoparticles that may overcome some of these difficulties.[63] They can be easily tagged with antibodies and photonic crystals,[64,65] which by careful manufacture can provide reliable substrates that can be tuned to specific resonance with excitation wavelengths. The use of antibody tags to enable molecule specific detection of disease has been demonstrated. Further developments have included the use of SERRS, a resonance SERS technique pioneered by Graham *et al.*, which is able to provide equivalent detection limits to fluorescence labelled dyes.[66] In the cancer environment, tagged nanoparticles enhancing specific signals from malignant markers are either being used *in vivo* (safety issues to be resolved)[67-69] or as molecular specific stains for histopathology;[70-72] with the possibility of numerous multiplexed SERS/SERRS stains providing hyperspectral images of locations of molecules of interest from the same spectral acquisition and tissue section.[73] The application of SERS for intracellular imaging and for monitoring drug distribution at physiological conditions has also been demonstrated.[74]

References

1. C. V. Raman and K. S. Krishnan, *Nature*, 1928, **121**, 501.
2. G. Landsberg and L. Mandelstam, *Naturwissenschaften*, 1928, **16**, 557.
3. M. Cardona, X. Zhou and T. Strach, in *Proc. 10th Anniv. HTS Workshop Phys. Mater Appl.*, ed. B. Battlog, World Scientific Singapore, 1996, p. 72–75.
4. C. Adjouri, A. Elliasmine and Y. Le Duff, *Spectroscopy*, 1996, **44**, 46.
5. R. C. Lord and N. T. Yu, *J. Mol. Biol.*, 1970, **20**, 509–24.
6. M. C. Tobin, *Science*, 1998, **161**, 68–69.
7. A. G. Walton, M. J. Deveney and J. L. Koenig, *Calc. Tiss. Int.*, 1970, **6**, 162.
8. N. T. Yu, B. H. Jo, R. C. C. Chang and J. D. Huber, *Arch. Biochem. Biophys.*, 1974, **160**, 614.
9. G. J. Puppels and J. Breve, in *"Biomedical Applications of Spectroscopy"*, eds. R.H.J. Clark and R.E. Hester, **vol. 25**, John Wiley and Sons, New York, 1996, Advances in Spectroscopy.
10. M. Gniadecka, H. C. Wulf, O. F. Nielsen, D. H. Christensen and J. Hercogova, *Photochem. Photobiol.*, 1997, **66**(4), 418–423.
11. C. M. Krishna, G. D. Sockalingum, L. Venteo, R. A. Bhat, P. Kustagi, M. Pluot and M. Manfait, Evaluation of the suitability of ex-vivo handled ovarian tissues for optical diagnosis by Raman microspectroscopy, *Biopolymers*, **5**, 269–276.
12. J. Smith, C. Kendall, A. Sammon, J. Christie-Brown and N. Stone, *Technol. Cancer Res. Treat.*, 2003, **2**(4), 327–331.
13. A. Molckovsky, L.M.W.K. Song, M. G. Shim, N. E. Marcon and B. C. Wilson, *Gastroint. Endosc.*, 2003, **57**(3), 396–402.
14. E. B. Hanlon, R. Manoharan, T.-W. Koo, K. E. Shafer, J. T. Motz, M. Fitzmaurice, J. R. Kramer, I. Itzkan, R. R. Dasari and M. S. Feld, *Phys. Med. Biol.*, 2000, **45**, R1–R59.
15. P. J. Caspers, G. W. Lucassen, R. Wolthuis, H. A. Bruining and G. J. Puppels, *Biospectroscopy*, 1999, **4**, S31–39.
16. U. Utzinger, A. Mahadevan-Jansen, D. Hinzelman, M. Follen and R. Richards-Kortum, *Appl. Spectrosc.*, 2001, **55**(8), 955.
17. N. Stone and P. Matousek, "Advanced Transmission Raman Spectroscopy – a promising tool for breast disease diagnosis", *Cancer Res.*, 2008, **68**, 4424–4430.
18. N. B. Colthup, L. H. Daly and S. E. Wiberley, *"Introduction to infrared and Raman spectroscopy"*, Academic Press, New York, 1975.
19. E. Smith and G. Dent, *"Modern Raman Spectroscopy: A Practical Approach"*, John Wiley and Sons, New York, 2005.

20. I. R. Lewis and H. G. M. Edwards, *"Handbook of Raman Spectro-scopy: From the Research Laboratory to the Process Line"*, CRC Press, 2001.

21. D. A. Long, *"The Raman Effect A Unified Treatment of the Theory of Raman Scattering by Molecules"*, John Wiley and Sons, New York, 2002.

22. F. S. Parker, *"Applications of Infrared, Raman, and Resonance Raman Spectroscopy in Biochemistry"*, Springer-Verlag, Heidelberg, 1983.

23. A. Lewis, J. Spoonhower, R. A. Bogomolni, R. H. Lozier and W. Stoeckenius, *Proc. Natl. Acad. Sci. USA*, 1974, **71**, 4462–4466.

24. R. J. H. Clark and T. J. Dines, *Angewandte Chemie*, 1985, **25**, 131–158.

25. M. Fleischmann, P. J. Hendra and A. J. McQuillan, *Chem. Phys. Lett.*, 1974, **26**, 163.

26. D. L. Jeanmaire and R. P. van Duyne, *J. Electroanal. Chem.*, 1977, **84**, 1–20.

27. I. Chourpa, F. H. Lei, P. Dubois, M. Manfait and G. D. Sockalingum, *Chem. Soc. Rev.*, 2008, **37**(5), 993–1000.

28. S. Nie and S. R. Emory, *Science*, 1997, **275**, 1102.

29. W. M. Tolles, J. W. Nibler, J. R. McDonald and A. B. Harvey, *Appl. Spectrosc.*, 1977, **31**(4), 253–339.

30. G. Eckhardt, D. P. Bortfeld and M. Geller, *Appl. Phys. Lett.*, 1963, **3**, 137.

31. L. D. Ziegler, *J. Raman Spectrosc.*, 2005, **21**, 769–779.

32. E. Ly, O. Piot, A. Durlach, P. Bernard and M. Manfait, *Appl. Spectrosc.*, 2008, **62**, 1088–1094.

33. J. Hutchings, C. Kendall, B. Smith, N. Shepherd, H. Barr and N. Stone, *J. Biophot.*, 2009, **2**, 91–103.

34. F. M. Lyng, E. Ó Faoláin, J. Conroy, A. Meade, P. Knief, B. Duffy, M. Hunter, J. Byrne, P. Kelehan and H. J Byrne, *Exp. Molec. Pathol.*, 2007, **82**, 121–129.

35. T. R. Chowdhury and J. R. Chowdhury, *Acta. Cytol.*, 1981, **25**(5), 557–565.

36. F. Draux, P. Jeannesson, A. Beljebbar, A. Tfayli, N. Fourré, M. Man-fait, J. Sulé-Suso and G. D. Sockalingum, *Analyst*, 2009, **134**(3), 542–548.

37. A.D. Meade, C. Clarke, F. Draux, G.D. Sockalingum, M. Manfait, F. M. Lyng, H. J. Byrne. *Anal. Bioanal. Chem.* Jan 20. [Epub ahead of print] (2010).

38. D. A. Moss, M. Keese and R. Pepperkok, *Vib. Spectrosc.*, 2005, **38**(1–2), 185–191.

39. F. Draux, P. Jeannesson, C. Gobinet, J. Sule-Suso, J. Pijanka, C. Sandt, P. Dumas, M. Manfait and G. D. Sockalingum, *Anal. Bioanal. Chem.*, 2009, **395**(7), 2293–301.

40. M. K. Kuimova, K. L. Chan and S. G. Kazarian, *Appl. Spectrosc.*, 2009, **63**(2), 164–71.

41. M. Isabelle, N. Stone, H. Barr, M. Vipond, N. Shepherd and K. Rogers, *Spectroscopy*, 2008, **22**, 97–104.

42. B. Mohlenhoff, M. Romeo, M. Diem and B. R. Wood, *Biophys. J.*, 2005, **88**, 3635–3640.
43. P. Bassan, H. J. Byrne, F. Bonnier, J. Lee, P. Dumas and P. Gardner, *Analyst*, 2009, **134**, 1586–1593.
44. P. Bassan, H. J. Byrne, J. Lee, F. Bonnier, C. Clarke, P. Dumas, E. Gazi, M. D. Brown, N. W. Clarke and P. Gardner, *Analyst*, 2009, **134**, 1171–1175.
45. P. Bassan, A. Kohler, H. Martens, J. Lee, H. J. Byrne, P. Dumas, El. Gazi, M. Brown, N. Clarke and P. Gardner, *Analyst*, 2010, **135**, 268–277.
46. V. Mazet, C. Carteret, D. Brie, J. Idier and B. Humbert, *Chemomet. Intell. Lab. Syst.*, 2005, **76**, 121.
47. C. A. Lieber and A. Mahadevan-Jahnsen, *Appl. Spectrosc.*, 2003, **57**, 1363.
48. J. Zhao, H. Liu, D. I. McLean and H. Zeng, *Appl. Spectrosc.*, 2007, **61**, 1225.
49. E. B. Hanlon, R. Manoharan, T.-W. Koo, K. E. Shafer, J. T. Motz, M. Fitzmaurice, J. R. Kramer, I. Itzkan, R. R. Dasari and M. S. Feld, *Phys. Med. Biol.*, 2000, **45**, R1–R59.
50. R. Na, I.–M. Stender, M. Henriksen and H.- C. Wulf, *J. Invest. Dermatol.*, 2001, **116**, 536.
51. G. J. Puppels, J. H. F. Olminkhof, G. M. J. Segers-Nolten, C. Otto, F. F. M. de Mul and J. Greve, *Exp. Cell Res.*, 1991, **195**, 361.
52. P. P. Calmettes and M. W. Berns, *Proc. Natl. Acad. Sci. USA*, 1983, **80**, 7197–7199.
53. D. O. Lapotko and V. P. Zharov, *Lasers in Surgery and Medicine*, 2005, **36**, 22–30.
54. Y. Liu, D. K. Cheng, G. J. Sonek, M. W. Berns, C. F. Chapman and B. J. Tromberg, *Biophysical Journal*, 1995, **68**, 2137–2144.
55. M. Bass, E.W. Van Stryland, D.R. Williams, W.L. Wolfe Eds, *Handbook of Optics Volume 1, Fundamentals, Techniques and Design*, 2nd edition, McGraw-Hill, New York, 1995.
56. N. Stone, C. Kendall, J. Smith, P. Crow and H. Barr, *Faraday Discuss.: Applications of Spectroscopy to Biomedical Problems*, 2004, **126**, 141–157.
57. M. D. Duncan, J. Reintjes and T. J. Manuccia, *Opt. Lett.*, 1982, **7**, 350.
58. A. Volkmer, *J. Phys. D*, 2005, **38**, R59–R81.
59. J.-X. Cheng and X. S. Xie, *J. Phys. Chem. B.*, 2004, **108**, 827–840.
60. M. ller and A. Zumbusch, *Chem. Phys. Chem.*, 2007, **8**, 2156–2170.
61. M. Fleischmann, P. J. Hendra and A. J. McQuillan, *Chem. Phys. Lett.*, 1974, **26**, 163–6.
62. K. Kniepp, *Phys. Rev. Lett.*, 1997, **78**, 1667.
63. W. E. Doering, M. E. Piotti, M. J. Natan and R. G. Freeman, *Adv. Mater.*, 2007, **19**, 3100–3108.
64. S. Cintra, M. E. Abdelsalam, P. N. Bartlett, J. J. Baumberg, T. A. Kelf, Y. Sugawara and A. E. Russell, *Faraday Discuss.*, 2006, **132**, 191–199.
65. S. Mahajan, M. Abdelsalam, Y. Suguwara, A. Russell, J. Baumberg and P. Bartlett, *Phys. Chem. Chem. Phys.*, 2007, **9**(1), 104–109.

66. G. Sabatte, R. Keir, M. Lawlor, M. Black, D. Graham and W. E. Smith, *Anal. Chem.*, 2008, **80**(7), 2351–2356.
67. A. Pal, N. R. Isola, J. P. Alarie, D. L. Stokes and T. Vo-Dinh, *Faraday Discuss.*, 2006, **132**, 293–301.
68. S. Keren, C. Zavaleta, Z. Cheng, A. de la Zerda, O. Gheysens and S. S. Gambhir, *Proc. Natl. Acad. Sci. USA*, 2008, **105**, 5844–5849.
69. X. M. Qian, X. H. Peng, D. O. Ansari, Q. Yin-Goen, G. Z. Chen, D. M. Shin, L. Yang, A. N. Young, M. D. Wang and S. M. Nie, *Nat. Biotech.*, 2008, **26**, 83–90.
70. L. Sun, K.-B. Sung, C. Dentinger, B. Lutz, L. Nguyen, J. Zhang, H. Qin, M. Yamakawa, M. Cao, Y. Lu, A. J. Chmura, J. Zhu, X. Su, A. A. Berlin, S. Chanand and B. Knudsen, *Nano Lett.*, 2007, **7**, 351–356.
71. B. Lutz, C. Dentinger, L. Sun, L. Nguyen, J. Zhang, A. J. Chmura, A. Allen, S. Chan and B. Knudsen, *Histochem. Cytochem.*, 2008, **56**(4), 371–379.
72. Y. N. Liu, Z. O. Zou, Y. O. Liu, X. X. Xu, G. Yu and C. Z. Zhang, *Spectrosc. Spect. Anal.*, 2007, **27**, 2045–2048.
73. K. Faulds, R. Jarvis, W. E. Smith, D. Graham and R. Goodacre, *Analyst*, 2008, **133**, 1505–1512.
74. I. Chourpa I, F. H. Lei, P. Dubois, M. Manfait and G. D. Sockalingum, *Chem. Soc. Rev.*, 2008, **37**(5), 993–1000.

Addendum A – Raman Calibration Procedure

(a) **Spectral Calibration**

Generally instrument software is based on a linear calibration and thus two reference points are sufficient. In the software, adjustment of the Zero (straight line intercept) and Co-efficient (straight line slope) is possible.

The grating should be tuned to the zero order and the zero point adjusted to ensure agreement. The grating should be tuned to the laser wavelength and the software calibration adjusted accordingly (in cases where the laser line is fixed rather than tuneable, this step is performed upon installation).

The instrument should be focused on a reference sample and the Raman response recorded. Silicon is often recommended as the reference material because it is stable and has a single strong narrow Raman mode at $520.7\,\mathrm{cm}^{-1}$ (see Figure 4.2). The Co-efficient parameter should be adjusted to ensure agreement. This can be performed most accurately by fitting a Gaussian/Lorentzian band to the peak.

These steps can be repeated a number of times to ensure a good calibration.

It is recommended at this point also to measure a multiline spectrum from a reference sample (PET, 1,4 Bis (2-methylstyryl) benzene, neon light) so that any nonlinearities of the calibration can be adjusted for in the final data.

(b) **Intensity Calibration**

The intensity calibration can be achieved via the following steps.

Record the dark current response of the detector over the region of interest.

Record the spectrum of a broadband intensity standard. Although white light sources such as halogen or tungsten lamps can be employed, the illumination geometry does not mimic that of the Raman collection well as it does not act as a point source. Thus the use of fluorescent standards (where they exist) such as those provided by the American NIST is recommended, although there are sometimes point to point variations in the signal found from these standards.

Section 2
Technical Aspects

Preparation of Tissues and Cells for Infrared and Raman Spectroscopy and Imaging

FIONA LYNG,[a] EHSAN GAZI[b] AND PETER GARDNER[b]

[a] Radiation and Environmental Science Centre, Focas Institute, Dublin Institute of Technology, Kevin St, Dublin 8, IRELAND; [b] School of Chemical Engineering and Analytical Science, Manchester Interdisciplinary Biocentre, University of Manchester, 131 Princess Street, Manchester, M1 7DN, UK

5.1 Introduction

Fourier transform infrared (FTIR) and Raman spectroscopy can provide important structural information on the molecular composition of a sample as well as relative quantification of lipids, proteins, carbohydrates and a variety of phosphorylated biomolecules within cells or tissue. However, sample preparation is the key to realizing the full potential of these technologies. This element of the experimental design can have a significant impact on the interpretation of spectra for their biochemical relevance and on the spatial distribution of biomolecules in imaging studies.

This chapter contains two main parts, the first part (section 5.2) deals with sample preparation for tissues, and the second part (section 5.3) deals with sample preparation for cells. In section 5.2, tissue preparation methods for spectroscopic analysis such as cryopreservation (section 5.2.2), fixation (section 5.2.3) and embedding (section 5.2.4) are described and a number of studies investigating the

RSC Analytical Spectroscopy Monographs No. 11
Biomedical Applications of Synchrotron Infrared Microspectroscopy
Edited by David Moss

effects of these tissue processing steps are presented and discussed. In section 5.3, cell preparation methods such as drying and fixation (section 5.3.2) are described and again studies investigating the effects of these processing steps are discussed. Cell preparation methods for biomechanistic studies (section 5.3.3) and the effects of cell culture growth media and substrates on spectroscopic analysis of cells (section 5.3.4) are also included in this section. The preparation of live cells for FTIR and Raman spectroscopy (section 5.3.5) is also discussed.

5.2   Tissue Preparation

5.2.1   Introduction to Tissue Preparation Methods

Surgically excised tissue may undergo one of two commonly used methods of preservation for long-term storage: paraffin embedding or flash-freezing. The choice between these two methods is based on the specific purpose of the resected tissue.

Currently, formalin fixation and paraffin preservation (FFPP) is the preferred source for the histological examination of tissue sections. This method involves immersing tissue in an aqueous formalin solution. Hydrated formalin (methylene glycol, OH–CH–OH) is a coagulative protein fixative, which cross-links the primary and secondary amine groups of proteins,[1] but preserves some lipids by reacting with the double bonds of unsaturated hydrocarbon chains.[2] Following formalin fixation, the tissue is dehydrated through consecutive immersions in increasing grades of ethanol up to 100% ethanol. Displacement of water with ethanol preserves the secondary structure of proteins but denatures their tertiary structure. Furthermore, formalin or ethanol induces coagulation of the globular proteins present in the cytoplasm, which can result in the loss of structural integrity of organelles such as mitochondria. Another disadvantage is that ethanol precipitates lipid molecules that are not preserved through the primary fixation step. However, stabilization of intercellular proteins by formalin and ethanol localizes associated glycogen. Following dehydration, the alcohol is replaced by an organic solvent such as xylene, which is miscible with both alcohol and molten paraffin wax. The tissue is then immersed in and permeated by molten paraffin wax. The infiltration of the wax into the intracellular spaces is promoted by the previous ethanol dehydration step that created pores in the plasma membrane of the cells. The tissue sample is then cooled to room temperature, which solidifies the wax. This process provides a physical support to the sample enabling thin sections (usually 2–7 μm) to be cut without deformation to the cellular structure or architecture.

Snap-freezing of fresh tissue is generally preferred for molecular based studies since this method avoids the use of organic solvents that cause degradation or loss of some cellular components. In particular, frozen sections are used to study enzymes and soluble lipids. Furthermore, this method is used to conduct immunohistochemical analysis since some antigens may be affected by extensive cross-linking chemical fixatives that denature their tertiary structure.

In the case of snap-freezing, the water contained within the cells acts as the supporting medium. Fresh tissue is snap-frozen in liquid nitrogen cooled isopentane ($-170\,^{\circ}$C) to promote vitreous ice formation and to prevent ice crystal damage, since the latter can produce holes in the tissue and destroy cellular morphology and tissue architecture. The hardened tissue can then be embedded in mounting medium such as OCT (optimal cutting temperature compound) for sectioning within a cryostat maintained at $-17\,^{\circ}$C and then subsequently stained.

Optimal cutting temperature compound is a viscous solution at room temperature, consisting of a resin–polyvinyl alcohol, an antifungal agent, benzalkonium chloride and polyethylene glycol to lower the freezing temperature.[3] Turbett *et al.* suggested that snap-frozen tissues should be stored without any medium since it was found that amplification of deoxyribonucleic acid (DNA), extracted from these tissues, was significantly affected for segments of greater than 300 base-pairs.[3] However, ribonucleic acid (RNA) was found to be unaffected.

Although the methods outlined above represent the gold standard tissue processing techniques in most pathology laboratories, some researchers have recently reported alternative tissue preparation protocols with the view of optimizing the assessment of specific biomolecular domains. Gillespie *et al.* conducted a comparative molecular profiling study in clinical tissue specimens that were fixed for long-term storage with widely used techniques (snap-frozen and formalin fixed paraffin embedded) and a less common method of 70% ethanol fixation and paraffin embedding.[4] The researchers found that although the total protein quantity was decreased in fixed and embedded tissues compared to snap frozen tissue, two-dimensional polyacrylamide gel electrophoresis (2D-PAGE) analysis of proteins from ethanol-fixed, paraffin-embedded prostate shared 98% identity with a matched sample from the same patient that was snap-frozen. This indicated that the molecular weights and isoelectric points of the proteins were not disturbed by the tissue processing method. The general quality and the quantity of the proteins in the ethanol-fixed samples were found to be superior to formalin-fixed tissue. Furthermore, Gillespie *et al.* reported that recovery of messenger RNA (mRNA) and DNA was more pronounced in ethanol-fixed specimens compared with formalin-fixed samples.[4] Thus, further improvements to tissue processing methodologies will play a key role in ultimately determining the complete molecular anatomy of normal and diseased human cell types.

Researchers working in the field of FTIR and Raman tissue diagnostics have employed a variety of methods for tissue preparation. Early FTIR spectroscopic studies have been carried out using ground samples of snap-frozen tissue for the bulk analysis of chemical composition.[5,6] Using this method, Andrus *et al.* reported an increasing ratio of peak areas corresponding to bands at $1121\,\text{cm}^{-1}$ and $1020\,\text{cm}^{-1}$ (attributed to RNA/DNA), which were associated with increased aggressiveness of malignant non-Hodgkin's lymphomas.[5] Takahashi *et al.* used bulk tissue analysis to study glycogen levels in tissues obtained from colorectal tumours, regions adjacent to the tumour and regions of normal colorectum.[6] The results indicated that there was a statistically significant difference in glycogen levels (peak area ratio $1045\,\mathrm{cm}^{-1}/1545\,\mathrm{cm}^{-1}$) between cancer tissue and the other two regions.[6]

The use of ground tissue provides indiscriminate and composite measurement of both epithelial and stromal tissue compartments. However, care must be taken with this type of analysis when molecular assignments are made for discriminatory bands. In the study by Andrus *et al.*, the influence of collagen absorbance was discussed.[5] However, other confounding variables exist in stromal tissue, namely a variety of cell types such as endothelial cells of blood vessels, fibroblasts, ganglia and erythrocytes, in addition to possible bisecting nerves and muscle. This was clearly demonstrated by Fernandez *et al.* who classified several prostate tissue components for diagnostic FTIR imaging.[7]

A number of papers have been published that compare the effects of different sample preparation protocols for FTIR and Raman microspectroscopic studies,[8–11] and these are discussed in the following sections.

5.2.2 Fresh and Cryopreserved Tissue

More than a decade ago, Shim and Wilson[8] demonstrated that dehydration at room temperature of fresh tissue specimens (subcutaneous fat, smooth muscle, cheek pouch epithelium, oesophagus) resulted in Raman spectra with a decrease in intensity of the $930\,\mathrm{cm}^{-1}$ (C–C stretch of proline and valine) peak relative to the peaks at $1655\,\mathrm{cm}^{-1}$ and $1450\,\mathrm{cm}^{-1}$. Although this may be indicative of protein denaturing, the authors did not observe any shifts in the amide I peak. However, an increase in the lipid:protein signal was observed with increasing drying times, providing evidence that the protein vibrational modes were perturbed by dehydration. Interestingly, Shim and Wilson found that Raman spectra obtained from OCT-freeze stored, snap-frozen tissue in phosphate buffered saline (PBS) were comparable to spectra obtained from fresh tissue in PBS.[8] Additional peaks observed at lower frequency ($764\,\mathrm{cm}^{-1}$ and $795\,\mathrm{cm}^{-1}$) in the spectra of snap-frozen adipose tissue were attributed to the coagulation of erythrocytes.

Faoláin *et al.* also conducted a comparative study of frozen and fresh tissue using parenchymal tissue from the placenta (Figure 5.1).[9] Here, the authors did find significant differences in the Raman spectra, a reduction in the intensity of bands at $1002\,\mathrm{cm}^{-1}$ (C–C aromatic ring stretching), $1447\,\mathrm{cm}^{-1}$ (CH_2 bending mode of proteins and lipids) and $1637\,\mathrm{cm}^{-1}$ (Amide I band of proteins) in the frozen tissue compared with fresh tissue. Additionally, frozen tissue exhibited a new peak at $1493\,\mathrm{cm}^{-1}$, which was not found to be OCT contamination but was attributed to an artefact of the freezing process. It was suggested that this artefact was due to depolymerization of the actin cytoskeleton, resulting in an increased contribution of the NH_3^+ deformation mode.[9] Using FTIR microspectroscopy, the frozen tissue was found to display an overall reduction in intensity, while shifts in the Amide I and II bands indicated changes in protein conformation in the frozen tissue section.[9]

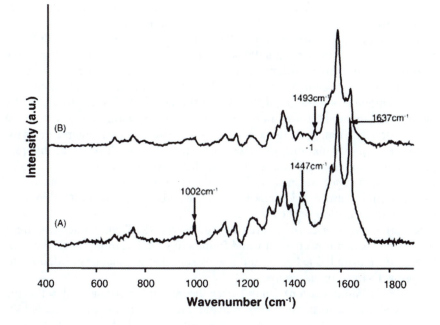

Figure 5.1 Raman spectra of (A) fresh tissue compared with (B) frozen tissue section. (Reproduced from reference [9].)

It is important to note that in the study by Faoláin *et al.*,[9] the comparison of fresh and frozen tissue was carried out with prior mounting onto MirrIR slides on which the frozen tissue had been thawed before analysis. Hence, depolymerization of proteins can also result from post-thawing of the frozen tissue, whereby the undesirable transition of vitreous ice into ice crystals could affect the integrity of the cytoskeletal proteins. It is well known within the structural cell biology community that this can be prevented by the application of freeze-drying. However, Shim and Wilson's investigations suggest that the spectral changes in protein vibrational modes, caused by heat induced denaturing of thawed frozen tissue, can be circumvented by thawing in PBS (maintained at room temperature).[8] Another molecular change associated with tissue thawing and dehydration was found to include a change in the relative intensities of the Amide I and methyl bending modes.[8]

The extent of protein depolymerization that has been observed in Raman spectra of freeze-dried tissue can be reduced by using an appropriate cryogen for the initial snap-freezing of the resected tissue specimen. Higher freezing rates are achieved using propane or mixtures of propane and isopentane in preference to isopentane.[12] Also, since ice crystals develop in a temperature range of 0 to $-140\,^{\circ}\text{C}$,[12] it is advantageous to maintain cryopreserved tissues at temperatures lower than $-140\,^{\circ}\text{C}$ during microtomy and freeze-drying. Finally, the size of the specimen also dictates the extent of ice crystal damage; in liquid nitrogen cooled propane, it has been reported that $0.5\,\text{mm}^3$ is the critical size that separates complete crystallization from partial vitrification.[12] Nevertheless, Stone *et al.* have demonstrated on a number of different tissue types that freeze-thawed tissue, without PBS, can be used to differentiate different pathologies with greater than 90% sensitivity and specificity.[13] The same method of sample preparation was used to demonstrate the biochemical basis for tumour progression of prostate and bladder cancers by determining the relative amounts of a number of tissue constituents.[14] This was carried out by obtaining Raman spectra of pure standards and correlating these with tissue spectra derived from each disease state using ordinary least-squares analysis. The biological explanations for these constituent changes through the different pathological states could be associated with known tumour behaviour. Thus, freeze–thawed tissue warmed to room temperature is not only diagnostically useful but may also provide relative quantifications and qualitative biomolecular characterization of the sample.

Recently, Wills *et al.* compared cryopreserved (banked) tissue with fresh tissue using Raman spectroscopy and found a high correlation, suggesting that frozen tissue from tissue banks could be used to develop diagnostic algorithms for use with fresh tissue samples.[15]

For FTIR investigations, snap-frozen tissue has been analysed following thawing and subsequent air-drying to avoid interference from water bands.[16–19] However, this dehydration process can result in undesirable perturbations to cellular chemistry, and some researchers have reported using cryosections dried under a stream of nitrogen gas to reduce oxidative and surface tension effects.[20,21] Both protocols have successfully been applied to the spectral classification of tissue pathologies,[16–21] and have been shown to generate detailed biospectroscopic maps that localize tumour lesions within oral[20] and brain tissues.[21] Additionally, using a univariate analytical approach to process tissue maps, Wiens *et al.* reported the use of dried snap-frozen sections to investigate the early appearance and development of scar tissue using FTIR signals corresponding to lipids, sugars and phosphates as well as collagen and its fibril orientation.[19]

5.2.3   Chemical Fixation of Tissue

It is important to note that the process of fixation is not instantaneous and two important properties of the fixative are its penetration rate and binding time. Medawar was the first to demonstrate that fixatives obey diffusion laws, whereby the depth of penetration was proportional to the square root of time.[22] The importance of fixative binding time was highlighted by Fox *et al.*, who investigated the binding of formaldehyde to rat kidney tissue, in which 16 μm thin sections were used so that penetration would not be considered a factor in the kinetics of the reaction.[23] They found that the amount of methylene glycol that covalently bound to this tissue increased with time until equilibrium was reached at 24 h. Thus, binding time is the limiting factor for tissue stabilization. These aspects of chemical fixation (penetration rate and binding time) may be a

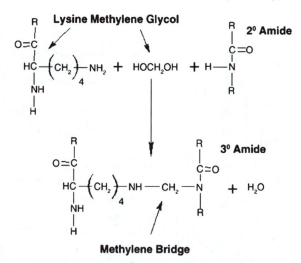

Figure 5.2 Scheme illustrating the formation of a methylene bridge during formalin fixation. (Reproduced from reference [9].)

potential source of biomolecular variance in pathological samples, since there exists a time lag in fixative exposure and binding between cells located within the core of the tissue compared with those at the extreme dimensions of the block.[23] In fact, Fox *et al.* reported that cells at the periphery of the tissue exhibit different morphological properties from cells that are a few tenths of a millimetre further within the specimen.[23]

Compared with fresh tissue, Raman spectra of formalin fixed tissue, following 24 h fixation, exhibit a significantly reduced intensity of the Amide I peak as well as the appearance of a peak at 1490 cm^{-1}.[9] The reduced intensity of the Amide I peak is attributed to the formation of tertiary amides (and loss of secondary amide), resulting from the reaction of methylene glycol (in formalin) cross-linking the nitrogen atom of lysine with the nitrogen atom of a peptide linkage (Figure 5.2).[9] The peak at 1490 cm^{-1} is thought to arise as a result of protein unravelling and increased activity of the NH_3^+ deformation mode. However, it has also been suggested[9] that the coupling of the C–N stretching vibration with the in-phase C–H bending in amine radical cations, which may be present in the methylene bridge following formalin fixation involving lysine (see Figure 5.3, below), could result in a peak at 1490 cm^{-1}. Thus, although previous FTIR based studies of formalin fixation of isolated proteins have demonstrated no measurable effects on protein secondary structure (the arrangement of the polypeptide backbone),[24] the data above suggest that it does have an effect on protein intensity. However, contrary to previous findings on isolated proteins,[24] Faoláin *et al.* found that formalin fixation of tissue produced a notable shift of 10 cm^{-1} in the Amide I and II bands in FTIR spectra.[9]

Formalin fixation was also found to preserve lipid signals in Raman spectra of lipid rich tissue such as adipose tissue and white matter of the brain.[8] Both

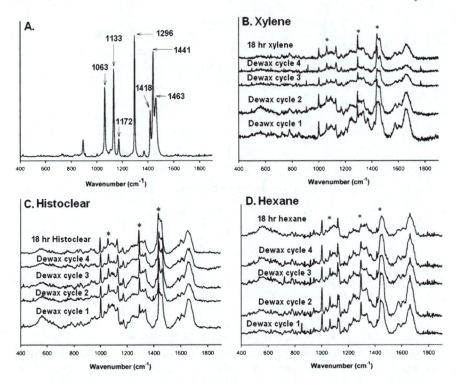

Figure 5.3   (A) Raman spectrum of paraffin wax; Raman spectra of paraffin embedded tissue following treatment with repeated cycles of deparaffinization
agents (B) xylene, (C) Histoclear and (D) hexane. * Marker peaks corresponding to paraffin residue at $1063 \, cm^{-1}$, $1296 \, cm^{-1}$ and $1441 \, cm^{-1}$.
Modified with permission and reproduced from reference [28].

Huang et al.[10] and Shim and Wilson[8] reported that a direct spectral con-
taminant from formalin in the Raman spectrum is a weak peak appearing at
$1040 \, cm^{-1}$. However, this artefact can be successfully removed through copious
washes with PBS.[9]

Pleshko et al.[11] have shown that 70% ethanol fixation of 35-day-old
embryonic rat femur gave rise to FTIR spectra that exhibited Amide I and II
peaks that were shifted to lower frequencies ($1647 \, cm^{-1}$ and $1546 \, cm^{-1}$,
respectively) compared with unfixed rat femur, which exhibited Amide I and II
peaks at $1651 \, cm^{-1}$ and $1550 \, cm^{-1}$. It was concluded that evaluation of protein
structure in this tissue should be limited to snap-frozen samples, or formalin-
fixed tissues where there were no observable shifts in the Amide I and II peaks.
Although this does not conform to the observations made by Faoláin et al.,[9]
this difference may be due to the different types of tissue used in each study,
calcified tissue in Pleshko et al.[13] and non-calcified tissue in Faoláin et al.[9]

Further studies of the effects of sample preparation on bone tissue
were undertaken by Aparicio et al.[25] and Yeni et al.[26] In the study by Aparicio

et al.,[25] seven different fixatives, absolute ethanol, 70% ethanol, glycerol, formaldehyde, EM fixative, and formalin in cacodylate or phosphate-buffered saline, and six different commonly used embedding media, Araldite, Epon, JB-4, LR White, PMMA and Spurr, were assessed by FTIR microspectroscopy. The FTIR spectra were compared with those from unprocessed ground tissue and with cryosections of unfixed tissue, fast-frozen in polyvinyl alcohol (5% PVA). Non-aqueous fixatives and embedding in LR White, Spurr, Araldite and PMMA were found to have the least effect on the spectral parameters measured (mineral to matrix ratio, mineral crystallinity and collagen maturity) compared with cryosectioned and non-fixed, non-embedded tissue.

In the study by Yeni *et al.*,[26] the effects of two fixatives, ethanol and glycerol, and six embedding media, Araldite, Eponate, Technovit, glycol methacrylate, polymethyl methacrylate and LR White, on the Raman spectral properties of bone were investigated. Non-fixed, freeze-dried and fixed but not embedded tissue samples were also included. Significant effects of fixation and embedding were observed and no single combination of fixative and embedding medium that left all spectral features unaltered was found. As found for the FTIR study by Aparicio *et al.*,[25] careful selection of a fixation/embedding combination for bone tissue is recommended to allow measurement of the spectral features of interest.

5.2.4 Paraffin Embedded Tissue

As mentioned in section 5.2.1, fixed tissue is immersed in xylene prior to impregnation with paraffin wax. Faoláin *et al.* have shown that Raman spectra of formalin fixed tissue exposed to xylene produce a number of strong peaks, associated with the aromatic structure, at $620 \, \text{cm}^{-1}$ (C–C twist of aromatic rings), $1002 \, \text{cm}^{-1}$ (C–C stretching of aromatic rings), $1032 \, \text{cm}^{-1}$ (C–C skeletal stretch of aromatic rings), $601 \, \text{cm}^{-1}$ (C–C in plane bending of aromatic rings) and $1203 \, \text{cm}^{-1}$ (C–C_6H_5 stretching mode of aromatic rings).[9] As mentioned in section 5.2.3, formalin fixation was found to reduce the intensity of the Amide I band. Interestingly, upon xylene exposure, the Amide I band was observed to reappear with appreciable intensity. It was suggested that the cross-linking of proteins by methylene glycol is reversed upon xylene treatment so that the Amide I band reverts back to the secondary amide.[9] As expected, the FTIR spectrum of xylene-treated formalin-fixed tissue demonstrated a loss of the lipid ester (C=O) band at $1740 \, \text{cm}^{-1}$, due to significant removal of cellular lipids.

Presently in the field of FTIR or Raman spectroscopy, there is lack of consensus with regard to a standard protocol for deparaffinization of paraffin embedded sections, and several approaches have been used. For example, Fernandez *et al.* deparaffinized prostate tissue sections by immersing in hexane at 40 °C with continuous stirring for 48 h.[7] During this period, the vessel was emptied every 3–4 h, rinsed thoroughly with acetone followed by hexane, and after thorough drying refilled with fresh hexane to promote dissolution of paraffin embedded in the tissue. The disappearance of a peak at $1462 \, \text{cm}^{-1}$ in the FTIR spectrum was used to ensure complete deparaffinization.

Sahu et al. deparaffinized samples of colon tissue using xylene and alcohol. The 10 μm paraffin embedded sections were washed with xylol for 10 min (three changes) with mild shaking.[27] Following this, the slide was washed with 70% alcohol for 12 h. To evaluate the efficacy of this procedure, FTIR spectra from the tissue were collected at each stage of the deparaffinization process: before deparaffinization, at each xylol washing step (following air-drying), and following alcohol treatment. It was reported that following two washes with xylol, a third xylol wash did not produce any significant changes to the lipid spectral regions (2800–3000 cm^{-1} and 1426–1483 cm^{-1}). Further treatment with alcohol produced changes to the region 900–1185 cm^{-1}, which was speculated to be the result of residual xylene removal. Alcohol treatment also showed a further reduction in lipid hydrocarbon signals in the spectral region 2800–3000 cm^{-1}. The authors observed that haematoxylin and eosin (H&E) sections of these deparaffinized tissues exhibited clear outlines for the cells that indicated the preservation of lipids in complex forms (membranes).

Faoláin et al. deparaffinized parenchymal tissue sections by immersing in two baths of xylene for 5 and 4 min, respectively, followed by two baths of absolute ethanol for 3 and 2 min and a final bath of industrial methylated spirits 95% for 1 min.[9] This method was found through Raman microspectroscopy to be inefficient at removing all of the paraffin since a number of strong signals from C–C and CH$_2$ vibrational modes were observed.[9]

Contrary to Fernandez et al.[7] and Sahu et al.[27] Faoláin et al. suggested that complete removal of paraffin from the tissue section is not possible to assess accurately with FTIR spectroscopy using the ∼ 1462 cm^{-1} signal.[9] This was concluded following analysis of Raman and FTIR spectra obtained from deparaffinized tissue sections (using an identical deparaffinization protocol for each mode of analysis) in which spectral peaks characteristic of paraffin were clearly resolved in Raman spectra (strong sharp bands), compared with FTIR spectra.[9] A follow-up study by Faoláin et al. comprehensively evaluated the efficiency of different deparaffinization agents to remove paraffin wax from cervical tissue sections.[28] In this study, one deparaffinization cycle involved two baths of deparaffinization agent (5 min and 4 min, respectively), followed by immersion in two baths of ethanol (3 min and 2 min, respectively) and a final bath in industrial methylated spirits 95%. This Raman based investigation demonstrated that paraffin signals at 1063 cm^{-1} (C–C stretch), 1296 cm^{-1} (CH$_2$ deformation) and 1441 cm^{-1} (CH$_2$ bending) (Figure 5.3A) were not completely removed by xylene and Histoclear even after 18 h of treatment (Figure 5.3B–C). However, hexane was found to be a superior deparaffinization agent, removing nearly all of the paraffin 18 hours post treatment (Figure 5.3D). These important findings were found to have practical implications for immunohistochemical staining. It was found that the intensity of positive staining in tissue sections treated with hexane for 18 h was 28% greater than in adjacent sections treated with xylene for the same period.[28] A more recent study confirmed that hexane was the most efficient dewaxing agent compared with xylene, histolene, cyclohexane, diethylether and 1.7% dishwashing soap, and showed conclusively that full deparaffinization by hexane resulted in improved immunohistochemical staining in rat oesophagus, skin, kidney and liver tissue.[29]

In the light of these findings from Faoláin *et al.*,[28] the protocol used by Fernandez *et al.*, where prostate tissue sections were treated with hexane for 48 h,[7] provides the most efficient method of paraffin removal used in an FTIR study to date. In fact, using this sample preparation protocol high levels of accuracy ($\geq 90\%$) were achieved for classifying a number of different tissue components for FTIR chemometric imaging of prostate tissue microarrays. Nevertheless, it has been shown by Gazi *et al.*,[30] who used a less rigorous method of deparaffinization involving Citroclear, which is less toxic than xylene, that the spectral region between $1481 \, cm^{-1}$ and $999 \, cm^{-1}$ could be used to discriminate and classify the different pathological grades of prostate cancer. A statistically significant distinction between tumours localized to the prostate gland and those showing extracapsular penetration could also be made. Moreover, the time efficient method of less rigorous deparaffinization is suitable for other FTIR based diagnostic parameters that do not include spectral regions that overlap with paraffin signals, such as the peak area ratio of the $1030 \, cm^{-1}$ (glycogen) to $1080 \, cm^{-1}$ (phosphate) bands, which differentiated malignant from benign prostate tissues in imaging studies.[31]

A recent study, by Meuse and Barker,[32] used FTIR spectroscopy to assess paraffin removal based on CH stretching bands near 2850, 2918 and $2956 \, cm^{-1}$. Hexane, xylene and limonene were compared as dewaxing agents and two human breast cancer cell lines were used as model tissues. With all three dewaxing agents, at least 97% of paraffin was found to be removed, with xylene found to be the most effective. The difference from the study by Faoláin *et al.*[28] is more than likely due to the fact that cell cultures were used, which would not have the complex intracellular components of tissue samples.

It may be argued that, for FTIR and Raman studies, the removal of paraffin is not necessary at all as only discrete frequency ranges corresponding to the lipid hydrocarbon modes are affected.

A number of studies have shown that non-dewaxed FFPP sections can be used successfully for skin and colon tissues.[33–37] For FTIR imaging, extended multiplicative signal correction (EMSC) was used to correct the spectral contribution from paraffin and PCA was used to improve the EMSC efficiency.[36,37] For Raman spectroscopy, independent component analysis (ICA) was used to obtain model sources of paraffin, and a spectral unmixing method, the non-negativity constrained least squares (NCLS) method, was used to eliminate the paraffin contribution from each individual Raman spectrum.[33–35] Untereiner *et al.* used FTIR imaging to compare FFPP sections without any dewaxing to frozen sections of peritoneal tissues and showed that sample preparation did not affect the discriminative potential of the technique.[38]

However, visualization of the unstained tissue's anatomical features has been shown to be severely hampered if the paraffin is not removed. Sahu *et al.* reported that colonic crypts in a 10 μm paraffin embedded tissue section appeared as circular entities when viewed under light microscopy.[27] Moreover, even between adjacent microtomed sections, tissue components can vary

Table 5.1   Summary of the main FTIR and Raman bands affected by standard tissue processing protocols used for histopathology.

	Main FTIR bands affected	*Main Raman bands affected*
Drying		Decrease in intensity of the 930 cm^{-1} peak, and an increase in the lipid:protein signal[20]
Cryopreservation	Overall reduction in intensity, shifts in the Amide I and II bands[19]	Reduction in the intensity of 1002[1], 1447 and 1637 cm^{-1} bands[19]
Formalin fixation	Shift of 10 cm^{-1} in the Amide I and II bands[19]	Reduced intensity of the Amide I peak,[19] and appearance of a peak at 1490 cm^{-1}
Paraffin embedding	1465 cm^{-1} band[19] 2850, 2918 and 2956 cm^{-1} bands[32]	1063, 1130, 1296 and 1436 cm^{-1} bands[19]

significantly, which in turn prevents the positioning of the beam upon a specific tissue location by comparison with an H&E section.

In Table 5.1 a summary of the main findings from section 2 is presented. The main FTIR and Raman bands affected by tissue drying, cryopreservation, formalin fixation and paraffin embedding are listed as these are the standard tissue processing methods used for histopathology.

5.3   Cell Preparation

5.3.1   Introduction to Cell Preparation Methods

Cells are naturally present in a hydrated form, whereby water molecules are bound to macromolecules such as proteins, phospholipids and carbohydrates and this contributes to their structural integrity and function. A review of the early literature concerning the application of FTIR to cell analysis for diagnostic or imaging purposes reveals that cells had generally been prepared by direct culture upon infrared (IR) transparent substrates, followed by removal from the culture medium and air-drying.[39–44] However, the air-drying process causes delocalization of biomolecules as a result of large surface tension forces associated with passing the water–air interface. Other researchers in the field have prepared cells by removing them from culture medium and then centrifugation,[45] drying under nitrogen gas,[46] or cytospinning,[47] with the view of minimizing the effects of these surface tension forces by increasing the rate of dehydration.

The removal of cells from pH buffered growth medium and subsequent air-drying can influence the osmotic pressure within these cells, resulting in cell shrinkage or swelling with the latter resulting in membrane rupture and leaching of intercellular components. In addition, drying of living cells can initiate autolytic processes whereby intracellular enzymes contained within lysosomes cause denaturing of proteins and dephosphorylation of mono-nucleotides, phospholipids and proteins. Furthermore, autolysis involves chromatin compaction, nuclear fragmentation (involving RNA and DNA nucleases) and cytoplasmic condensation and fragmentation. Thus, in FTIR based biomechanistic studies, where researchers are interested in identifying the metabolites formed as a result of the cell's response to specific stimuli, the effects of autolysis as a consequence of inappropriate cell preparation may obscure these investigations.

In cell biology, a critical and fundamental step in any investigation is "fixation". This is used to quench autolysis, minimize leaching of biomolecular constituents, whilst at the same time using optimized dehydration protocols to bypass surface tension distortions and preserve the structural and functional chemistry of biomolecules for analysis.

Another fundamental aspect of sample preparation that can influence cellular biochemistry is the surface on which they are grown. The surface can induce changes in cell adhesion and motility, in their proliferation and differentiation and in gene expression. It is desirable for *in vitro* cultures to mimic the *in vivo* environment as closely as possible and, in this context, progress has recently been made in modelling cellular systems in two-dimensional cultures. Studies have also been carried out detailing the use of biomaterial surfaces (Matrigel™, fibronectin, laminin, gelatin) for this type of cell culture. The influence of these surfaces on cell morphology and the spectral information obtained is also discussed.

5.3.2 Chemical Fixation of Cells

As mentioned in section 5.3.1, to avoid potential confounding variables from autolytic processes initiated by cells during air-drying, it is important that the cells are appropriately fixed to maintain localizations of biomolecular species. To this end, Gazi *et al.* studied the use of three chemical fixation methods for spectroscopically mapping single prostate cancer cells with synchrotron (SR) based FTIR microspectroscopy.[48] The cells were cultured directly onto MirrIR reflection substrates for transflection mode analysis. In the first method, the cells were fixed in 4% formalin (in PBS) for 20 min at room temperature with a brief rinse in doubly deionized water (3 s) before being air-dried. Water rinsing was found to be an important step for removing residual PBS from the surface of the cells so that a clear distinction could be made between the nuclear and cytoplasmic compartments (Figure 5.4A–B). This also reduced light-scattering artefacts during analysis. Although trypan blue staining of these cells demonstrated loss of plasma membrane integrity, which is likely to be due to the 1% methanol present in the fixative, SR-FTIR images (collected with $7 \mu m \times 7 \mu m$ sampling-aperture-size and $3 \mu m$ step-size) revealed localizations of lipid [v_s(C=O)] and phosphate [v_{as}(PO$_2$)] domains (Figure 5.4C). Spectral subtraction of the neat formalin spectrum from the FTIR spectrum of the formalin

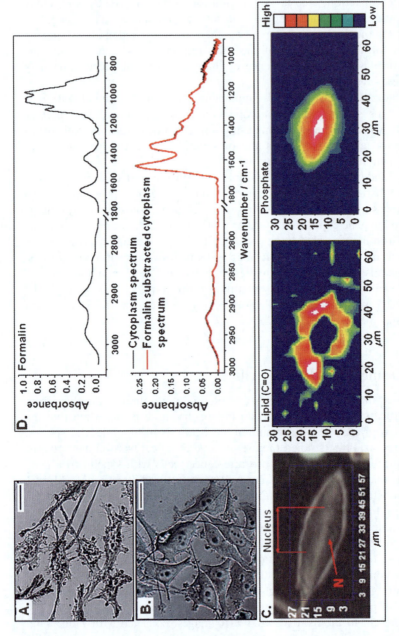

Figure 5.4 Photomicrographs of formalin-fixed prostate cancer cells, (A) without subsequent rinsing in deionized water and (B) with 3 s rinse in deionized water to remove residual PBS from the surface of the cells. Scale bar in all photomicrographs = 50 μm. (C) Optical image of a single, formalin-fixed, PC-3 cell. The cell's nucleus and nucleolus (N) are identified. SR-FTIR images depicting the intensity profiles of lipid ester $v_s(C=O)$ (1752–1722 cm^{-1} peak area) and phosphate $v_{as}(PO_2)$ (1280–1174 cm^{-1} peak area). (D) The FTIR spectrum of formalin and overlay of the FTIR spectrum of the cytoplasm with the same spectrum processed to remove their theoretical formalin content. (Reproduced from reference [48].)

fixed cell resulted in negligible differences in the intensities of peaks across the frequency range 3000–1100 cm^{-1} (Figure 5.4D). This was performed following normalization of the spectrum of formalin to the intensity value of the peak at 1000 cm^{-1} in the formalin fixed cell spectrum, since this frequency gives rise to the most intense peak in the spectrum of formalin.

In the second method, cells were formalin fixed and subsequently critical point dried (CPD). The CPD process involves first displacing gradually the intercellular water molecules in the formalin-fixed cells with increasing concentrations of ethanol. The ethanol is then displaced by acetone, which is miscible with liquid CO_2. The acetone within the cells is then displaced by liquid CO_2, within a chamber. The chamber is heated with a simultaneous rise in pressure as liquid CO_2 enters the vapor phase. At a specific temperature and pressure, the density of the vapor equals the density of the liquid, the liquid–vapor boundary disappears, and the surface tension is zero. Thus, this method reduces any residual distortions that may occur in the pre-fixed cell as a result of air-drying. Since the formalin fixed CPD dried cells were exposed to significant lipid-leaching reagents (ethanol and acetone), the cells were positive to trypan blue staining and the SR-FTIR spectrum of these cells demonstrated loss of the lipid ester v_s(C=O) peak.

In the third fixation method cells were fixed with glutaraldehyde and osmium tetroxide (OsO_4) prior to CPD. Glutaraldehyde polymerizes in solution, where dimers and trimers are the most abundant polymers.[49] The aldehyde groups of glutaraldehyde react with the amino groups of proteins to form imines in an irreversible reaction. The commonly used post-fixative to glutaraldehyde is OsO_4, which preserves unsaturated lipids by the formation of cyclic esters and is also an irreversible reaction.[49] With the exception of a weak peak at 960 cm^{-1}, it also does not give rise to absorption bands within the spectral region of interest in the mid-IR range where most biomolecules absorb. CPD–glutaraldehyde–OsO_4 fixation of cells preserves fine structure as observed in electron microscopy studies. The cells were found to be negative to trypan blue, indicating good preservation of plasma membrane lipid molecules. Various SR-FTIR images of cells fixed using this method (optical image in Figure 5.5A) are shown in Figures 5.5B–C. The neat FTIR spectrum of glutaraldehyde is shown in Figure 5.5D and a spectrum obtained from the SR-FTIR image of the cell is also presented. The localizations of phosphate and lipid ester v_s(C=O) signals were consistent with SR-FTIR maps of formalin fixed cells (see Figure 5.4). Although no significant spectral markers from glutaraldehyde were detected in the SR-FTIR spectrum of these cells, compared with formalin fixed cells (see Figure 5.4D), a reduction in the intensity of peaks between 1500 cm^{-1} and 1000 cm^{-1} was observed and the lipid ester v_s(C=O) signal appeared as a less resolved shoulder on the amide I band (Figure 5.5D).

Harvey *et al.* compared formalin fixation and alcohol fixation (SurePath preservative for liquid based cytology) for prostate cells in culture.[50] Formalin fixed cells showed a better diagnostic performance. This was, however, primarily due to the Raman tweezers experimental set-up. A much lower laser power was used for SurePath fixed cells because of problems with cells escaping

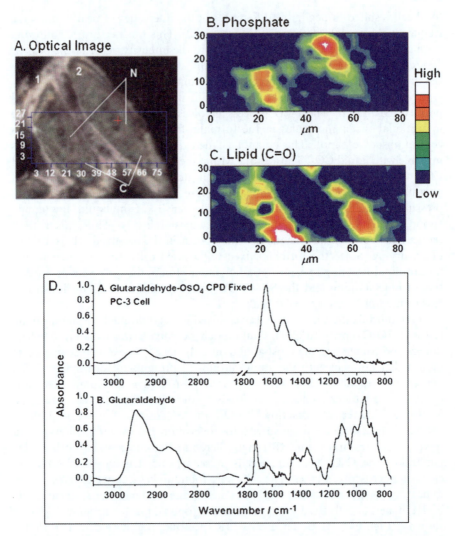

Figure 5.5 (A) Optical image of two glutaraldeyde–osmium tetroxide critical point dried (CPD) fixed PC-3 cells (labelled 1 and 2). N designates the nucleus, C designates the cell cytoplasm. Localizations and intensity profiles of (B) phosphate (1271–1180 cm^{-1} peak area) and (C) lipid ester (C=O) (1756–1722 cm^{-1} peak area). (D) FTIR spectrum of a glutaraldehyde–osmium tetroxide fixed PC-3 cell; the IR spectrum of neat glutaraldehyde. (Reproduced from reference [48].)

the laser trap. This resulted in much lower signal to noise ratios for SurePath fixed cells, leading to better prediction rates for formalin fixed cells.

Mariani *et al.* compared the effects of light fixation in 2% formalin, air drying and desiccation for 1 h each on two different cell lines using Raman spectroscopy.[51] For human keratinocyte cells (HaCaT cell line), air dried samples provided the best intensity for all components. Desiccation produced a slightly less intense spectral profile when compared with air-drying. Both air-drying and desiccation maintained significant spectral features particularly in the $1600–1500\,cm^{-1}$ region. Formalin fixation produced the weakest overall spectrum with many of the less intense features becoming largely unidentifiable, particularly in the $1600–1500\,cm^{-1}$ region. However, formalin fixation was found to be the most consistent method of cell preservation, with air-drying found to be the most inconsistent method. For activated human peripheral macrophages, C.A.MM6 cells, the most intense spectra were from formalin fixed and desiccated samples. Lower intensity bands detected in the spectra of desiccated and formalin fixed samples from $1253\,cm^{-1}$ to $1127\,cm^{-1}$ specific for mostly lipid and protein constituents could not be identified in the air-dried sample. Despite the preservation of many important strong and weaker spectral features in the C.A.MM6 cells with formalin fixation, overall reproducibility of this fixation method was found to produce significant inconsistencies. These inconsistencies resulted from the differential distribution of cell surface proteins and lipid conformations present. This was found to be particularly important for C.A.MM6 cells because activated cells express more proteins and undergo significant plasma membrane changes compared with non-activated cells. Structural component organization and lipid conformations are also known not to be identical between individual cells of the same lineages. These inter-lineage differences can vary in response to fixation, leading to varied methylene protein cross-linking. Air-drying and desiccation were found to be more consistent methods of cell preservation for these cells. In general, desiccation was found to provide the best compromise between signal intensity and preservation of cellular components.

A more recent study made the important step of using live cells as a reference for evaluation of fixation effects.[52] This study compared the effects of three commonly used fixatives, formalin, Carnoy's fixative (60% absolute ethanol, 30% chloroform, 10% glacial acetic acid) and methanol:acetic acid (3:1), on three cell lines using Raman spectroscopy. The level of fixation effect was found to be cell line dependent but, in general, all the fixatives were observed to affect the vibrational modes of lipid, protein, nucleic acid and carbohydrate moieties. On examination of the difference spectra and the spectral loadings and the results of k-means cluster analysis, it was concluded that formalin fixation produces a cellular spectrum that is globally more similar to that of the live cell, and therefore best preserves cellular integrity, out of the fixatives studied. Interestingly, Carnoy's fixative and methanol:acetic acid were found to affect nucleic acid spectral contributions significantly, despite being recommended for preservation of nucleic acids (Figure 5.6).

Chemical fixation was investigated for the preparation of adipocytes for FTIR analysis.[53] Adipocytes are specialized for the synthesis and storage of fatty acids (FAs) as triacylglycerides (TAGs) as well as for FA mobilization through lipolysis. Figure 5.7A shows the appearance of adipocytes in growth medium and illustrates the presence of numerous lipid droplets contained within their cytoplasm. Figure 5.7B shows adipocytes, prepared for FTIR

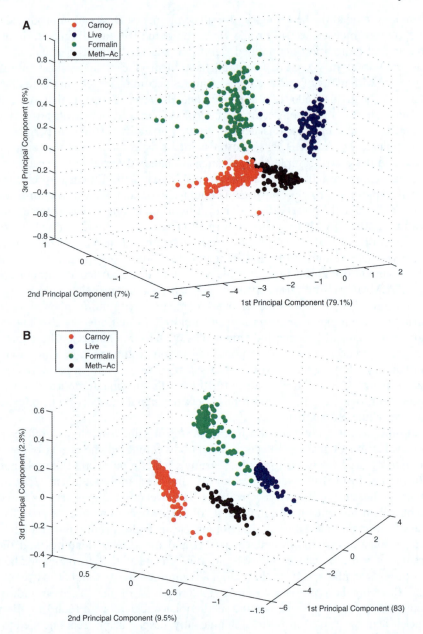

Figure 5.6 Principal components analysis (PCA) score plot for Raman spectra of (A) A549, (B) BEAS2B and (C) HaCaT cell lines. Percentage labels on each axis denote the variance described by that PC. There is clear separation of cellular spectra fixed with Meth-Ac and Carnoy's fixative relative to live and formalin fixed spectra. A degree of similarity between the spectral content of formalin fixed and live cell spectra is implied by the proximity of their clusters. (Reproduced from reference [52].)

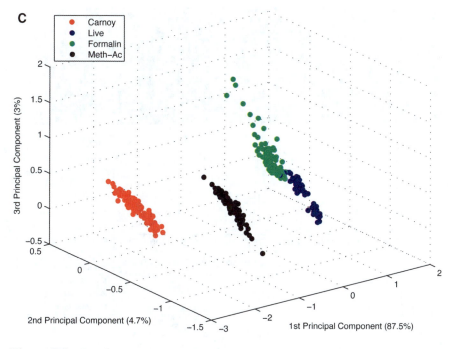

Figure 5.6 Continued.

analysis, following fixation in 4% formalin (in PBS), a brief water rinse to remove residue salts and air-drying at ambient conditions. Although this fixation protocol was appropriate for the preservation of prostate cancer cells for SR-FTIR analysis, it resulted in the collapse of intracellular lipid droplet structures into an unordered lipid deposit in the adipocytes (Figure 5.7B).

An FTIR spectrum of the intracellular lipid droplet of the formalin fixed air-dried adipocyte [Figure 5.7D(ii)] exhibits a lipid ester $v_s(C=O)$ peak at $1744\,cm^{-1}$, which is the same frequency of absorption as the lipid ester $v_s(C=O)$ peak in the reference TAG spectrum [Figure 5.7D(i)]. Several other characteristic peaks of the glycerol moiety of TAG are also observed in the lipid deposit of the formalin-fixed adipocyte at frequencies $>1500\,cm^{-1}$ and these are identified in Figure 5.7D(ii) as peaks labelled 8–12. However, the peaks in this spectrum [Figure 5.7D(ii)] are broader than the same peaks in the reference TAG spectrum [Figure 5.7D(i)]. This is due to collapse of the lipid droplets that give rise to a range of bonding strengths with neighbouring molecular species for those functional groups absorbing at frequencies $>1500\,cm^{-1}$. Figure 5.7C shows adipocytes containing well-preserved lipid droplets following para-formaldehyde (PF) fixation with OsO_4 post-fixation and CPD.

The advantages of this sample preparation over formalin fixation, water rinsing and air-drying are: (a) formalin in PBS contains methanol, which permeates the plasma membrane and results in a faster fixation compared with PF that does not contain methanol. However, methanol extracts intracellular

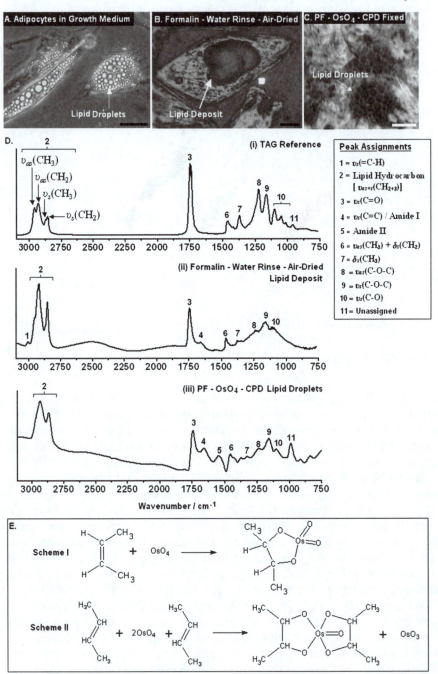

A. Adipocytes in Growth Medium — Lipid Droplets

B. Formalin - Water Rinse - Air-Dried — Lipid Deposit

C. PF - OsO₄ - CPD Fixed — Lipid Droplets

D.

(i) TAG Reference

$\nu_{as}(CH_3)$
$\nu_{as}(CH_2)$
$\nu_s(CH_3)$
$\nu_s(CH_2)$

(ii) Formalin - Water Rinse - Air-Dried Lipid Deposit

(iii) PF - OsO₄ - CPD Lipid Droplets

Wavenumber / cm⁻¹

Peak Assignments

1 = ν_s(=C-H)
2 = Lipid Hydrocarbon [$\nu_{as+s}(CH_{2+3})$]
3 = ν_s(C=O)
4 = ν_s(C=C) / Amide I
5 = Amide II
6 = ν_{as}(CH₃) + δ_s(CH₂)
7 = δ_s(CH₃)
8 = ν_{as}(C-O-C)
9 = ν_s(C-O-C)
10 = ν_s(C-O)
11 = Unassigned

E.

Scheme I

Scheme II

lipids, which is inappropriate for adipocyte fixation. (b) The OsO_4 post-fixative preserves lipids. (c) The 3-dimensional structure of the adipocyte is retained since the sample is dried without surface-tension effects, through CPD, and the localization of intracellular lipid droplets of the adipocyte is persevered. The disadvantage of this fixation protocol is that the mode of action by which OsO_4 preserves lipids is through complexation reactions within the double bonds of lipid hydrocarbon chains or complexation and cross-linking between unsaturated hydrocarbon chains (Figure 5.7E). Thus, the $v_s(=C–H)$ signal from unsaturated hydrocarbons is present in the lipid-deposit spectrum of the formalin fixed, water rinsed, air-dried adipocyte [Figure 5.7D(ii)], but is not observed in the lipid-droplet spectrum of the $PF–OsO_4–CPD$ adipocyte [Figure 5.7D(iii)]. Additionally, both methods of fixation (formalin–water rinse–air-dried and $PF–OsO_4–CPD$) result in a decrease in peak resolution of the $v_{as}(CH)_2$ and $v_{as}(CH)_3$ modes and the $v_s(CH)_2$ and $v_s(CH)_3$ modes.

The $PF–OsO_4–CPD$ protocol outlined above was used to preserve samples of prostate cancer cells (PC-3 cell line; prostate cancer cells derived from bone metastases) that were co-cultured with adipocytes pre-loaded with deuterated palmitic acid (D-PA_{31}).[53] This specimen was used in an FTIR tracing experiment to determine whether PC-3 cells could uptake the fatty acids stored within adipocytes. Figure 5.8A shows an optical image of a $PF–OsO_4–CPD$ fixed adipocyte surrounded by PC-3 cells and stroma cells. In this figure the adipocytes are visualized as large dark bodies (designated by Adp in Figure 5.8A), whereas PC-3 cells (1–4) are lighter in appearance and possess lamellipodia/pointed processes. The dark stain results from the binding of OsO_4 to the lipids. The boxed area was mapped using FTIR microspectroscopy, and the $v_{as}(C-D)_{2+3}$ signal intensity distribution is shown in Figure 5.5B, above. As expected, there was localization of the $v_{as}(C-D)_{2+3}$ signal with high intensity to the adipocyte; however it was also found that this signal illuminated the PC-3 cells (Figure 5.8B). Given that the only source of $v_{as}(C-D)_{2+3}$ signal in the PC-3 cells is through incorporation of D_{31}-PA released by the adipocytes, these data unequivocally demonstrate the translocation of D_{31}-PA between these cell types without cell isolation or external labelling. Appropriate fixation was necessary in this experiment, since delocalization/bleeding of lipid molecules

---

Figure 5.7 Optical photomicrographs showing (A) adipocytes in growth medium with prominent intracellular droplets (scale bar $= 10 \,\mu m$); (B) adipocyte following formalin fixation, water rinsing and air-drying (scale bar $= 20 \,\mu m$); (C) adipocytes fixed in paraformaldehyde (PF) and osmium tetroxide (OsO_4) and critical point dried (CPD) (scale bar $= 30 \,\mu m$); (D) typical FTIR spectrum of (i) triacylglyceride (TAG) reference with C_2–C_{10} saturated hydrocarbon chains, (ii) the lipid deposit from a formalin fixed air-dried adipocyte, and (iii) the lipid droplets from a paraformaldehyde (PF) and osmium tetroxide (OsO_4) CPD adipocyte; (E) OsO_4 reaction with unsaturated hydrocarbon chains to form cyclic esters. **Scheme I**. Reaction with a single unsaturated hydrocarbon chain. **Scheme II**. Cross-linking reaction with adjacent unsaturated hydrocarbon chains. (Reproduced from reference [53].)

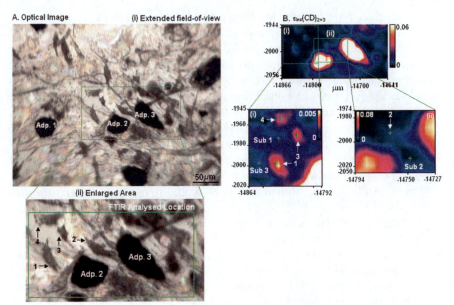

Figure 5.8 (Ai) Optical photomicrograph showing PF–OsO$_4$–CPD fixed adipocytes (Adp. 1–3) surrounded by prostate cancer cells and stroma cells. **(Aii)** Magnified region of Adp. 2 and 3 with surrounding prostate cancer cells (Labelled 1–4). The boxed area was analysed by imaging FTIR microspectroscopy. (B) FTIR spectral maps depicting the intensity distribution of the v_{as}(C-D)$_{2+3}$ signal. The boxed areas (i) and (ii) were expanded and the colour intensity threshold changed to provide better contrast of the v_{as}(C-D)$_{2+3}$ signal in cells relative to the substrate. (Reproduced from reference [53].)

from adipocytes in the adipocyte–PC-3 cell co-culture system could result in false-positive results concerning PC-3 uptake of adipocyte-derived D$_{31}$-PA.

A summary of the main findings from section 5.3.2 is presented in Table 5.2. The main FTIR and Raman bands affected by the commonly used cell preservation methods of drying and fixation are listed.

5.3.3 Cell Preparation for Biomechanistic Studies

In the FTIR study by Tobin *et al.*,[45] drying cells by centrifugation was used to investigate the response of cervical cancer cells to epidermal growth factor (EGF). Cells were incubated with EGF with increasing incubation times. Changes in protein conformation (noted by shifts in Amide I peak position) as a result of phosphorylation by EGF (monitored by the peak area of the phosphate monoester vibration at 970 cm^{-1}), at consecutive time points, were observed. Such a time-course experiment required a rapid method of sample preparation because of the short interval between sampling time points. Gazi *et al.* used formalin fixation to study the temporal fluctuations in phosphate,

Table 5.2 Summary of the main FTIR and Raman bands affected by standard cell processing protocols.

	Main FTIR bands affected	Main Raman bands affected
Drying		
Air drying Desiccation		Loss of features in 1253–1127 cm⁻¹ region[51] Less intense bands at 1612 cm⁻¹ and 1180 cm⁻¹ compared with formalin fixation[51]
Chemical fixation		
Formalin	No significant effects[38,83]	Formalin fixation closest to live cell[52] but cell line dependent[51] Nucleic acid bands (804–831, 1167, 1102–1200, 1300 cm⁻¹); protein bands (1669, 1450, 1345–1256, 1102–1200 cm⁻¹) and lipid bands (1170, 1284–1220, 1339–1305 cm⁻¹)[52] Loss of features in 1600–1500 and 1448–1127 cm⁻¹ regions[51]
Formalin and critical point drying	Loss of lipid ester peak (1756–1722 cm⁻¹) due to exposure to acetone and ethanol[38]	
Glutaraldehyde–osmium tetroxide and critical point drying	Reduction in intensity of 1500–1000 cm⁻¹ Region and less resolved lipid ester signal (1756–1722 cm⁻¹)[48]	
Carnoy's fixative (60% absolute ethanol, 30% chloroform, 10% glacial acetic acid)		Nucleic acid bands (670–788, 788–840, 870, 895–898, 928, 1060–1095 cm⁻¹; protein bands (1668–1661, 1450, 1339–1305, 1247–1305, 1176 cm⁻¹) and lipid bands (1045, 1065–1062, 1070, 1176, 1300, 1339–1247, 1453–1447 cm⁻¹)[52]
Methanol:acetic acid (3:1)		Similar to Carnoy's fixative[52]
Acetone	Loss of bands associated with cellular lipids (1740, 2925, 2854 cm⁻¹) and modifications to Amide I and II bands[83]	

protein secondary structures and endogenous non-isotopically labelled lipid signals following stimulation with different concentrations of D_{31}-PA and deuterated arachidonic acid (D_8-AA).[54] It was found that the shortest practical time interval between sampling points during which the cells could be fixed was 15 min. As an example, Figure 5.9A shows these biochemical fluctuations for

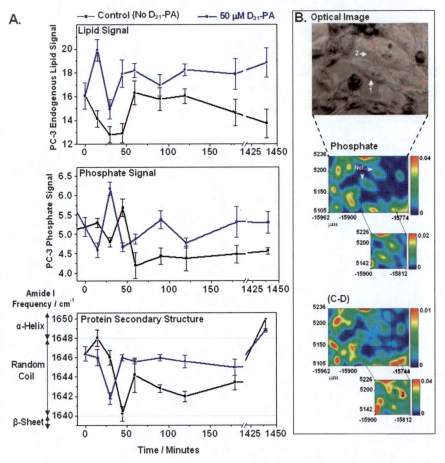

Figure 5.9   (A) Temporal fluctuations in various biomolecular domains probed by FTIR, for PC-3 cells exposed to 50 μM D_{31}-PA or no D_{31}-PA (control). Endogenous mean lipid hydrocarbon peak area intensities ($\pm$SE); mean phosphate diester peak area [$v_{as}(PO_2)$] intensities ($\pm$SE) and Amide I frequency shifts ($\pm$SE). (B) Optical image of PC-3 cells following incubation with D_{31}-PA for 24 h. This area was analysed using imaging FTIR microspectroscopy. Infrared biospectral maps show the intensity distributions of phosphate [nuclei are labelled (Ncl)] and $v_{as+s}(CD_{2+3})$ peak area (D_{31}-PA or its metabolites). In each image, cells 1 and 2 (see optical image) are magnified to demonstrate the intensity of IR signals in greater detail. FTIR spectra were obtained from points 1 (nucleus) and 2 (cytoplasm) in this image. (Reproduced from reference [54].)

PC-3 cells incubated with 50 μm D_{31}-PA in serum-free culture media, compared with control (PC-3 cells incubated in identical conditions but without D_{31}-PA). The endogenous lipid signal in the control PC-3 cells initially fell and is induced by metabolic/cytokine/growth factor imbalance resulting from the exchange of media to serum-free RPMI media at the zero minute time-point. Conversely, cells incubated with D_{31}-PA showed an initial rise in endogenous lipids. Since the incubation medium (RPMI) contains no FAs, this increase in lipid content must be due to *de novo* biosynthesis. This initial rise in endogenous lipid signal was followed by a fall, attributed to metabolic breakdown into adenosine triphosphate (ATP), which is a major product of lipid metabolism. This notion is supported by a phosphate spike at 30 min accompanied by a significant shift in the Amide I frequency, indicating protein phosphorylation.

The time-efficient formalin fixation method not only suitably preserved biomolecular composition so that lipid metabolism and protein phosphorylation could be measured, but also preserved the subcellular localizations of biomolecules for imaging studies. Figure 5.9B shows an optical photomicrograph of PC-3 cells on MirrIR substrate, following exposure to 50 μM D_{31}-PA for 24 h. This area was analysed by imaging FTIR microspectroscopy and the resulting distribution of the integrated intensity of the phosphate diester $[\upsilon_{as}(PO_2); 1274\text{--}1181\ \text{cm}^{-1}]$ peak area is shown. As expected, for cells 1 and 2 in the optical image, it can be seen that the most intense phosphate signals localize at the nucleus. In contrast, the most intense $\upsilon_{as+s}(CD_{2+3})$ signal localized at the cytoplasm, suggesting that the subcellular localization of D_{31}-PA or its metabolites is predominately in the cytoplasm.

Another FTIR based dose–response study has been undertaken where prior to spectroscopic examination, drug induced cells had been removed from culture media, washed in PBS and air-dried.[55] This study reports spectroscopic changes (ratios of peaks) that could be associated with exposure of the cells to increasing doses of the chemotherapeutic drug. Spectroscopic changes were correlated with cell sensitivity to the drug measured using the MTT [(3-(4,5-dimethylthiazol-2-yl)-2,5-diphenyltetrazolium bromide] assay. Thus, there is evidence to suggest that spectroscopic changes associated with drug exposure can be determined and that this is in fact dominant over metabolite perturbations resulting from autolysis during the drying process.

5.3.4 Growth Medium and Substrate Effects on Spectroscopic Examination of Cells

A number of studies have investigated the use of FTIR or Raman microspectroscopy as diagnostic tools to differentiate and classify cell lines, *in vitro*, based on their pathological state[56,57] or resistance to drugs.[58] Interestingly, we find that some researchers have grown their different cell lines in the same culture media,[56–58] whereas others have used different media for each cell type.[59,60] The European Collection of Cell Cultures (ECACC) provides standard protocols for the optimum growth of different cell lines. In some instances, cell culture media may be different for cells of the same epithelial origin, for example ECACC suggest that PC-3 cells (a prostate cancer epithelial cell line derived from bone metastases) should be grown in Ham's F-12, whereas LNCaP (a prostate cancer epithelial cell line derived from lymph node metastases) should be grown in RPMI 1640. Both RPMI 1640 and Ham's F-12 are complex mixtures consisting of a range of inorganic salts, amino acids, vitamins, nucleotides and glucose as well as small-molecule precursors. However, differences among media can exist with respect to the relative concentrations of each component as well as compositional differences such as the presence or absence of a major biomolecular class, for instance RPMI 1640 contains no fatty acids, unlike Ham's F-12, which contains the ω6-FA, linoleic acid (LA).

In a recent study by Harvey *et al.*,[60] reflection mode FTIR photoacoustic spectroscopy (PAS) was used to obtain spectra from four different formalin fixed prostate cell lines (BPH = benign prostatic hyperplasia; LNCap-FGC = prostate cancer epithelial cells derived from lymph node metastases; PC-3 = prostate cancer epithelial cells derived from bone metastases; PNT2-C2 = normal prostate epithelial cells immortalized by transfection with the genome of the SV40 virus). Unsupervised principal component analysis (PCA) of this spectral data set yielded separation of clusters corresponding to each of these cell lines (Figure 5.10A). Two of these cell lines were grown in the same media (LNCaP and PNT2-C2), whereas two were grown in different media (BPH and PC-3). Importantly, the two cell lines that were grown in identical media (LNCaP and PNT2-C2) showed significant separation, realized by anticorrelation on PC-2.

In a follow-up study, Harvey *et al.* acquired conventional FTIR spectra from PC-3 cells and LNCaP cells, each grown separately in their optimum culture medium (as advised by ECACC protocols) and a "foreign medium".[61]

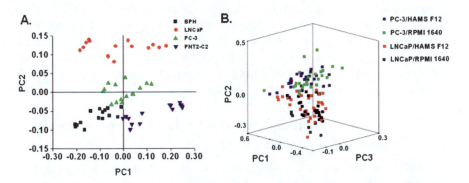

Figure 5.10 (A) Principal components analysis (PCA) scores plot of the background-subtracted, vector normalized first derivative FTIR-PAS spectra of four different prostate cell lines (BPH, LNCaP, PC-3 and PNT2-C2). (Reproduced from reference [60].) (B) PCA scores plot of vector normalized, first derivative FTIR spectra of PC-3 and LNCaP cell lines, grown in their "optimum" culture medium or "foreign" culture medium.

Unsupervised PCA of these spectra demonstrated clustering of the two cell lines that was independent of the culture medium in which they were grown, but was principally dependent on the cell type (Figure 5.10B). Thus, it may be concluded that at least for the prostate cell lines PC-3 and LNCaP, the influence of the basic media under investigation in this study (RPMI 1640 and Ham's F-12) on the cell metabolites is not the primary influence on spectroscopic measurements. However, it must be acknowledged that cells are a function of their environment. Thus, the same cell line grown in two different media with relatively larger compositional differences (or containing potent stimuli) will affect spectroscopic classification.

If the cell is exposed to an environment that does not sustain its optimum growth and down-regulates the expression of biomolecular features (such as cell surface antigens, hormone receptors, protein expression), which characterize that cell type *in vivo*, then this may ultimately convert the cell to a new class. *In vivo*, it is well known that stromal–cell interactions are particularly important in cancers such as of the breast, where the stromal compartment plays a critical role in directing proliferation and functional changes in the epithelium.[62] Moreover, environmental stimuli directing cell phenotype have been recently studied with imaging FTIR by Krafft *et al.*[63] In this study, human mesenchymal stem cells were treated with osteogenic stimulatory factors that induced their differentiation. Differentiation was detected by FTIR through changes in the Amide I band shape (indicative of protein composition/structural changes) and phosphate levels (indicative of the expression of calcium phosphate salts).

In a similar manner to compositional differences in the growth media that may or may not elicit changes in cell biochemistry and thus its spectra, substrates can also induce morphological as well as functional changes in the cell. Meade *et al.* studied the influence of a range of substrates on the normal human epithelial keratinocyte cell line (HaCaT) using FTIR and Raman spectroscopy.[64] The substrate extracellular matrix (ECM) coatings under evaluation were two glycoproteins, fibronectin and laminin, and one protein, gelatin (derived from thermal denaturing of collagen). Gelatin was coated onto MirrIR slides for FTIR experiments or quartz slides for Raman experiments, by incubation for 24 h at 4 °C. Laminin and fibronectin were coated onto these substrates by incubation for 4 h and 40 min, respectively, at room temperature. For the fibronectin and laminin coated slides, excess solution was aspirated from the substrates and washed in PBS prior to cell deposition. For the gelatin coated slides, excess solution was aspirated and cells were deposited for culture without prior washing in PBS.

Fluorescence assays were conducted at three days post-seeding, as well as fixation using 4% formalin (in PBS) with water rinse for FTIR and Raman investigations. Cellular proliferation, viability and protein content were found to be down-regulated when cells were grown on uncoated quartz compared with uncoated MirrIR substrates. However, increases in proliferation and viability were more pronounced when cells were grown on coated quartz than coated MirrIR substrates. Additionally, it was found that quartz coated with all three ECM coatings generated significantly enhanced proliferation compared to the control (uncoated quartz). However, this was not the case for MirrIR, which resulted in a significant increase in proliferation only for the laminin coated slide. Viability was significantly increased when cells were grown on laminin and gelatin coated quartz, whereas for MirrIR substrate, viability was only significantly increased when this was coated with gelatin. The authors suggested that gelatin provides a coating with similar proliferation effects on quartz and MirrIR and increases viability, which is desirable for long-term cultures.

The FTIR and Raman spectroscopic analyses of the coated slides demonstrated that the gelatin coating did not give rise to sufficiently high signals to significantly influence FTIR or Raman spectra of the cells cultured upon them.[64] First derivative FTIR spectra and Raman spectra of cells on gelatin, fibronectin and laminin demonstrated spectral changes on each of these substrates that were associated with nucleic acid, lipid and protein expression. Raman spectra provided further insight and relative quantification, since it was found that coatings that promoted proliferation gave rise to increases in spectral regions associated with DNA, RNA and proteins, with a decrease in lipids. This has been attributed to an increase in the sustained production of signalling proteins, as a result of integrin binding to the coating, that promotes cellular proliferation. Supporting this, the authors also found through FTIR that the ratio of protein (sum of integral absorbencies corresponding to the Amide I, II and III bands) to lipid (integral absorbance 1370–1400 cm^{-1}) gave rise to values that could be significantly correlated with an increase in proliferation (as measured by fluorescence assay).

The experimental set-up described above consisted of thin layers of ECM that were barely detectable in the FTIR or Raman spectrum; however Lee *et al.* studied prostate cancer cells that had been cultured onto relatively thicker layers of ECM.[65] In their investigation, Matrigel was used as the artificial ECM. The major constituents of Matrigel are collagen type IV (a protein), heparin sulfate (a proteoglycan), laminin and entactin (glycoproteins). Figure 5.11A shows an optical photomicrograph of prostate cancer cells on Matrigel. This area was analysed using imaging FTIR microspectroscopy. As expected, the lipid hydrocarbon signal demonstrates high intensity at the cell locations, relative to the Matrigel surroundings, due to the cumulative absorption of lipid containing biomolecules in the cells and Matrigel (Figure 5.11B). Since the lipid background signal is nearly homogeneous, it suggests that the Matrigel is of constant thickness within the analysis field-of-view. However, the protein background exhibits a heterogeneous distribution of intensity (Figure 5.11C), which is likely to be due to concentration differences when taking into consideration the lipid intensity image. As expected, cells adhered to the low protein concentration exhibited a higher protein intensity signal than the surrounding layer, whereas those on a high protein concentration or thick surface revealed an unexpected lower protein intensity signal. This is illustrated in the protein cross-section in Figure 5.11D, which was plotted with values taken from a region of high protein intensity to a region of low protein intensity and bisecting the cells.

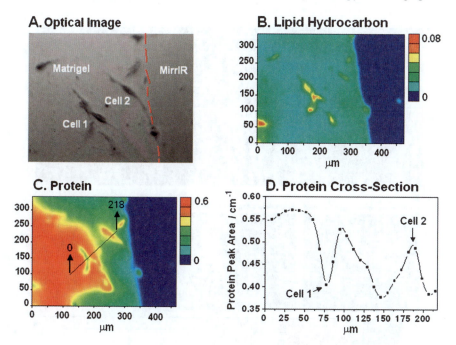

Figure 5.11 (A) Optical image of PC-3 cells on Matrigel. The red dotted line denotes the Matrigel (left)–MirrIR™ substrate (right) interface. FTIR spectral maps depicting the intensity distribution of the (B) lipid hydrocarbon and (C) protein peak area signals. (D) A cross-section through the protein intensity map is displayed as a graphical plot depicting the protein peak area values from a region of high concentration of Matrigel (at 0 μm) to one of lower concentration (towards 218 μm) and bisecting cells 1 and 2. (Reproduced from reference [65].)

Local proteolysis of the Matrigel matrix was investigated using time-lapse video microscopy. The final frame of the time-lapse video and the brightfield image of the same area after fixation, relocated for FTIR-microspectroscopic imaging, are shown in Figures 5.12B (i) and (ii), respectively. These optical images demonstrate that the morphology of the cells (elongated and rounded), when in culture, is suitably retained by the formalin fixation procedure. In Figure 5.12A, the video frame captured at the start of the time-lapse recording shows that the cells at internal locations on Matrigel display a rounded morphology. The cell marked with a red arrowhead was stationary throughout the course of time-lapse recording and retained its rounded appearance. It is reasonable to assume that the low protein intensity at this cellular location [Figure 5.12B(iii)] would be indicative of local proteolysis or mechanical degradation of the Matrigel. However, between 5 h and the point of termination of the time-lapse study (22 h, 22 min), the cell marked with the green arrowhead migrated, several times, towards and away from the cells marked with the blue and white arrowheads. Since these cells (green, blue and

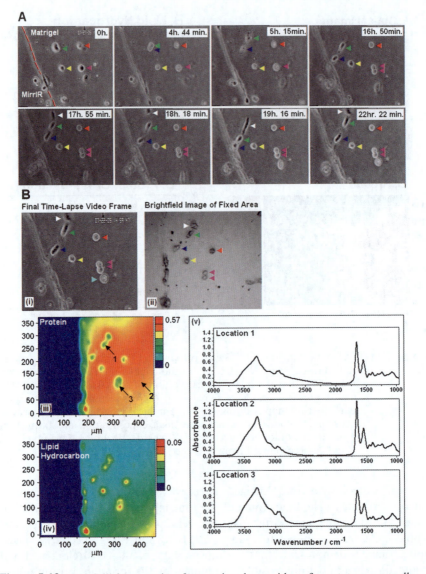

Figure 5.12   (A) Still frames taken from a time-lapse video of prostate cancer cells on Matrigel. The red dotted line at zero hours designates the Matrigel–MirrIR interface. Individual cells are labelled with coloured arrowheads for the ease of tracking cell migration across consecutive time frames and following figures. (B) Optical images of the (i) final frame of the time-lapse video (at 22 h, 22 min) and (ii) brightfield image of the same area, post fixation. Note that the image of the fixed area is rotated 22° compared with the image of the final frame such that the Matrigel–MirrIR interface is approximately vertical. The fixed area was analysed by FTIR microspectroscopy and the intensity distribution of the (iii) protein and (iv) lipid hydrocarbon domains are shown. (v) Representative raw IR spectra taken from three locations [1 and 3 (on cell) and 2 (on Matrigel)], from the protein intensity map, are also displayed. (Reproduced from reference [65].)

white arrowheads) were motile throughout the time-lapse recording, it is unlikely that there was local digestion of the Matrigel just prior to termination of the time-lapse recording *via* matrix metalloproteinases produced by the prostate cancer cells. Moreover, if proteolytic digestion was a dominant mechanism by which the green, blue and white arrowhead cells transversed the Matrigel, then one would expect low protein signals to arise from the entire path occupied by these cells. It was concluded from this, that light-scattering artefacts influenced the protein intensity maps of these cells on Matrigel, giving the illusion of protein degradation at the cell locations.[65]

A model was produced, which showed that a switch from a higher than background signal to a lower than background signal will occur at a given thickness or concentration of protein within the Matrigel layer. Importantly, the model includes light that is directly back-scattered into the microscope collection optics. These findings have a fundamental impact on research in the field of FTIR spectroscopy concerning cells on two-dimensional matrices.

5.3.5   Preparation of Living Cells for FTIR and Raman Studies

5.3.5.1   FTIR Studies

A number of studies concerning the analysis of living cells by FTIR have been performed with synchrotron radiation sources.[66-68] An early study by Holman *et al.* reported spectral changes in HepG2 cells (human hepatocellular carcinoma) treated with increasing doses of an environmental toxin.[66] In this study, post-treated cells were detached from culture substratum using trypsin, followed by two washes in PBS then kept as a suspension at 4 °C and measured with SR-FTIR within 24 h. Although the cool temperature minimizes the enzymatic effects of autolysis, without fixation there may be biochemical differences between cells at the zero time point compared with cells stored in PBS for 24 h, particularly for glycogen stores, since the cells were in a nutrient deficient environment. Nevertheless, it has been shown that spectra from these cells showed spectroscopic changes in the ratio of peak intensities ($1082 \, cm^{-1}$/ $1236 \, cm^{-1}$) that could be correlated with increasing doses of toxin exposure. In a situation where the effect of time on cell biochemistry has not been assessed, one must be careful when associating spectral changes to the direct result of a condition administered to the cell. However, if spectral discrimination is achieved following randomized sampling of cells exposed to each of the different conditions, then this may evaluate whether live cell spectra are significantly influenced by their duration in nutrient deficient media.

More recently, specialized equipment for maintaining live T-1 cells (aneuploid cells from human kidney tissues) on gold-coated slides for *in-situ* SR-FTIR analysis has been investigated by Holman *et al.*[68] A mini-incubator system was used to sustain cell viability by maintaining a humidified environment, so as to retain a thin layer of growth medium around the cell during SR-FTIR measurements. The mini-incubator was temperature controlled at 37 °C *via* circulating water from a water bath, and IR transparent CaF_2 windows on the top cover were separately temperature controlled to avoid condensation. Using this incubator, the authors investigated any possible cytotoxic effects that may be elicited in the cell by exposure to the SR-IR radiation. Using the Alcian blue exclusion assay, it was found that the cells showed negative staining 24 h after exposure to 20 min of SR-IR radiation, which indicated that the cell membranes remained intact. The effects of 20 min SR-IR exposure on cell metabolism were assessed using the MTT assay. This confirmed that both control cells (not exposed to SR radiation) situated near to exposed cells and exposed cells produced mitochondrial dehydrogenases, which are associated with glycolysis and indicate negligible effects on this metabolic pathway. Finally, colony-forming assays demonstrated that there was no long-term damage as a result of SR-IR exposure.

Although these assays could not have been carried out *in situ* within the mini-incubator, it is encouraging to find that the length of time (20 min) that these cells were placed in the incubator had no short-term or long-term effects. Furthermore, the researchers reported that consecutive SR-FTIR spectra obtained at 10 min intervals for 30 min exhibited an unchanging IR spectrum to within 0.005 AU across the entire mid-IR spectral range. This provides supporting evidence for the justification of this experimental set-up for measuring single-point SR-FTIR spectra from single living cells. Since the experiment was carried out for only 30 min, changes in the spectrum over an extended time period, which is required to obtain cell maps, is unknown.

However, using a different experimental design for the sample compartment, Miljkovic *et al.* reported no spectral changes in spectra collected from live cells when data were collected every 30 min for 3 h.[69] Miljkovic *et al.* collected FTIR images at $6.25 \mu m \times 6.25 \mu m$ pixel resolution of living HeLa cells (cervical cancer cell line) using the linear array detector Spotlight microspectrometer equipped with a glowbar source.[69] In this study, different approaches were used to prepare cells for transflection and transmission mode analysis. For transmission mode, live cells in growth medium were placed into a $6 \mu m$ pathlength CaF_2 liquid cell. This preparation resulted in the compression and rupturing of larger cells, however the smaller cells were left intact (Figure 5.13B). For transflection measurements, cells in buffered saline solution were placed as a drop onto a MirrIR slide and a CaF_2 or BaF_2 coverslip was placed on top. This preparation also involved the use of a $5 \mu m$ Teflon spacer to prevent the coverslip touching the MirrIR slide. Raw spectra obtained from FTIR images of HeLa cells using both modes of analysis showed an unusual Amide I to II ratio (Figure 5.13A and C), which was more apparent in transflection mode spectra (Figure 5.13A). This was attributed to the longer path length ($10 \mu m$) in the transflection mode measurement. The origin of this distorted Amide I to Amide II ratio was determined to be due to overcompensation of the water background from the cell spectrum, since the cell contains less water than the surrounding medium or buffer.

The authors suggested that correction of the Amide I and II peaks could be carried out by visually fitting a scaled buffer spectrum to the raw cell spectrum, until the resulting corrected spectrum shows a normal Amide A envelope. Although subjective, Miljkovic *et al.* demonstrated that it was possible to

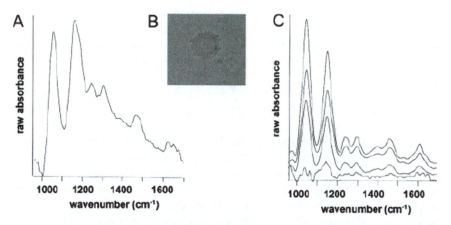

Figure 5.13 (A) Reflection/absorption spectrum and (B) visual image of a HeLa cell in BSS buffer, (C) Absorption spectra taken from a HeLa cell in growth medium. All spectra were collected with 128 scans at $4\,cm^{-1}$ spectral resolution. (Adapted from reference [69].)

obtain a protein intensity image in which the HeLa cell displayed the expected high protein intensity, centered at its nucleus.[69]

In contrast to using cells in suspension as in the study by Miljkovic *et al.*,[69] cells used by Moss *et al.* were cultured directly onto CaF_2 plates.[67] This plate was placed into a liquid cell consisting of a $15\,\mu m$ Teflon spacer, providing a pathlength of $11–12\,\mu m$ and maintained at $35\,°C$. A constant flow of cell culture medium was passed through the cell at a rate of $230\,\mu l/h$. As in the study of Miljkovic *et al.*,[69] a background spectrum of growth medium was collected in a cell-free region of the sample and the ratios with the cell spectrum estimated. There was high reproducibility between SR-FTIR spectra obtained from 10 individual fibroblast cells when a spectrum of each cell was acquired every 24 min for 2 h. Although intrasampling differences were observed between cells, these were very much smaller than the standard deviation of repeated measurements for each cell. Moss *et al.* provide further support for the low spectral variance observed for live cell FTIR spectra, when collected within the first few hours of transfer to the sample analysis chamber.[67]

In agreement with the study by Miljkovic *et al.*,[69] a distorted Amide I to Amide II intensity ratio was observed by Moss *et al.*[67] However, since the spectrum of background water is different from that of water bound to macromolecules, it was suggested that it is not possible to eliminate this background absorbance accurately. The authors also suggest that if the goal of the experiment is to obtain a spectrum from the same position of the exact same cell, before and after administration of a stimulus, then the difference between the spectra can be resolved even in the presence of background water. Additionally, it was found in this study that non-confluent cells could migrate out of the measuring SR beam. Moss *et al.* suggest that this could be minimized by placing the cell in a well.[67]

A recent study by Draux *et al.* investigated different optical substrates, quartz, calcium fluoride and zinc selenide, for preservation of cell integrity for Raman spectroscopy of single living cells.[70] Quartz was found to be the most appropriate based on cell morphology and proliferation rate.

5.3.5.2   Raman Studies

(See also Chapter 4, sections 4.7 and 4.8.) The spatial resolution of Raman spectroscopy is inherently higher than that of FTIR owing to the shorter wavelength of the excitation radiation (the diffraction limit is generally given as $\lambda/2$). An image obtained by Raman microspectroscopy requires raster scanning of a focused laser beam across the cell. Using this mode of data collection, an increase in spatial resolution, which is a function of step size and beam diameter, also increases the time for chemical mapping. Although in FTIR studies it has been shown that, for up to 3 h, spectral changes are not observed at the *whole-cell level* (see section 5.3.5.1), previously reported Raman maps of living cells have required ≥ 3 h collection times.[71] At the *subcellular level* one might expect that biochemical changes could occur within this period for a given sampling point. However, it has been shown that localization and spectral distinction between cytoplasmic and nuclear compartments in living cells was not affected in Raman maps of two different cell types (human osteogenic sarcoma cell and human embryonic lung epithelial fibroblast) that required long collection times (up to 20 h).[71] In this study by Krafft *et al.*,[71] difference spectra of the cytoplasm and nucleus identified the important discriminatory variables that distinguished these compartments: nucleic acids and lipids. In fact, analysis of living cells grown on quartz and analysed in media provided Raman spectra containing features of subcellular components that were more pronounced than those obtained from frozen-hydrated cells. This was attributed to conformational changes and aggregation of biomolecular constituents caused by the freeze-drying process and which are not present when the cells are analysed hydrated.

The acquisition time for a Raman map of a cell can be improved by increasing the sensitivity of the technique. Kneipp *et al.* have demonstrated that enhanced Raman signals (10–14 orders of magnitude) for the native constituents of a cell can be achieved by incorporating colloidal gold particles into the cell.[72,73] The gold nanoparticles give rise to surface-enhanced Raman scattering (SERS) where Raman molecules close to the vicinity of the nanoparticles experience electronic interaction with enhanced optical fields due to resonances of the applied optical fields with the surface plasmon oscillations of the metallic nanostructures. This process results in an increase in the scattering cross-section of the Raman molecules, which enabled Raman maps to be collected at 1 μm lateral resolution (1 s for one mapping point)[72] where each spectrum in the map consisted of the spectral region 400–1800 cm^{-1}.

Delivery of the nanoparticles into the cell interior can be carried out in two ways, sonication or fluid-phase uptake.[72] The fluid-phase uptake method involves supplementing the culture medium with colloidal gold suspensions (60 nm in size), 24 h prior to experiments. The cells internalize the nanoparticles through endocytosis and without further induction (Figure 5.14A).[73] This can result in the formation of colloidal aggregates inside the cell that may be 100 nm to a few micrometers in size.[72] The cells are washed in buffer to remove non-incorporated nanoparticles and replaced in fresh buffer for SERS analysis. The second method of delivering nanoparticles into the cell is by sonication, where rupture of the cell membrane enables an influx of nanoparticles before self-annealing within a few seconds. However, in low-intensity ultrasound mediated gene transfection, it has been found that sonication can induce stress responses in the cell,[74] and so it should be carried out 24 h prior to experiment to allow enough time for the cell to repair any damage. Incorporation of the nano-particles into the cell using the fluid-phase uptake method was not found to yield any visible changes in growth characteristics such as signs of apoptosis or cell detachment when compared to a control cell culture.[72] Raman spectra obtained from different locations in the nanoparticle doped cell gave rise to very different spectral profiles, illustrating the biochemical heterogeneity of the cell (Figure 5.14B).

Another signal enhancement technique, CARS (coherent anti-Stokes Raman spectroscopy) uses multiple lasers to excite coherent anti-Stokes Raman scattering. CARS microscopy is a rapid vibrational imaging technique and has been used to image mitosis and apoptosis in living cells,[75] for real-time visualization of organelle transport in living cells,[76] and for quantitative imaging of intracellular lipid droplets.[77] Stimulated Raman scattering, which has a significantly greater sensitivity than spontaneous Raman scattering, has been used to monitor the uptake of omega-3 fatty acids by living human lung cancer cells.[78] No plasma membrane blebbing was observed following repeated imaging of the same cells indicating that no photodamage had occurred.

As well as the application of Raman microspectroscopy to live cell imaging, the technique has also been applied for the phenotypic typing of live cells. Krishna and colleagues collected Raman spectra of two different cell lines and their respective drug resistant analogues: breast cancer cell line MCF7 and its subclone resistant to verapamil (MCF7/VP) and promyelocytic leukaemia HL60S cell line and its multidrug-resistant phenotypes (HL60/DOX: resistant to doxorubicin; HL60/DNR: resistant to daunorubicin).[58] PCA analyses of these Raman spectra were able to generate score plots that showed clustering and separation for each cell line and its drug-resistant clone. The authors also carried out these experiments using FTIR and found that classification, discrimination and reproducibility were greater using this method. However, with the view of translating this type of analysis to clinical application, it would be desirable for the chosen method to incorporate minimal sample preparation for high-throughput screening. For the Raman study, a cell pellet consisting of 1×10^6 cells (washed in 0.9% NaCl) was used directly for spectroscopic analysis, whereas for the FTIR experiments a time-limiting step was required that consisted of drying a cell suspension under mild vacuum onto a zinc selenide sample wheel.

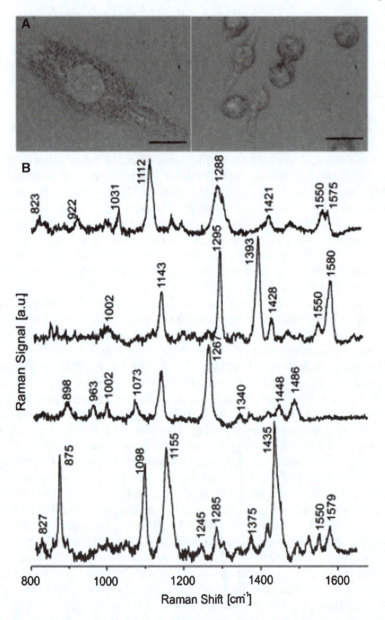

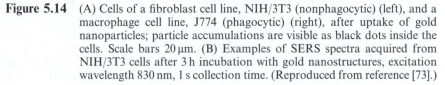

Figure 5.14   (A) Cells of a fibroblast cell line, NIH/3T3 (nonphagocytic) (left), and a macrophage cell line, J774 (phagocytic) (right), after uptake of gold nanoparticles; particle accumulations are visible as black dots inside the cells. Scale bars 20 μm. (B) Examples of SERS spectra acquired from NIH/3T3 cells after 3 h incubation with gold nanostructures, excitation wavelength 830 nm, 1 s collection time. (Reproduced from reference [73].)

The sample preparation method used by Krishna *et al.* for the Raman study requires fast data acquisition times, since live cell pellets surrounded by a thin layer of aqueous buffer may undergo biochemical changes over-time.[58] In their study, 25 spectra were collected for each pellet, where one spectrum took 4.5 min to collect. Thus, between the first and final spectrum there was a time-lag of 1 h and 53 min. If biochemical changes did occur during this period, then it may have contributed to the lower discriminatory power achieved using Raman spectra in this study. In contrast, the FTIR spectra were obtained from dried cells, providing perhaps a background interference that is constant over all cells and so differences due to MDR or drug sensitive phenotypes could be more readily resolved. This provides further evidence that possible artefacts from the drying process do not hamper spectroscopic differentiation between cells of differing phenotypes (as mentioned in section 5.3.2).

More recent phenotyping studies by Chan *et al.*[79–81] and Harvey *et al.*[82] have used Laser Tweezers Raman Spectroscopy (LTRS) to optically trap single living cells suspended in PBS.

5.4 Conclusions

Sample preparation is a key aspect of any experimental design and particularly so for spectroscopic analysis of cells and tissues. The continuing developments in tissue preservation for optimum detection of specific biomolecules using emerging bioanalytical approaches will shape the tissue repositories of the future. These developments will also have an impact on biomedical vibrational spectroscopy, since this technology can play an important role in determining the biochemical basis underpinning disease progression. Nevertheless, it is apparent that existing tissue banks have proven adequate for FTIR and Raman studies of tissue pathologies, providing high classification power. This is despite the fact that spectral artefacts exist as a result of sample processing. These spectral artefacts can be due to protein depolymerization or a change in the lipid to protein ratio for dried cryosections, or in the case of deparaffinized sections, due to residual paraffin, coagulation of proteins and loss of lipids.

As discussed in section 5.2, some of these artefacts can be minimized. Protein depolymerization of freeze-dried/thawed cryosections can be reduced by careful attention to the cryogen used for initial tissue snap-freezing as well as cryo-microtomy and freeze-drying environmental temperatures. Other artefacts such as residual paraffin can now be confidently removed in the light of work carried out by Faoláin *et al.*[28] It appears that deparaffinization using hexane for >20 h is an appropriate method for this purpose and this has wider implications in immunohistochemical pathology. However, this protocol can be time limiting and so less rigorous protocols may be sufficient where spectroscopic markers for pathological assessment do not overlap with paraffin signals.[30,31]

The early work of Fox *et al.*,[25] investigating the binding time of formalin to tissue, may be of significance to those vibrational spectroscopists presently using formalin fixed cells in imaging or biomechanistic studies, since the effects of formalin binding time on cell spectra have not been assessed. Although cell line dependent, formalin fixation has been shown to provide cell preservation similar to the live cell state.[52]

Tailored chemical fixation protocols for vibrational spectroscopy may however be necessary in some instances. For example, as described in section 5.3.2, the lipid component of adipocytes requires fixation with OsO_4.[53] Coupled to paraformaldehyde, a well-preserved sample can be obtained for spectroscopic analysis. The lengthy procedure of OsO_4–paraformaldehyde with CPD is appropriate where experiments are capturing a cellular event at time-frames that are far apart. However, for shorter time frames (intervals of 15 min), faster fixation methods are required and formalin has so far been proven to be adequate.[54] Interestingly, it has also been demonstrated that air-dried cells following exposure to a pharmacological drug or stimulus can also produce spectral changes that may be associated with response to the condition administered.[55]

There have been encouraging results reported within the context of live cell experiments using FTIR. The collective demonstration of biochemical stability for different cell types (T-1, HeLa and fibroblasts) by the various research groups working within this field,[67–69] together with the suggestion of Moss *et al.*[67] that correction for water absorbance may not be necessary, suggests that future FTIR studies may be able to measure early biochemical responses of

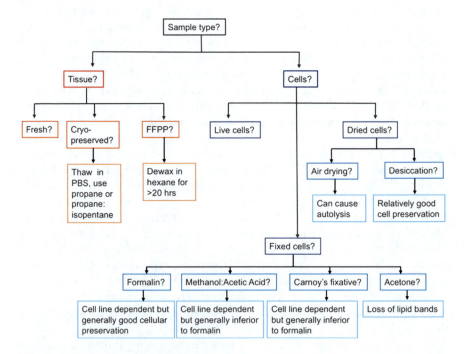

Figure 5.15   Sample preparation choices for tissues and cells.

single living cells to stimuli. Raman based live cell studies have shown excellent prospects for cell phenotyping as well as probing the distributions of native biomolecules of a cell with high sensitivity and spatial resolution and without the requirement for exogenous labelling.

In summary, for spectroscopic analysis of tissue samples, fresh or carefully prepared frozen tissue would be considered ideal, but FFPP tissue can be used successfully if deparaffinized using hexane. Other dewaxing agents and protocols can also be used successfully if the paraffin signals do not overlap with the spectral regions of interest. For analysis of cell samples, live cells would be considered ideal, but in cases where this is not possible or practical, fixed cells can be used. Formalin has been shown to provide good cellular preservation for spectroscopic analysis in many studies but some experimentation may be necessary to find the optimum fixative for each particular cell sample. For this optimization, fixed cell samples should be compared with live cell samples where possible.

Figure 5.15 provides an overview of common sample preparation choices for tissue and cell samples together with some recommendations based on the studies discussed in this chapter.

Acknowledgements

Support was received from the Association for International Cancer Research (AICR Grant number 04-518), The Prostate Cancer Foundation (PG and EG) and from the NBIPI programme, funded by the Irish Government's Programme for Research in Third Level Institutions, Cycle 4, National Development Plan 2007–2013, supported by the European Union Structural Fund and the Technology Sector Research (Strand 3) programme of the Irish Higher Education Authority (FL) during the writing of this article and some of the experiments described within it. We are grateful to Dr Stephen Murray (Paterson Institute for Cancer Research, UK) for use of the time-lapse video microscope.

References

1. J. A. Kiernan, Formaldehyde, formalin, paraformaldehyde and glutaraldehyde: What they are and what they do, *Microsc. Today*, 2000, **00–1**, 8–12.
2. D. Jones, *Introduction*, In: *Fixation in Histochemistry*, P. J. Stoward, (eds). Chapman and Hall, London, 1973, 2–7.
3. G. R. Turbett and L. N. Sellner, The use of optimal cutting temperature compound can inhibit amplification by polymerase chain reaction, *Diagn. Molec. Pathol.*, 1997, **6**, 298–303.
4. J. W. Gillespie, Evaluation of Non-Formalin Tissue Fixation for Molecular Profiling Studies, *Am. J. Pathol.*, 2002, **160**, 449–457.

5. P. G. L. Andrus and R. D. Strickland, Cancer grading by Fourier Transform Infrared Spectroscopy, *Biospectroscopy*, 1998, **4**, 37–46.
6. S. Takahashi, A. Satomi, K. Yano, H. Kawase, T. Tanimizu, Y. Tuji, S. Murakami and R. Hirayama, Estimation of glycogen levels in human colorectal cancer tissue: relationship with cell cycle and tumour outgrowth, *J. Gastroenterol.*, 1999, **34**, 474–480.
7. D. C. Fernandez, R. Bhargava, S. M. Hewitt and I. W. Levin, Infrared spectroscopic imaging for histopathological recognition, *Nature Biotechnol.*, 2005, **23**, 469–474.
8. M. G. Shim and B. C. Wilson, The effects of ex vivo handling procedures on the near-infrared Raman spectra if normal mammalian tissues, *Photochem. Photobiol.*, 1996, **63**, 662–671.
9. E. O. Faoláin, M. B. Hunter, J. M. Byrne, P. Kelehan, M. McNamara, H. J. Byrne and F. M. Lyng, A study examining the effects of tissue processing on human tissue sections using vibrational spectroscopy, *Vib. Spectrosc.*, 2005, **38**(1–2), 121–127.
10. Z. W. Huang, A. McWilliams, S. Lam, J. English, D. I. McLean, H. Lui and H. Zeng, Effect of formalin fixation on the near-infrared Raman spectroscopy of normal and cancerous human bronchial tissues, *Int. J. Oncol.*, 2003, **23**, 649–655.
11. N. L. Pleshko, A. L. Boskey and R. Mendelsohn, An FT-IR microscopic investigation of the effects of tissue preservation on bone, *Calcif. Tissue Int.*, 1992, **51**, 72–77.
12. J. L. Stephenson, Ice crystal growth during the rapid freezing of tissues, *J. Biophys. Biochem. Cytol.*, 1956, **2**, 45–52.
13. N. Stone, C. Kendall, J. Smith, P. Crow and H. Barr, Raman spectroscopy for identification of epithelial cancers, *Faraday Discuss.*, 2003, **126**, 141–157.
14. N. Stone, M. C. H. Prieto, P. Crow, J. Uff and A. W. Ritchie, The use of Raman spectroscopy to provide an estimation of the gross biochemistry associated with urological pathologies, *Anal. Bioanal. Chem.*, 2007, **387**, 1657–1668.
15. H. Wills, R. Kast, C. Stewart, R. Rabah, A. Pandya, J. Poulik, G. Auner and M. D. Klein, Raman spectroscopy detects and distinguishes neuroblastoma and related tissues in fresh and (banked) frozen specimens, *J. Pediatr. Surg.*, 2009, **44**, 386–391.
16. M. Jackson, J. R. Mansfield, B. Dolenko, R. L. Somorjai, H. H. Mantsch and P. H. Watson, Classification of breast tumours by grade and steroid receptor status using pattern recognition analysis of infrared spectra, *Cancer Detect. Prevent.*, 1999, **23**, 245–253.
17. M. Meurens, J. Wallon, J. Tong, H. Noel and J. Haot, Breast cancer detection by Fourier transform infrared spectrometry, *Vib. Spectrosc.*, 1996, **10**, 341–346.
18. G. Muller, W. Wasche, U. Bindig and K. Liebold, IR-Spectroscopy for tissue differentiation in the medical field, *Laser Phys.*, 1999, **9**, 348–356.

19. R. Wiens, M. Rak, N. Cox, S. Abraham, B. H. J. Juurlink, W. M. Kulyk and K. M. Gough, Synchrotron FTIR microspectroscopic analysis of the effects of anti-inflammatory therapeutics on wound healing in laminectomized rats, *Anal. Bioanal. Chem.*, 2007, **387**, 1679–1689.

20. C. P. Schultz, The potential role of Fourier transform infrared spectroscopy and imaging in cancer diagnosis incorporating complex mathematical methods, *Technol. Cancer Res. Treat.*, 2002, **1**, 95–104.

21. C. Beleites, G. Steiner, M. G. Sowa, R. Baumgartner, S. Sobottka, G. Schackert and R. Salzer, Classification of human gliomas by infrared imaging spectroscopy and chemometric image processing, *Vib. Spectrosc.*, 2005, **38**, 143–149.

22. P. B. Medawar, The rate of penetration of fixatives, *J. Roy. Microsc. Soc.*, 1941, **61**, 46.

23. C. H. Fox, F. B. Johnson, J. Whiting and P. P. Roller, Formaldehyde fixation, *J. Histochem. and Cytochem.*, 1985, **33**, 845–853.

24. T. J. Mason and T. J. O'Leary, Effects of formaldehyde fixation on protein secondary structure: a calorimetric and infrared spectroscopic investigation, *J. Histochem. Cytochem.*, 1991, **39**, 225–229.

25. S. Aparicio, S. B. Doty, N. P. Camacho, E. P. Paschalis, L. Spevak, R. Mendelsohn and A. L. Boskey, Optimal methods for processing mineralized tissues for Fourier transform infrared microspectroscopy, *Calcif. Tissue Int.*, 2002, **70**, 422–429.

26. Y. N. Yeni, J. Yerramshetty, O. Akkus, C. Pechey and C. M. Les, Effect of Fixation and Embedding on Raman Spectroscopic Analysis of Bone Tissue, *Calcif. Tissue Int.*, 2006, **78**, 363–371.

27. R. K. Sahu, S. Argov, A. Salman, U. Zelig, M. Huleihel, N. Grossman, J. Gopas, J. Kapelushnik and S. Mordechai, Can Fourier transform infrared spectroscopy at higher wavenumbers (mid IR) shed light on biomarkers?, *J. Biomed. Opt.*, 2005, **10**, 05017–05027.

28. E. O. Faolain, M. B. Hunter, J. M. Byrne, P. Kelehan, H. A. Lambkin, H. J. Byrne and F. M. Lyng, Raman spectroscopic evaluation of efficacy of current paraffin wax section dewaxing agents, *J. Histochem. Cytochem.*, 2005, **53**, 121–129.

29. K. Poon, I. Lydon, H. Lambkin, N. Rashid, H. J. Byrne, F. M. Lyng, Investigation of dewaxing agents and protocols on formalin-fixed paraffin-embedded tissues by Raman spectroscopy, *Manuscript in preparation*.

30. E. Gazi, M. Baker, J. Dwyer, N. P. Lockyer, P. Gardner, J. H. Shanks, R. S. Reeve, C. A. Hart, M. D. Brown and N. W. Clarke, A Correlation of FTIR Spectra Derived From Prostate Cancer Tissue with Gleason Grade and Tumour Stage, *Eur. Urol.*, 2006, **50**, 750–761.

31. E. Gazi, J. Dwyer, P. Gardner, A. Ghanbari-Siahkali, A. Wade, J. Miyan, N. P. Lockyer, J. C. Vickerman, N. W. Clarke, J. H. Shanks, L. J. Scott, C. Hart and M. Brown, Applications of FTIR -Microspectroscopy to Benign Prostate and Prostate Cancer, *J. Pathol.*, 2003, **201**, 99–108.

32. C. W. Meuse and P. E. Barker, Quantitative Infrared Spectroscopy of Formalin-fixed, Paraffin-embedded Tissue Specimens, Paraffin Wax

Removal With Organic Solvents, *Appl. Immunohistochem. Mol. Morphol.*, 2009, **17**(6), 547–52.

33. A. Tfayli, O. Piot, A. Durlach, P. Bernard and M. Manfait, Discriminating nevus and melanoma on paraffin-embedded skin biopsies using FTIR microspectroscopy, *Biochim. Biophys. Acta.*, 2005, **1724**, 262–269.

34. A. Tfayli, C. Gobinet, V. Vrabie, R. Huez, M. Manfait and O. Piot, Digital Dewaxing of Raman Signals: Discrimination Between Nevi and Melanoma Spectra Obtained from Paraffin-Embedded Skin Biopsies, *Appl. Spectrosc.*, 2009, **63**(5), 564–70.

35. C. Gobinet, V. Vrabie, A. Tfayli, O. Piot, R. Huez and M. Manfait, Pre-processing and source separation methods for Raman spectra analysis of biomedical samples, *Conf. Proc. IEEE Eng. Med. Biol. Soc.*, 2007, 6208–11.

36. E. Ly, O. Piot, R. Wolthuis, A. Durlach, P. Bernard and M. Manfait, Combination of FTIR spectral imaging and chemometrics for tumour detection from paraffin-embedded biopsies, *Analyst*, 2008, **133**, 197–205.

37. E. Ly, O. Piot, A. Durlach, P. Bernard and M. Manfait, Differential diagnosis of cutaneous carcinomas by infrared spectral micro-imaging combined with pattern recognition, *Analyst*, 2009, **134**, 1208–1214.

38. V. Untereiner, O. Piot, M. D. Diebold, O. Bouché, E. Scaglia and M. Manfait, Optical diagnosis of peritoneal metastases by infrared micro-scopic imaging, *Anal. Bioanal. Chem.*, 2009, **393**, 1619–1627.

39. L. Chiriboga, P. Xie, H. Yee, V. Vigorita, D. Zarou, D. Zakim and M. Diem, Infrared spectroscopy of human tissue. I. Differentiation and Maturation of epithelial cells in the human cervix, *Biospectroscopy*, 1998, **4**, 47–53.

40. H. Y. N. Holman, M. C. Martin, E. A. Blakely, K. Bjornstad and W. R. McKinney, IR spectroscopic characteristics of a cell cycle and cell death probed by synchrotron radiation based Fourier transform IR spectro-microscopy, *Biopolymers (Biospectroscopy)*, 2000, **57**, 329–335.

41. P. Lasch, M. Boese, A. Pacifico and M. Diem, FT-IR spectroscopic investigations of single cells on the subcellular level, *Vib. Spectrosc.*, 2002, **28**, 147–157.

42. P. Lasch, A. Pacifico and M. Diem, Spatially resolved IR microspectro-scopy of single cells, *Biopolymers (Biospectroscopy)*, 2002, **67**, 335–338.

43. D. Yang, D. J. Castro, I. E. El-Sayed, M. A. El-Sayed, R. E. Saxton and N. Y. Zhang, A Fourier-transform infrared spectroscopic comparison of cultured human fibroblast and fibrosacroma cells: A new method for detection of malignancies, *J. Clin. Laser Med. Surg.*, 1995, **13**, 55–59.

44. A. Salman, J. Ramesh, V. Erukhimovitch, M. Talyshinksy, S. Mordechai and M. Huleihel, FTIR microspectroscopy of malignant fibroblasts transformed by mouse sarcoma virus, *J. Biochem. Biophys. Meth.*, 2003, **55**, 141–153.

45. M. J. Tobin, M. A. Chesters, J. M. Chalmers, F. J. M. Rutten, S. E. Fisher, I. M. Symonds, A. Hitchcock, R. Allibone and S. Dias-Gunasekara, Infrared microscopy of epithelial cancer cells in whole tissues and in tissue culture, using synchrotron radiation, *Faraday Discuss.*, 2004, **126**, 27–38.

46. H. P. Wang, H. C. Wang and Y. J. Huang, Microscopic FTIR studies of lung cancer cells in pleural fluid, *Sci. Total Environ.*, 1997, **204**, 283–287.

47. N. Jamin, P. Dumas, J. Moncutt, W.-H. Fridman, J.-L. Teillaud, G. L. Carr and G. P. Williams, Highly resolved chemical imaging of living cells by using synchrotron infrared microspectrometry, *Proc. Natl. Acad. Sci. USA*, 1998, **95**, 4837–4840.

48. E. Gazi, J. Dwyer, N. P. Lockyer, P. Gardner, J. Miyan, C. A. Hart, M. D. Brown and N. W. Clarke, Fixation Protocols for subcellular imaging using synchrotron based FTIR-microspectroscopy, *Biopolymers*, 2005, **77**, 18–30.

49. J. A. Kieran, *Chapter 2 Fixation*, In: *Histological and Histochemical Methods: Theory & Practice*, Pergamon Press, Oxford, UK, 1990, 10–35.

50. T. J. Harvey, C. Hughes, A. D. Ward, E. C. Faria, A. Henderson, N. W. Clarke, M. D. Brown, R. D. Snook and P. Gardner, Classification of fixed urological cells using Raman tweezers, *J. Biophoton.*, 2009, **2**(1–2), 47–69.

51. M. M. Mariani, P. Lampen, J. Popp, B. R. Wood and V. Deckert, Impact of fixation on in vitro cell culture lines monitored with Raman spectroscopy, *Analyst.*, 2009, **134**(6), 1154–61.

52. A. D. Meade, C. Clarke, F. Draux, G. D. Sockalingum, M. Manfait, F. M. Lyng, H. J. Byrne, Studies of chemical fixation effects in human cell lines using Raman microspectroscopy, *Anal. Bioanal. Chem.*, Epub, Jan 20, 2010.

53. E. Gazi, P. Gardner, N. P. Lockyer, C. A. Hart, N. W. Clarke and M. D. Brown, Probing Lipid Translocation Between Adipocytes and Prostate Cancer Cells with Imaging FTIR Microspectroscopy, *J. Lipid Res.*, 2007, **48**, 1846–1856.

54. E. Gazi, T. J. Harvey, P. Gardner, N. P. Lockyer, C. A. Hart, N. W. Clarke and M. D. Brown, A FTIR Microspectroscopic Study of the Uptake and Metabolism of Isotopically Labelled Fatty Acids by Metastatic Prostate Cancer, *Vib. Spectrosc.*, 2009, **50**, 99–105.

55. J. Sule-Suso, D. Skingsley, G. D. Sockalingum, A. Kohler, G. Kegelaer, M. Manfait and A. J. El Haj, FT-IR microspectroscopy as a tool to assess lung cancer cells response to chemotherapy, *Vib. Spectrosc.*, 2005, **38**, 179–184.

56. P. Crow, B. Barrass, C. Kendell, M. Hart-Prieto, M. Wright, R. Persad and M. Stone, The use of Raman spectroscopy to differentiate between different prostatic adenocarcinoma cell lines, *Br. J. Cancer*, 2005, **92**, 2166–2170.

57. C. M. Krishna, G. D. Sockalingum, G. Kegelaer, S. Rubin, V. B. Kartha and M. Manfait, Micro-Raman spectroscopy of mixed cancer cell populations, *Vib. Spectrosc.*, 2005, **38**, 95–100.

58. M. C. Krishna, G. Kegelaer, I. Adt, S. Rubin, V. B. Kartha, M. Manfait and G. D. Sockalingum, Characterisation of uterine sarcoma cell lines exhibiting MDR phenotype by vibrational spectroscopy, *Biochim. Biophys. Acta*, 2005, **1726**, 160–167.

59. E. Gazi, J. Dwyer, P. Gardner, A. Ghanbari-Siahkali, A. Wade, J. Miyan, N. P. Lockyer, J. C. Vickerman, N. W. Clarke, J. H. Shanks, L. J. Scott, C. Hart and M. Brown, Applications of FTIR-Microspectroscopy to Benign Prostate and Prostate Cancer, *J. Pathol.*, 2003, **201**, 99–108.

60. T. J. Harvey, E. Gazi, N. W. Clarke, M. D. Brown, E. C. Faria, R. D. Snook and P. Gardner, Discrimination of Prostate Cancer Cells by FTIR Photo-Acoustic Spectroscopy, *Analyst.*, 2007, **132**, 292–295.

61. T. J. Harvey, E. Gazi, A. Henderson, R. D. Snook, N. W. Clarke, M. Brown and P. Gardner, Factors influencing the Discrimination and Classification of Prostate Cancer Cell Lines by FTIR Microspectroscopy, *Analyst*, 2009, **134**, 1083–1091.

62. S. Z. Haslam and T. L. Woodward, Host microenvironment in breast cancer development: Epithelial-cell–stromal-cell interactions and steroid hormone action in normal and cancerous mammary gland, *Breast Cancer Res.*, 2003, **5**, 208–215.

63. C. Krafft, R. Salzer, S. Seitz, C. Ern and M. Schieker, Differentiation of individual human mesenchymal stem cells probed FTIR microscopic imaging, *Analyst*, 2007, **132**, 647–653.

64. A. D. Meade, F. M. Lyng, P. Knief and H. J. Byrne, Growth substrate induced functional changes elucidated by FTIR and Raman spectroscopy in in-vitro cultured human keratinocytes, *Anal. Bioanal. Chem.*, 2007, **387**, 1717–1728.

65. J. Lee, E. Gazi, J. Dwyer, N. P. Lockyer, M. D. Brown, N. W. Clarke and P. Gardner, Optical Artifacts in Transflection Mode FTIR Microspectro-scopic Images of Single Cells on a Biological Support: Does Rayleigh Scattering Play a Role?, *Analyst*, 2007, **132**, 750–755.

66. H.-Y. N. Holman, R. Goth-Goldstein, M. C. Martin, M. L. Russell and W. R. McKinney, Low-dose responses to 2,3,7,8-tetrachlorodibenzo-*p*-dioxin in single living human cells measured by synchrotron infrared spectromicroscopy, *Environ. Sci. Technol.*, 2000, **34**, 2513–2517.

67. D. Moss, M. Keese and R. Pepperkok, IR microspectroscopy of live cells, *Vib. Spectrosc.*, 2005, **38**, 185–191.

68. H. Y. N. Holman, M. C. Martin and W. R. McKinney, Synchrotron-based FTIR spectromicroscopy: Cytotoxicity and heating considerations, *J. Biomed. Phys.*, 2003, **29**, 275–286.

69. M. Miljkovic, M. Romeo, C. Matthaus and M. Diem, Infrared micro-spectroscopy of individual human cervical cancer (HeLa) cells suspended in growth medium, *Biopolymers*, 2004, **74**, 172–175.

70. F. Draux, P. Jeannesson, A. Beljebbar, A. Tfayli, N. Fourre, M. Manfait, J. Sulé-Suso and G. D. Sockalingum, Raman spectral imaging of single living cancer cells: a preliminary study, *Analyst*, 2009, **134**(3), 542–8.

71. C. Krafft, T. Knetschke, A. Siegner, R. H. W. Funk and R. Salzer, Mapping of single cells by near infrared Raman microspectroscopy, *Vib. Spectrosc.*, 2003, **32**, 75–83.

72. K. Kneipp, A. S. Haka, H. Kneipp, K. Badizadegan, N. Yoshizawa, C. Boone, K. E. Shafer-Peltier, J. T. Motz, R. R. Dasari and M. S. Feld,

Surface-enhanced Raman spectroscopy in single living cells using gold nanoparticles, *Appl. Spectrosc.*, 2002, **56**, 150–154.

73. K. Kneipp, H. Kneipp and H. Kneipp, Surface-enhanced Raman scattering in local optical fields of silver and gold nanoaggregates – From single-molecule Raman spectroscopy to ultrasenstive probing in live cells, *Acc. Chem. Res.*, 2006, **39**, 443–450.

74. L. B. Feril, T. Kondo, Y. Tabuchi, R. Ogawa, Q.-L. Zhao, T. Nozaki, T. Yoshida, N. Kudo and K. Tachibana, Biomolecular effects of low-intensity ultrasound: apoptosis, sonotransfection, and gene expression, *Japan. J. Appl. Phys.*, 2007, **46**, 4435–4440.

75. J. X. Cheng, Y. K. Jia, G. Zheng and X. S. Xie, Laser-scanning coherent anti-Stokes Raman scattering microscopy and applications to cell biology, *Biophys. J.*, 2002, **83**, 502–509.

76. X. Nan, E. O. Potma and X. S. Xie, Nonperturbative chemical imaging of organelle transport in living cells with coherent anti-stokes Raman scattering microscopy, *Biophys J.*, 2006, **91**(2), 728–35.

77. H. Rinia, K. N. Burger, M. Bonn and M. Muller, Quantitative label-free imaging of lipid composition and packing of individual cellular lipid droplets using multiplex CARS microscopy, *Biophys. J.*, 2008, **95**, 4908–4914.

78. C. W. Freudiger, W. Min, B. G. Saar, S. Lu, G. R. Holtom, C. He, J. C. Tsai, J. X. Kang and X. S. Xie, Label-free biomedical imaging with high sensitivity by stimulated Raman scattering microscopy, *Science*, 2008, **322**(5909), 1857–61.

79. J. W. Chan, D. S. Taylor, T. Zwerdling, S. M. Lane, K. Ihara and T. Huser, Micro-Raman spectroscopy detects individual neoplastic and normal hematopoietic cells, *Biophys. J.*, 2006, **90**(2), 648–56.

80. J. W. Chan, D. S. Taylor, S. M. Lane, T. Zwerdling, J. Tuscano and T. Huser, Nondestructive identification of individual leukemia cells by laser trapping Raman spectroscopy, *Anal. Chem.*, 2008, **80**(6), 2180–7.

81. J. W. Chan and D. K. Lieu, Label-free biochemical characterization of stem cells using vibrational spectroscopy, *J. Biophoton.*, 2009, **2**(11), 656–68.

82. T. J. Harvey, E. Correia Faria, E. Gazi, A. D. Ward, N. W. Clarke, M. D. Brown, R. D. Snook and P. Gardner, The Spectral Discrimination of Live Prostate and Bladder Cancer Cell Lines Using Raman Optical Tweezers, *J. Biomed. Opt.*, 2008, **13**, 064004.

83. G. Hastings, R. Wang, P. Krug, D. Katz and J. Hilliard, Infrared microscopy for the study of biological cell monolayers. I. Spectral effects of acetone and formalin fixation., *Biopolymers*, 2008, **89**, 921–930.

CHAPTER 6

Data Acquisition and Analysis in Biomedical Vibrational Spectroscopy

PETER LASCH[a],* AND WOLFGANG PETRICH[b]

[a] Robert Koch-Institut (P25), Nordufer 20, 13353 Berlin, Germany; [b] Faculty of Physics and Astronomy, University of Heidelberg, Albert-Überle-Str. 3-5, 69120 Heidelberg, Germany and Roche Diagnostics GmbH, Sandhofer Str., 116, 68305, Mannheim, Germany

6.1 Introduction

Present data acquisition and/or analysis tools such as OPUS (Bruker Optics), Perkin Elmer's Spotlight software (Perkin Elmer), Resolutions Pro (Varian), Grams (Thermo Fisher Scientific), The Unscrambler (CAMO), CytoSpec (www.cytospec.com) and various Matlab (The MathWorks) toolboxes allow for the easy recording and evaluation of infrared (IR) spectra. However, care has to be taken concerning the particular choice of data acquisition parameters, pre-treatment of spectra and data analysis procedures. With the attempt to move biomedical IR spectroscopy "from bench top to bedside" further questions of reproducibility and standardisation arise.

This chapter is intended to assist the procedure of biomedical vibrational spectroscopy in the context of synchrotron IR spectroscopy. It starts with the recording of spectra in standardised settings and continues through the assessment of spectral (and system) quality, spectral pre-processing up to data analysis (both quantitative and classification analysis). For the latter part, the main emphasis is placed on the role of independent validation.

RSC Analytical Spectroscopy Monographs No. 11
Biomedical Applications of Synchrotron Infrared Microspectroscopy
Edited by David Moss
© Royal Society of Chemistry 2011
Published by the Royal Society of Chemistry, www.rsc.org

6.2 Standardisation of the Infrared Spectral Measurements

The migration of research results of IR spectroscopy to practical applications requires a high level of spectral quality and reproducibility. Standardisation and optimisation of the measurement conditions and measurement parameters are therefore considered to be necessary prerequisites for further technological progress. In this section we will discuss systematic factors, which potentially impact the spectral quality and reproducibility, such as sample preparation, the optical set-up, the problem of water vapour absorption, and aspects of spectral and spatial resolution.

Sample preparation (see also Chapter 5): Reproducible sample acquisition and preparation is a key step in biomedical infrared spectroscopy and a number of sample preparation techniques for tissues, cells, and biofluids, that are compatible with mid-IR, have been suggested.[1] One of the major technical problems of biomedical IR spectroscopy is the presence of water: water exhibits very strong absorption features in the mid-IR region that is relevant to bio-medicine, which may mask the absorption signals of proteins, lipids, amino acids or other important compounds. One of the simplest solutions to over-come the water absorption problem is to dry the sample. Tissues, for example, can be snap frozen, and studied after cryo-sectioning and drying. Sample preparation of eukaryotic cells is often performed by growing the cells directly on the optical substrate with rapid drying afterwards. A similar approach is used for serum analyses. Here, a small volume of serum is transferred onto a sample carrier and allowed to air dry prior to measurements being taken.

An alternative approach of tissue sample preparation for IR spectroscopy is based on the paraffin-embedding technique. When applying this technique, it is however important to realise that any treatment of the objects under study with xylol (for de-paraffinisation) or alcohol (for cell fixation) may alter the structure (protein aggregation) and/or chemical composition of the samples (removal of lipids). Furthermore, as these procedures are dependent on time and con-centration they represent an additional source of error and should be avoided if possible. Therefore cryogenic tissue samples are often preferred over paraffin-embedded samples for IR spectroscopic investigations. Tissue sections of 8 μm thickness are usually produced by cryo-sectioning. A thickness of 8 μm is advantageous because it allows for preparation of specimens of a relatively large lateral size, while the limited thickness results in well-suited absorbance values of approx. 0.8 AU in the amide I region in transmission-type experi-ments. In this regime a non-linear response of the mercury cadmium telluride (MCT) detector is avoided, while at the same time an adequate signal-to-noise ratio (SNR) is maintained.

Optical material: After sectioning, the tissue slices are mounted on IR transparent supportive substrates such as CaF_2 windows for transmission-type measurements. CaF_2 is a popular window material in IR spectroscopy because it is almost insoluble in water and thus allows post-staining of the sections by haematoxylin and eosin (H&E). Furthermore, compared with other window materials such as ZnSe or KRS-5 the refractive index of CaF_2 is relatively low, so that the extent of optical fringing and chromatic aberration introduced by the window material is minimised (see references [2,3] for details). Finally, CaF_2 is transparent and colourless in the visible spectral range, which allows for effective documentation of the post-stained specimens. On the other hand, the low-frequency cut-off at approximately $950\,cm^{-1}$ and costs of approximately 150 € per window are obvious disadvantages of CaF_2 as an IR window material. Less expensive supporting substrates such as polymers (PEN, polyethylene naphthalate, or PET, polyethylene terephtalate, Du Pont) are rarely used in biomedical vibrational spectroscopy since these materials exhibit sharp and intense absorption bands in the spectral region of interest which cannot be fully compensated.

Water vapour, carbon dioxide: Long optical pathlengths in IR microspectrometers and the high extinction coefficients of water vapour can result in a very low transmission of IR light at the water vapour absorption lines between 1350 and $1950\,cm^{-1}$. These effects are often overlooked due to the intrinsic low half-widths of water vapour bands and the low spectral resolution of $4\,cm^{-1}$ or $8\,cm^{-1}$ commonly used in biomedical IR spectroscopy. As an illustration the single channel spectra of Figure 6.1 demonstrate that the sharp absorption features of water vapour in the mid-IR spectral region between $1350–1950\,cm^{-1}$ and $3600–3900\,cm^{-1}$, respectively, may considerably contaminate the spectra and obscure many important details of the underlying spectra. Thus removing unwanted absorption features of atmospheric water vapour from the sample spectra is a challenge in biomedical IR spectroscopy that is both frequent and well known.

The effective minimisation of water vapour contamination of mid-IR spectra can be achieved by two means. Firstly, the instrumentation and sample area can be purged by dry air. Secondly, water vapour bands can be removed by subtracting a spectrum of pure water vapour from the sample spectrum. Spectral subtraction is not an easy task as it requires the water vapour spectrum to be subtracted, and to fit the water vapour bands in the sample spectrum. In practice this is rarely the case because the precise shape of the water vapour spectrum is highly variable and depends on many factors such as the temperature or the partial pressure in the instrument. Aside from these well-known facts, it is of particular importance that sharp water vapour absorption features are typically only partly resolved at spectral resolutions of $4\,cm^{-1}$, or larger (see Figure 6.1). This observation is a direct consequence of the fundamental Nyquist–Shannon sampling theorem, which states that the sampling rate must be at least twice the signal frequency to be detected. As a result, water vapour signals in typical biomedical low resolution spectra are strongly distorted. Another spectral effect of strong water vapour absorptions is associated with the fact that most of the IR spectrometers are single beam instruments which consecutively measure the radiant power transmitted (single beam sample spectrum, SC_{SA}) and radiant power incident (single channel reference, SC_{BG}). As absorption or transmission spectra are obtained on the basis of the ratio between these two spectra, regions with intense water vapour signals may

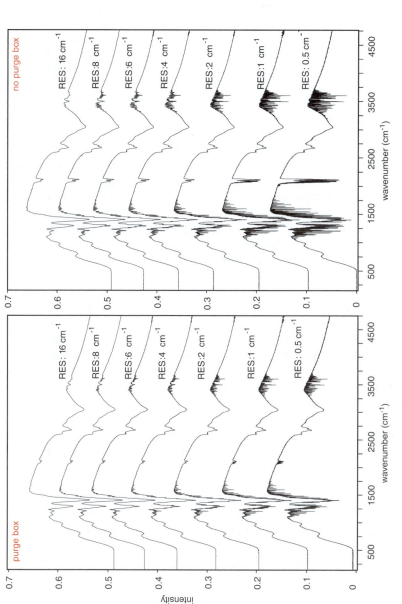

Figure 6.1 Dependence of water vapour absorption features in single channel mid-IR spectra with (left panel) and without (right panel) purging of the microscope sample compartment by dry air. Spectra were taken in transmission mode from real-world samples (cryo-sections of colon tissue mounted on CaF_2 windows). Note the sharp and intense water vapour absorption bands in the spectral regions 1350–1950 and 3600–3900 cm^{-1}, with nearly total absorption between 1500 and 1800 cm^{-1} (no purge box, RES = 0.5 cm^{-1}).

exhibit dramatically amplified noise, or – in the case of nearly total water vapour absorptions – essential discontinuities in the spectra (division by zero). It is therefore virtually impossible to computationally compensate water vapour contaminations in IR spectra acquired without purging.

As a basic principal in biomedical IR spectroscopy, the water vapour absorption problem should be thus addressed by effective purging of the instrumentation by dry air or nitrogen and/or by using appropriate desiccants. We consider purging/drying as an essential prerequisite for the successful computational removal of residual interfering water vapour. This need is illustrated by Figure 6.1, which shows single channel spectra of colon cryo-sections as a function of spectral resolution and purge conditions. These spectra have been acquired in transmission mode using a Fourier transform IR (FT-IR) spectrometer and an IR microscope, both purged by dry air. Spectra shown in the left panel were collected using an additional purge box that also encloses the sample compartment of the IR microscope. Interestingly, one can find almost total water vapour absorptions even in cases where the FT-IR instrument and the microscope are purged and only the short distance between the Cassegrain objective/condenser pair is accessible to atmospheric water vapour (see lower spectrum of the right panel with $RES = 0.5\,cm^{-1}$).

The absorption spectra obtained from two consecutive single beam measurements under the same conditions as described above are given in Figure 6.2. These spectra demonstrate amplified noise in water vapour regions under purge conditions (left panel) and noise plus uncompensated water vapour bands in the case where no purge box was present (right panel). The shape of the baseline in the spectra of the right panel indicates that a computational removal of unwanted atmospheric distortions in FT-IR microspectra cannot be achieved if the water vapour level exceeds a defined threshold. An example where purging of the instrumentation was completely absent is given in Figure 6.3. Here the spectral effects of total water vapour absorptions are shown in a single channel (left) and an absorption spectrum (right).

Consequently, the reduction of water vapour by purging and/or the use of desiccants along the complete optical path of IR light within IR instruments, including the sample compartment, and a stable concentration of carbon dioxide are considered to be essential prerequisites for the successful computational removal of water vapour residues in the IR spectra. It is thus recommended that adequate drying and/or purging of the instrumentation is ensured and that software algorithms for water vapour and carbon dioxide correction are applied for reproducible spectra acquisition.[4,5]

Spectral resolution: In FT-IR spectroscopy the spectral resolution is directly related to the maximum retardation of the moveable mirror and thus to the scan time and the data acquisition time: the lower the spectral resolution, the lower the time required to collect a spectrum. This fact becomes important in cases where large numbers of spectra are acquired. In IR imaging with single detector instrumentation for example, a compromise has to be found between spectral resolution and the required low scan time. The optimum choice will depend on the exact type of sample. In addition, the zero-filling factor (ZFF) has to be

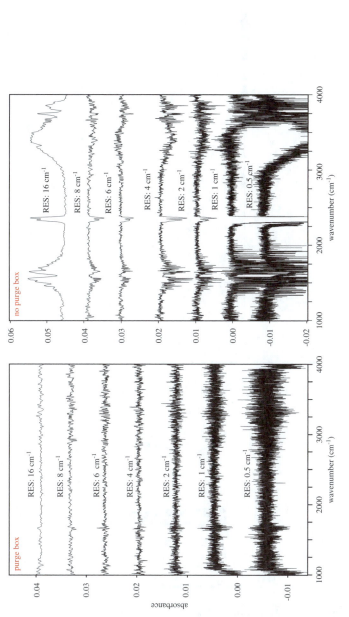

Figure 6.2 Spectral quality of mid-IR absorption spectra as a function of spectral resolution and purge conditions. Single channel sample and single background spectra were acquired consecutively from colon cryo-sections on CaF_2 by using a Bruker IRScope II microscope connected to an IFS28/B FT-IR spectrometer. Absorption spectra were obtained on the basis of the following equation: $ABS = -\log(SC_{SA}/SC_{BG})$. Further measurement parameters: detector: small-sized $100 \times 100\ \mu m$ N_2-cooled single element MCT detector. SR: approx. $25\ \mu m$ (aperture diameter: $900\ \mu m$, $36 \times$ Cassegrain objective), 128 scans, apodization function: Happ-Genzel.

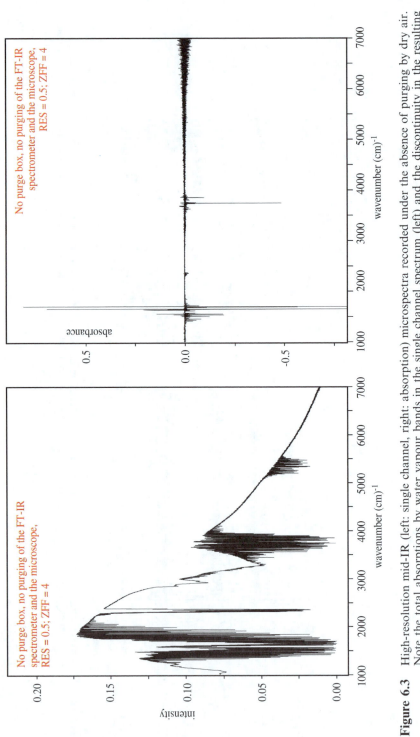

Figure 6.3   High-resolution mid-IR (left: single channel, right: absorption) microspectra recorded under the absence of purging by dry air. Note the total absorptions by water vapour bands in the single channel spectrum (left) and the discontinuity in the resulting absorption spectrum (right panel). See text and captions of Figures 6.1 and 6.2 for further details.

considered together with spectral resolution. The ZFF is a purely computational parameter; it defines the number of zero values, which are added to the side lobes of an interferogram prior to the Fourier transformation. Note that zero filling corresponds to an interpolation and does not add any structural information. We found, however, that the ZFF is important for subsequent Savitzky–Golay derivation, which is routinely applied in spectral pre-processing (*vide infra*).

In order to illustrate the impact of spectral resolution and zero filling in FT-IR microspectrometry of tissue sections, a series of test measurements with various settings for spectral resolution (RES) and ZFF was carried out. In these measurements spectra from a specific sample position of a human colorectal cryo-section were acquired. The sample was mounted on a CaF_2 window of 1 mm thickness and measured in transmission mode. In order to amplify spectral details, raw absorption spectra were pre-processed by applying a Savitzky–Golay first derivative filter with nine smoothing points. The results of the test (Figure 6.4) demonstrate that most of the "real" spectral features of an arbitrary tissue area are discernible at a combination of $RES = 6 \, cm^{-1}$ and $ZFF = 4$. This combination of measurement parameters ensures a good compromise between detectability of spectral details, measurement time and the SNR (*cf.* spectral features indicated by an arrow at approx. $1520 \, cm^{-1}$).

Apodisation: In FT-IR spectroscopy, the finite interferometer pathlength leads to the fact that the inverse Fourier transform, which is required to obtain the spectra from the interferogram, cannot be calculated from $-\infty$ to $+\infty$ as would be required by the mathematics of Fourier transformation. One way to mitigate the impact of this shortcoming is to introduce a so-called apodisation function. Numerous types of function have been suggested and a compromise has to be found between disguising existing spectral features and introducing artificial spectral features by apodisation. We found that the Happ–Genzel, Norton–Beer (NB) medium and Blackman–Harris (BH) 3-term apodisation appear to be well suited for the biomedical vibrational spectroscopy of dried specimens (see Figure 6.5).

Spatial resolution (SR): With wavelengths of light between 2.5 μm and 25 μm, Abbe's diffraction limit constrains mid-IR microspectroscopy to the observation of areas which are of similar size to the cellular compartments. While Abbe's criterion describes the principal limitation for the lateral spatial resolution (SR), many other factors restrict the practically achievable SR further. Furthermore, the choice of the size of the sample area to be measured depends on the problem and the type of biomedical samples. However, depending on the specific case, the maximum SR is often not required in MIR microspectroscopy.

Owing to the distinct levels of morphological heterogeneity in cells and tissues, the spatial resolution in a given IR imaging set-up strongly affects the local area of averaging and hence the IR spectral patterns obtained from the biomedical samples. This is of particular importance when spectral databases of reference microspectra are collected from well-characterised tissue structures.

Figure 6.6 exemplarily illustrates how the SR may affect the patterns of tissue microspectra. In this example the spatial resolution of an IR imaging data set

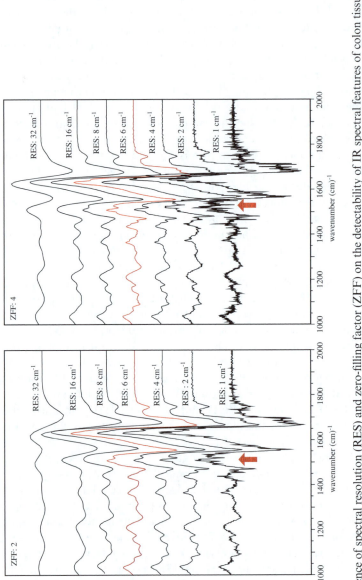

Figure 6.4 The influence of spectral resolution (RES) and zero-filling factor (ZFF) on the detectability of IR spectral features of colon tissue. In this example, identical positions of a tissue sample mounted on a CaF$_2$ window of a thickness of 1 mm were characterised by utilising a Bruker IR Scope II IR microscope. Transmission type IR spectra were recorded using a circular aperture of 900 μm diameter and a Cassegrain objective (36 ×, NA 0.5, SR *ca.* 25 μm). A Happ–Genzel apodization function and a first derivative Savitzky–Golay filter with nine smoothing points were applied to the spectra.

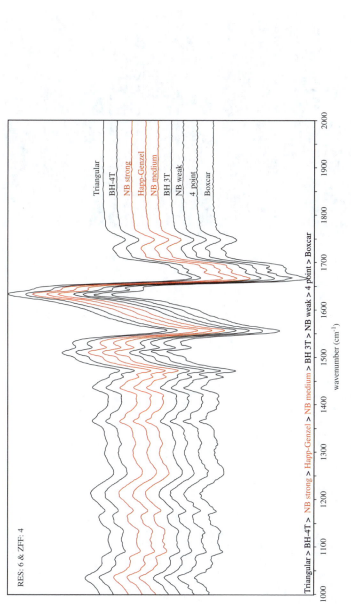
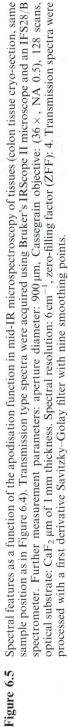

Figure 6.5   Spectral features as a function of the apodisation function in mid-IR microspectroscopy of tissues (colon tissue cryo-section, same sample position as in Figure 6.4). Transmission type spectra were acquired using Bruker's IRScope II microscope and an IFS28/B spectrometer. Further measurement parameters: aperture diameter: 900 μm, Cassegrain objective: (36 ×, NA 0.5), 128 scans, optical substrate: CaF_2 μm of 1 mm thickness. Spectral resolution: 6 cm^{-1}, zero-filling factor (ZFF): 4. Transmission spectra were processed with a first derivative Savitzky–Golay filter with nine smoothing points.

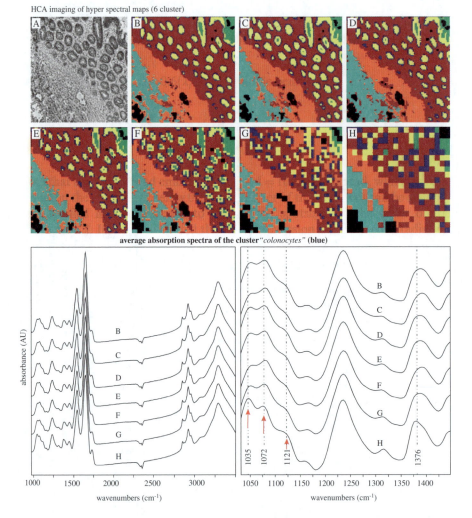

HCA imaging of hyper spectral maps (6 cluster)

average absorption spectra of the cluster *"colonocytes"* (blue)

Figure 6.6   Computational example for the variation of spatial resolution (SR) on mid-IR spectral patterns of tissues. The SR of an original map of IR point spectra (SR: 8.8 μm at 4000 cm^{-1}) was computationally reduced by pixel binning. The illustration shows class average spectra of the class "colonocytes" as a variable of the SR (traces B: 8.8 μm to G: 76 μm, see Table 6.1). Note the spectral "contamination" of the colonocyte spectra by absorption bands of carbohydrates, mostly due to mucin (see red arrows). (Adapted from reference 6.)

was computationally varied by pixel binning between 8.8 μm and 76 μm (*cf.* Table 6.1). Infrared spectral maps of panels C–H were produced on the basis of these binned microspectra and agglomerative hierarchical clustering (A: the Nomarski contrast image, B: cluster map of the original IR data). Furthermore, the mean cluster spectra of the cluster "colonocytes" are given (see lower panels

Table 6.1 Interpolation of an infrared spectral map in the spatial domains. Original map size: $1444 \times 1388\,\mu m^2$ corresponding to 232×223 individual pixel spectra (B). The number of pixel spectra was reduced by retaining the original image aspect ratio. The spatial resolution of the original map (B) was experimentally obtained with $8.8\,\mu m$ at $2.5\,\mu m$ wavelength ([a], estimated; [b], pixel size).

	Map size (in pixel spectra)	Spatial resolution at 2.5 μm wavelength [in μm]
B	232×223	8.8
C	160×154	10.0^{a}
D	133×128	14.0^{a}
E	80×77	18.3^{b}
F	60×58	24.5^{b}
G	40×38	37^{b}
H	20×19	76^{b}

of Figure 6.6). The analysis of these cluster means spectra indicates "contamination" of the high-resolution colonocyte spectra by carbohydrate absorption bands (red arrows, see also reference [6] for details). The carbohydrate signatures originate from mucin, a mixture of muco- and glycoproteins present as a precursor in goblet cells next to the colonocytes. Obviously, the variation of the spatial resolution affects not only the number of details which are resolved in the IR images but also affects the character and type of the IR spectral patterns. The computational example of Figure 6.6 thus illustrates the need to standardise the SR if spectral databases of reference spectra are collected.

In order to achieve standardisation of SR it is recommended to monitor the following measurement parameters or factors:

Geometry factors:
- in mapping experiments: step size in *x*- and *y*-directions, shape and diameter of the aperture
- multichannel detectors: detector size and spacing of elements of a focal plane array detector.

Optical factors:
- shape and geometry of Cassegrain objectives, mirrors and field stops
- transmission or transflection type measurements
- confocal or non-confocal set-up
- in transmission type measurements: thickness and refractive index of the supporting substrate
- the optical alignment of the IR microscope.

For a more detailed review of these and further factors, which affect the spatial resolution in IR microspectrometry, please consult the literature.[3,6]

6.3 Assessing the Quality of the Obtained Spectra

It is of course worthwhile confirming all of the issues mentioned in section 6.2 during a series of measurements as well as after the measurements. However, here we want to focus on particular issues of comparing multiple measurements and relating measurement-to-measurement variations to the variations within a single spectrum.

In an ideal set-up the repeated measurement of a sample would provide exactly the same spectrum. There are, however, two fundamentally different constraints: on the one hand, there is always a contribution caused by random error, *i.e.* noise, which can originate from, for example, the statistics of photons or from the electronics of signal amplification. Noise can be reduced, *e.g.* by cooling the detector, but based on fundamental laws of physics it cannot be zero. On the other hand, apart from noise there are small remaining spectral variations among spectra of identical samples, which are summarised under the expression systematic errors. They are caused, for example, by sample carrier inhomogeneity or differences in the drying process, and they also affect the reproducibility of the spectroscopy.

A method is required to analyse noise and reproducibility in biomedical vibrational spectroscopy. If spectra are compared at a single wavenumber only, the nomenclature and calculation procedures could readily follow the univariate methods used to evaluate single parameter measurements in standard analytical chemistry. The concepts of precision and bias are frequently introduced in analytical chemistry in order to distinguish between the random noise and the systematic error. While precision evaluates the deviation among repeated measurements and is therefore meant to address the random errors in the quantification of analytes, the bias enumerates the difference between the mean of the average results of repeated measurements and the reference value. Finally, the term accuracy covers both the precision and the bias and it is defined as the closeness of agreement between a test result and the accepted reference value. In analytical chemistry, another frequent distinction is made between the within-run precision ("repeatability") and the between-run precision ("reproducibility"). These terms are adequate for a univariate comparison of a measured value with a reference value.

There are numerous ways to define noise and reproducibility for the multivariate analysis of spectra. In analogy with univariate analysis – and keeping in mind that spectral noise is usually calculated from a single spectrum rather than multiple measurements at a given wavenumber – we derive the noise and reproducibility of the multivariate spectral data of biomedical vibrational spectroscopy as follows. The overall spectral signal can be represented as

$$S(\kappa) = S_t(\kappa) + N(\kappa) + E(\kappa) \qquad (6.1)$$

where S is the observed absorbance signal, S_t is the "true" absorbance signal in the absence of noise N and reproduction errors E, and κ is a running index number representing the wavenumber. N and E can be estimated by means of the formulas summarised in Appendix A. Simply speaking, for multivariate analysis we found it helpful to define the noise as an average over the standard deviations of $S–S_t$ within a given wavenumber interval, while we calculated the reproduction error as the standard deviation among the mean spectral signals of repeated measurements. Here, the repetition should be as close as possible to a clinical setting, *i.e.* the repetition should start with the same sample and go through the complete procedure of sample application, drying, spectroscopy, *etc.* again. In passing we would like to note that it is also interesting to assess the reproducibility for various stages along the line of sample handling, sample preparation and spectroscopic measurements.

To illustrate the usability of the distinction between random errors and systematic errors we compared the spectral contributions of repetitive measurements of identical aliquots of the same sample with the finding that reproducibility rather than signal-to-noise ratio is limiting the IR spectroscopy-based quantification of analytes in dried films of serum.[7] In turn, this finding should mean that additional noise should not hamper the accuracy of the quantification. Based on the spectra from an experiment on reagent-free quantification of analytes in serum,[8] we investigated the impact of noise, which here is artificially added onto the spectra measured, and evaluated the data as a function of SNR. Figure 6.7 shows the prediction error (root mean square error of prediction, RMSEP) for the prediction of the glucose concentration in serum as derived from IR spectroscopy as a function of SNR. While the SNR deteriorated from its original value of 3000 by more than an order of magnitude the quantification error remains low. Only for SNR ≤ 100 did we find a significant impact of added noise on the prediction accuracy. This result confirms the independent finding that reproducibility rather than noise is limiting the analysis of dried films of serum.[7]

A variety of further tests concerning the repeatability, reproducibility and quality of IR spectra obtained can be performed. In fact, tests for these three important spectral parameters were first introduced to biomedical IR spectroscopy in the late 1980's within the context of a project for FT-IR analyses of intact micro-organisms.[9] In these tests, the spectral "quality" is checked with regard to absorbance values of the raw spectral data, the signal-to-noise ratio (SNR), spectral contributions from water vapour, optical fringes and more. Quality tests have been adapted and expanded for IR microspectroscopic imaging and are comprised now of five independent quality checks:

1. A test for spectral contributions of water vapour,
2. A check for sample thickness (integrated intensity),
3. The test of the spectral signal-to-noise ratio (SNR),
4. A check called "test for an additional band" (tissue embedding medium)
5. A so-called "bad pixel" test to eliminate spectra from dead pixels of focal plane array detector measurements.

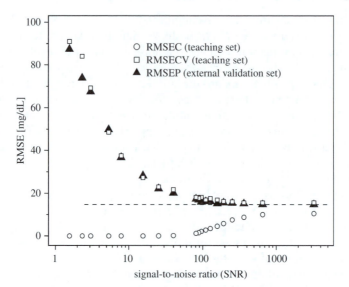

Figure 6.7 The root mean square error of calibration (RMSEC), leave-one-out cross validation (RMSECV) and prediction (RMSEP) are plotted as a function of the signal-to-noise ratio (SNR). While the intrinsic SNR amounts to 3000, random noise was artificially added to mid-IR spectra of 247 serum samples (which decreases the SNR) and the concentration of glucose was recalculated by means of partial least squares (PLS) based on the noisy spectra. The open symbols refer to assessing the quality of quantification within the teaching set, while the filled symbols relate to an external validation set. The data show that the noise can be increased by more than an order of magnitude before the prediction accuracy of the independent external validation set (RMSEP) is affected. In addition, it can clearly be observed that the RMSEC is a poor measure of accuracy since it suggests delivering seemingly better results for lower SNRs, while in fact the calibration simply tends to fit the noise for low values of SNR (see section 6.7).

Outliers, *i.e.* spectra that have failed one of the tests, are routinely removed from the hyperspectral imaging data sets and are thus not accepted for multivariate classification analysis.

6.4 Spectral Pre-processing

In classification analysis, pre-processing aims at removing quantitative aspects of the IR spectral information and at improving the accuracy and robustness of subsequent pattern analysis. Thus an optimal pre-processing workflow is the key to obtaining reliable classification and routinely includes tests for spectral quality, normalisation routines, spectral filtering and feature selection.

Spectral pre-processing for quantitative analysis (*vide infra*) requires alternative concepts. Here it is important to maintain the original absorbance information. Popular pre-processing routines in quantitative analysis such as baseline correction or the (extended) multiplicative signal correction (MSC/ EMSC) framework, aim at the separation of absorption and other optical effects of light propagation (scattering, dispersion). Calibration models developed on the basis of such pre-processed data generally perform better than models created with raw absorption spectra.[10]

Quality tests as integral parts of spectral pre-processing have been described in the previous sections. In the following we will therefore focus on pre-processing methods for classification analysis such as normalisation, spectral filtering (de-noising, derivatives) and feature selection.

Normalisation: According to Beer–Lambert's law, in a transmission type of experiment the absorbance A is defined as the product of the optical pathlength l, the concentration c of the substance under study and the respective absorption coefficient ε. While the last two variables facilitate qualitative analysis on structure and composition of the biological objects, a variable optical pathlength l does not add valuable information for classification analysis. In many instances variations of sample thickness may even hamper the identification of specific IR spectral patterns.

One possible approach to computationally reduce the effects of different optical pathlengths is vector normalisation. The result of vector normalisation is a spectrum in which the sum of all absorbance values squared equals 1. This transformation does not require reference spectra and is thus independent from so-called collective data sets. Another popular normalisation method, the min–max normalisation, scales and shifts spectra such that the minimum/maximum absorbance value occurring equals 0, or 1, respectively.

An example of how normalisation affects the differentiability of spectra from different tissue structures in classification analysis is given in Figure 6.8. This illustration shows representative FT-IR microspectra from nine different structures of the human colon: spectra of the submucosa (green), crypts (red), and of seven other pre-defined tissue structures (black). Panel A depicts the raw spectral data, while spectra in panel B have been offset corrected and vector-normalised using the spectral region of 1200–1770 cm^{-1}. In this simple case, characteristic spectral features of crypts (mucin features) and of the submucosa (features of collagen) are easily discernible in the pre-processed spectra, but not in the raw data. Furthermore, univariate analysis of the spectra in panel B would, in principle, allow for the introduction of certain thresholds suitable for the identification of both classes, namely at 1450 cm^{-1} for submucosa and at 1080 and 1733 cm^{-1} for the class crypts. The illustration demonstrates that normalisation and baseline correction are essential prerequisites for classification analysis.

Spectral filtering: In more complex classification tasks the utilisation of spectral (frequency) filters turned out to be essential. Popular filters are noise filters for de-noising, Savitzky–Golay derivative filters for "resolution enhancement" and various types of frequency filter in the Fourier space (Fourier self-deconvolution). A common aspect of these filters is that the person conducting the experiment must consider a trade-off between noise and the detectability of spectral fine structures, *i.e.* between SNR and the resolution.

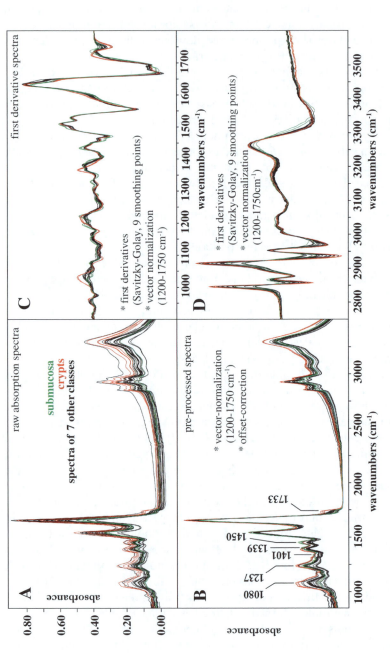

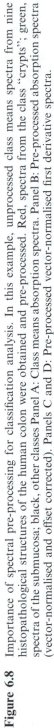

Figure 6.8 Importance of spectral pre-processing for classification analysis. In this example, unprocessed class means spectra from nine histopathological structures of the human colon were obtained and pre-processed. Red, spectra from the class "crypts"; green, spectra of the submucosa; black, other classes Panel A: Class means absorption spectra. Panel B: Pre-processed absorption spectra (vector-normalised and offset corrected). Panels C and D: Pre-processed vector-normalised first derivative spectra.

Derivative filters: The main advantage of derivative spectroscopy lies in the enhancement of the spectral fine structures combined with a reduction of broad baseline effects. Unfortunately, derivative spectroscopy requires a high SNR, which is sometimes hard to achieve, particularly if the IR microspectra are acquired with high spatial resolution. The example of Figure 6.8, panels C and D, demonstrates that the application of a first derivative Savitzky–Golay filter with nine smoothing points to the raw spectral data in combination with vector normalisation dramatically enhances the number of discriminative spectral features. We consider this combination to be the most effective and robust combination of pre-processing routines for classification analysis.[10]

6.5 Data Analysis: Quantitative Analysis

The quantification of analytes was one of the first applications of data analysis in IR spectroscopy. In its most simple form, the height of an absorbance peak is IR spectroscopy compared to the concentration of an analyte. For example, if the peak height increases linearly with analyte concentration, one could plot this linear dependence and would be able to predict the concentration of this analyte in a similar sample. In this example the peak height x correlates with the concentration y of the analyte and the degree of correlation could be enumerated by means of Pearson's correlation coefficient (see Appendix B). This correlation analysis would use a single spectral range (the peak) only and thus would be called a univariate analysis. Analysis would be restricted to simply finding that linear equation $y = ax + b$ which optimally fits the data. The aim of the fitting process is that the deviation between the fitted line and the data, *i.e.* the sum of squares of the deviations, is at a minimum. This procedure is called a least square fit.

In vibrational spectroscopy each peak can be conceptually assigned to a given vibration (or rotation) of a molecular bond. The large number of conceivable vibrations of a single biomolecule gives rise to a large number of peaks in that molecule's spectrum such that the molecular properties are revealed by the multitude of peaks rather than a single peak. For this reason, it is the multivariate analysis rather than the univariate analysis that is of particular importance in biomedical vibrational spectroscopy. Conceptually, multivariate analysis follows the univariate principle of "fitting a line" and there is a variety of multivariate methods for finding the optimum correlation between spectra and concentrations.

Introductions to multivariate data analysis are given, for example, in reference 11. In short, some of the major concepts are Classical Least Square (CLS) Regression, Principle Component Regression (PCR) and Partial Least Square (PLS) Regression. In CLS Regression, which is also called "multiple linear regression" or "ordinary least square regression", the univariate concept of minimising the squared error between the data and a regression line $y = ax + b$ is simply extended to fitting the function $y = a_1x_1 + a_2x_2 + a_3x_3 + \ldots + b$. Instead of using the x_i data directly, PCR first

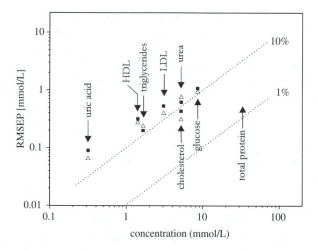

Figure 6.9   Root mean square error of prediction (RMSEP) of the concentration of analytes in serum (as derived from IR spectroscopy) as a function of average concentration. The data shown refer to an independent validation set of samples from 99 donors. (From reference 8.)

reduces the spectra to those components that are responsible for the highest variations in the data. A regression is subsequently performed between these components and the concentration values. Finally, in PLS regression the identification of the most important spectral components is not performed by merely investigating the variance in the spectra themselves, but by including the concentration values already in this identification analysis.

Figure 6.9 shows the result of the prediction of the concentration of analytes based on the mid-IR spectroscopy of dried serum samples.[8] Total protein content, lipids, lipoproteins, glucose as well as urea and uric acid can be quantified with an accuracy that is in the medically acceptable range. For example, the prediction error of 10% for the quantification of glucose is clearly within the range of acceptable accuracy as recommended for monitoring diabetes.

6.6   Data Analysis: Classification

6.6.1   Unsupervised Classification Analysis

The main goal of unsupervised classification analysis is to organise data (patterns) into meaningful or useful groups using some type of similarity measure. Unsupervised classification analysis is driven by spectral information only and it is particularly useful for extracting information from unclassified patterns, or during an exploratory phase of pattern recognition for assessing internal data similarities.[12] One particular form of unsupervised classification analysis is clustering, which encompasses a number of different algorithms for statistical data analysis. Clustering techniques aim at sorting different objects into groups or clusters on the basis of distance measures. Cluster analyses can be of the partitional or the hierarchical type. While the first form of algorithms determines all clusters at once by producing a unique assignment of each data object to a cluster, hierarchical algorithms find clusters in a series of partitions on the basis of successively established clusters. In hierarchical clustering one can furthermore distinguish between agglomerative or divisive forms. The top-down divisive strategy begins with the whole data set and separates the data objects successively into finer clusters. In contrast, agglomerative algorithms follow a bottom-up strategy and begin with each element as a separate cluster. Clusters are then merged stepwise to larger clusters.

The outcome of agglomerative hierarchical cluster analysis is a crisp cluster membership function, which can take only the values 0 (no membership) or 1 (membership). Other non-hierarchical clustering techniques such as k-means cluster (KMC)[13] analysis still follow this concept, whereas fuzzy C-means (FCM)[14] clustering returns fuzzy class memberships. The latter method thus departs from the classical (0 or 1) two-valued logic and uses soft linguistic system variables, *i.e.* degrees of class membership values varying between 0 and 1.

In biomedical IR spectroscopy, agglomerative hierarchical cluster analysis (HCA) enjoys great popularity, mostly due to its simplicity and ease of interpretation.

In the first step of HCA, a distance matrix is calculated that contains the complete set of interspectral distances. The distance matrix is symmetric along its diagonal and has the dimension $n \times n$, with n as the number of patterns. Spectral distance can be obtained in different ways depending on how the *similarity* of two patterns is calculated. Popular distance measures are Euclidean distances, including the city-block distance (Manhattan block distance), Mahalanobis distance,[15] and so-called differentiation indices (D-values, see also Appendix B)[16].

The second step of agglomerative HCA involves the clustering process. First, the two most similar spectra, *i.e.* spectra with the smallest inter-spectral distance, are determined and merged to form a new cluster. Then, according to the pre-defined linkage method, the spectral distances between the new cluster and the remaining spectra are evaluated. The process is repeated $n - 1$ times until all objects are combined into one single cluster. The sequence of fusions can be represented graphically as a dendrogram (see e.g. Figure 6.10.E).

In cluster imaging the cluster membership functions are utilised to visualise the spatial distribution of spectral patterns. Image assembly on the basis of IR microspectroscopy and cluster analysis follows the idea of assigning unique colours to spectra belonging to a given cluster. Pseudo-colour images are thus reassembled by plotting specifically coloured pixels as a function of the spectra's spatial coordinates. It is important to note that in FCM cluster imaging the colour intensity is scaled according to the fuzzy cluster membership value. This is different from the standard way of "crisp" cluster imaging (HCA or KMC) in which only colours with the intensities of 1 and 0 are used.

The computational demands of the various forms of cluster analysis vary significantly.[17] While small data sets with fewer than 10000 spectra can be

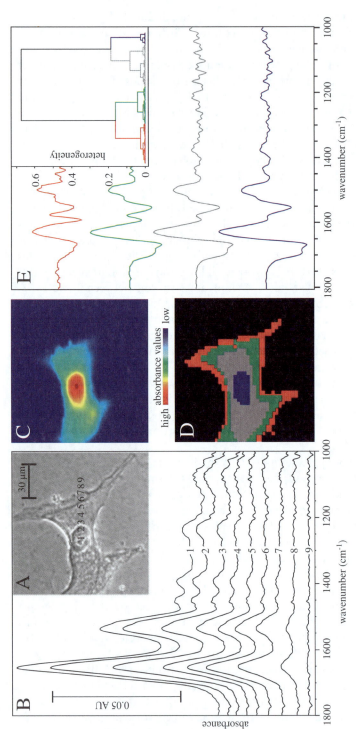

Figure 6.10 Example of the application of hierarchical cluster analysis (HCA) and HCA imaging in synchrotron IR microspectroscopy. Panel **A:** Nomarski contrast image of a benign human skin fibroblast grown directly onto a CaF$_2$ slide after enzymatic removal of phospholipids and RNA. Panel **B:** Single point spectra collected at numbered positions in panel A. Panel **C:** Chemical map obtained from baseline corrected absorption spectra in the amide I region (1620–1680 cm^{-1}). Panel **D:** HCA map produced from first derivative spectra on the basis of Euclidean distances and Ward's algorithm. Panel **E:** Dendrogram and class means spectra of the four-class-classification approach given in panel D. Spectral parameters: IR source, U10B beamline at NSLS; aperture of $10 \times 10\,\mu m^2$; spectral resolution: 8 cm^{-1}; number of scans: 128; step size: 3 µm; sampled area: $126 \times 126\,\mu m^2$ for 43×43 IR spectra. See text for further experimental details.

effectively partitioned by agglomerative HCA, this particular cluster approach turns out to be inefficient for larger hyperspectral data sets. This is for the most part due to the fact that the central processing unit (CPU) time in agglomerative HCA scales with the square of the number of spectra and not linearly (first order approximation) as for FCM or KMC cluster imaging (see reference [17] for details). Although the best correlation between histopathology and IR imaging is often found for agglomerative HCA, the high computational demands generally limit the usability of HCA in IR microspectroscopy. To circumvent this limitation and also to facilitate HCA imaging of large hyperspectral data sets, some software packages offer the option of "reduced HCA imaging". In this approach, HCA is initially performed with a reduced number of randomly selected microspectra. When the HCA has been completed, the cluster memberships of the remaining spectra can be computed on the basis of a distance measure such as D-values between these and cluster means spectra.[18]

An example of the application of unsupervised classification analysis to synchrotron IR microspectroscopic imaging is given in Figure 6.10. It shows in panel A a Nomarski contrast image of a human skin fibroblast grown directly on the CaF_2 supportive substrate. The cell was treated by ethanol and RNAse prior to characterisation by IR spectroscopy. The dark spot in the centre of the cell is the nucleus. Microspectra were collected at the U10B beam line of the election storage ring at National Synchrotron Light Source (NSLS) in Brookhaven, using a Nicolet Magna 860 step-scan FT-IR instrument coupled to a Continuum microscope. Owing to the highly collimated IR-beam of the synchrotron source, an aperture as small as $10 \times 10 \, \mu m^2$ was used with adequate SNR. The IR spectral map was collected with a spectral resolution of $8 \, cm^{-1}$ from a sample area of $126 \times 126 \, \mu m^2$ using a step size of $3 \, \mu m$ in x- and y-directions, respectively (43×43 spectra).

Individual microspectra displaying IR spectral properties of individual cell compartments are given in Figure 6.10B (*cf.* numbering of the spectra with the inset of panel A). These microspectra form the basis for IR spectral imaging of panels C and D. Panel C of Figure 6.10 depicts a chemical map obtained from baseline-corrected absorption spectra in the amide I region ($1620–1680 \, cm^{-1}$). It is quite obvious that this chemical map mainly reflects the thickness of the cell, which is at a maximum at the site of the nucleus. Qualitative spectral differences can be inferred by agglomerative HCA (panel D). This map was produced from first derivative spectra on the basis of Euclidean distances and Ward's algorithm.[19] The cluster map and the mean cluster spectra in panel E demonstrate the presence of distinct differences between spectra from the nucleus (blue) and the cytosol (grey, green; for a discussion see reference [20]).

Finally, one class of unsupervised methods is represented by self-organising maps (SOM), or Kohonen maps, named after the Finnish professor Teuvo Kohonen.[21] A SOM is a type of artificial neural network that needs to be trained but does not require labelling of the input vectors. Examples of classification analysis by SOMs in biomedical IR and Raman spectroscopy are given in references.[22,23]

6.6.2   Supervised Classification Analysis

The term "supervised", or "concept driven classification analysis" is used to describe a group of techniques in which a model is created that maps input objects (spectra) to desired outputs (class assignments).[24] Supervised classification analysis requires supervised learning, a machine learning technique for creating a classification function from training data. The following points are important when creating and applying supervised classification models in practice:

1. Determine the problem. Here it is of particular importance to define output categories (classes).
2. One should test for the internal data structure on the basis of unsupervised classification techniques (such as data clustering). This might be useful to determine whether the data principally allow the desired class assignments to be accomplished.
3. Gather the training data. Training data should be representative for the given problem and ideally contain a statistically relevant number of patterns of each pre-defined class. These patterns should cover intra-class and inter-class variances taken under realistic conditions.
4. Adequate data pre-processing. The accuracy of the function learned depends strongly on how the input patterns are represented. Data pre-processing intends to increase the robustness of classification, *i.e.* by normalisation, and also includes the steps of feature selection and feature reduction that reduce the dimensionality of the classification problem.
5. Selection of the supervised classification technique or the combination of techniques suitable for accomplishing the classification task. Popular supervised classifiers are Multi-Layer Perceptron Artificial Neural Networks (MLP-ANN), Support Vector Machines (SVM), k-Nearest Neighbours (k-NN), combinations of genetic algorithms (GA) for feature selection with Linear Discriminant Analysis (LDA), Decision Trees and Radial Basis Function (RBF) classifiers.
6. Teaching of the classifier on the gathered training data. In this phase, the performance of the classifier can be optimised by adjusting parameters on an internal validation subset. The final classification accuracy can then be measured on an external validation set, which should be kept totally separate from the training and internal validation subsets.

Supervised classifiers have been successfully applied to many problems in biomedical IR spectroscopy and imaging: ANNs or k-NNs in microbiology,[25,26,27] ANNs and LDA in serum analysis,[28,29,30] and ANNs in cancer diagnostics by IR spectroscopy,[31] or IR microspectroscopy and imaging.[32,33,34] The large number of successful studies carried out on the basis of ANNs is certainly due to the availability of the NeuroDeveloper,[35] a software package provided by Synthon Analytics.[36] This software package was specifically designed for the classification of spectroscopic data using artificial neural networks and comprises standard tools for spectral pre-processing, feature selection, network training and classification. Furthermore, a so-called *ModuleDeveloper* allows hierarchically organised (modular) neural networks of any degree of complexity to be created. These modular types of ANN allow a break-down of complex multi-class classification problems into small and independent (ideally two-class) classification tasks. Independent modular ANNs can be specifically optimised and thus offer a number of advantages. A discussion of these types of ANN can be found in the literature.[37,38]

The ANN classification of hyperspectral imaging data sets can be employed in a similar way to cluster imaging: the ANN class membership functions, that is the output activations, are converted to colour intensities and plotted as a function of the *xy*- coordinates of the pixel spectra in an imaging data set. Depending on the type of colour transfer function, the ANN images may be of the types "crisp" or "fuzzy". In the latter case the colour intensities scale linearly with the output activations. An example of hyperspectral imaging, using supervised ANN classifiers for image segmentation, is given in Figure 6.11. Infrared microspectra from a well-differentiated cryostat section of a rectal adenocarcinoma were collected in transmission mode using PerkinElmer's Spectrum Spotlight 300 IR imaging system. Spectra were acquired in the 4:1 imaging mode, which corresponds to a nominal pixel size of $6.25 \times 6.25\,\mu m^2$ and gives a lateral spatial resolution of approximately 12 μm at 6 μm wavelength.[6] The sampling area was $1737.5 \times 1625\,\mu m^2$ (279×261 microspectra).

For comparative purposes, panels A and B of Figure 6.11 show the photomicrograph of the H&E post-stained tissue area and a false colour map produced on the basis of vector normalised first derivative spectra and agglomerative HCA (10 classes), respectively. It is evident that the main tissue structures can be successfully differentiated by unsupervised data clustering. Fibrovascular tissues, for example, appear in this HCA map as yellow, light green or green regions while apical parts of neoplastic epithelia form the cluster encoded in orange.

The lower two panels of Figure 6.11 (C/D) show results of image segmentation by supervised multilayer perceptron artificial neural networks (MLP-ANNs). The crisp false colour image given in panel C was reassembled on the basis of a monolithic *toplevel* ANN with only five pre-defined spectral categories (see inset). Classification results of a modular *combined* ANN, a set of five hierarchically organised ANNs, are shown in Panel D. When comparing the false colour ANN images with the HCA map it is important to realise that HCA images were reassembled utilising only spectra of the current IR map, *i.e.* from one sample only. In contrast, the ANN classifiers were trained and optimised with microspectra from a tissue database which contained IR reference spectra from 12 relevant tissue structures and 28 patient samples.[34] Thus, while the HCA methodology is useful in an exploratory stage of spectral analysis, the ANN approach is applicable to image segmentation in a prospective clinical routine application (see reference[34] for details). Pre-conditions for the successful application of the IR based technique in histopathology are: (i) adequate data pre-treatment (quality tests, pre-processing), (ii) feature

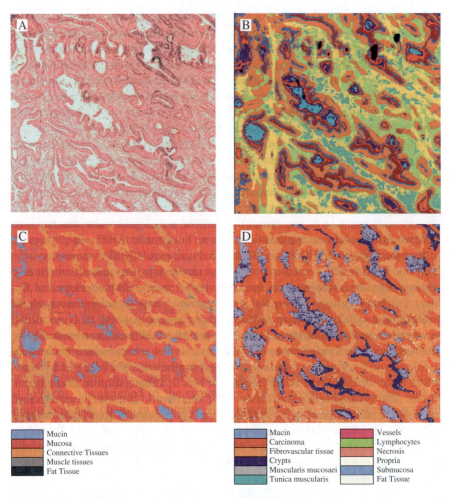

Mucin
Mucosa
Connective Tissues
Muscle tissues
Fat Tissue

Mucin   Vessels
Carcinoma   Lymphocytes
Fibrovascular tissue   Necrosis
Crypts   Propria
Muscularis mucosaei   Submucosa
Tunica muscularis   Fat Tissue

Figure 6.11 FT-IR microspectroscopic imaging of a cryostat section from a well
differentiated adenocarcinoma of the rectum. Panel **A**: Photomicrograph
of the H&E stained cryostat section. The sampling area was
$1737.5 \times 1625\,\mu m^2$ (279×261 microspectra). Panel **B**: HCA map produced from vector normalised first derivative spectra (ten-class classification approach). Panel **C**: Image segmentation by the toplevel artificial
neural network (ANN) with five predefined classes of IR microspectra.
Panel **D**: IR imaging based on the sublevel ANN with 12 classes (see text
and reference [34] for details).

selection and (iii) the utilisation of dedicated classification models such as
modular ANN classifiers.

6.6.3 The DPR Approach

Pattern recognition algorithms have been applied to a variety of tasks in biomedical vibrational spectroscopy. One specific application example is the direct

link between the IR spectrum of a blood (or plasma or serum) sample and the donor's health status. We dubbed this particular scenario "Diagnostics Pattern Recognition (DPR)". Instead of deconvoluting the biochemical parameters of the serum within the framework of quantitative spectroscopy (see section 6.5) and subsequently reassembling the biochemical information in order to enable a diagnosis, the DPR concept directly links the spectral pattern to the donor's state of health. This procedure has been exemplified in the fields of rheumatoid arthritis,[39] diabetes,[40] metabolic syndrome,[41] cardiovascular medicine,[42] and also in veterinary medicine,[29,30] namely bovine spongiform encephalopathy (BSE). In the latter studies, different classification methods have been combined to yield the overall classification. Table 6.2 provides an overview of the results of these studies. Typical sensitivities and specificities of 80% and above have been achieved.

6.7 The Role of Independent Validation

Given the powerful analysis tools which are readily available today, there is the danger of over- fitting the data and/or over-interpreting the results. While much of the early work in biomedical vibrational spectroscopy was purely explorative, it is important to note that nowadays, *i.e.* as the field is approaching application, any claim on suitability for diagnostic applications needs to be accompanied by an independent validation.

Figure 6.12 shows the process of analysis as it has been established over recent years in supervised analysis. Immediately after spectroscopy, the data are randomly split into a training, or teaching, data set and subsets for internal and external validation. For quantification as well as classification analysis, it is important to note that the independent external validation subset needs to be disregarded completely during the teaching process. Even more so, these validation data should be ideally blinded to the algorithm development team and remain blinded until the final algorithm has been established. Only after the teaching has been finished completely and the external validation spectra have been processed through the trained algorithm to give a predicted class assignment (or concentration) should this independent validation data be unblinded. In a clinical setting it is a prerequisite that the algorithm is established and new samples are acquired and processed afterwards. Obviously, particularly in this latter scenario the long-term stability of the set-up, standardisation and other issues mentioned in the foregoing sections are involved here.

Frequently the efforts are hampered by the lack of a sufficiently large number of samples. In our experience, it appears to be helpful if the number of samples exceeds the number of parameters (wavenumber intervals, principal components, latent variables, *etc.*) by at least a factor of five. This finding is supported by calculating this ratio between the number of teaching samples and the number of parameters both in the field of diagnostic pattern recognition (see the column "ratio "in Table 6.2) as well as in the quantitative analysis of serum.

Table 6.2 Diagnostic Pattern Recognition (DPR) in various applications (R-LDA, robust linear discriminant analysis; LDA, linear discriminant analysis; QDA, quadratic discriminant analysis; RDA, regularised discriminant analysis; ANN, artificial neural network; PCA, principal component analysis; SVM, support vector machine; N_{teach}, number of teaching samples; N_{para}, number of parameters used for classification (principal components *etc.*); ratio, N_{teach}/N_{para}; N_{val}, number of independent validation samples; SE, sensitivity; SP, specificity; LOO, leave-one-out validation. AMI, acute myocardial infarction; BSE, bovine spongiform encephalopathy).

Study	Method	N_{teach}	N_{para}	Ratio	N_{val}	SE	SP	Ref.
diabetes mellitus ⇔ "healthy"	R-LDA	80	8	10	LOO	73	75	40
	RDA	80	8	10	LOO	80	86	39
rheumatoid arthritis ⇔ "healthy"	R-LDA	258	8	32.2	126	84	88	41
metabolic syndrome ⇔ "healthy"	R-LDA	127	8	15.9	101	80	82	42
AMI ⇔ non-AMI	R-LDA	1008	8	126	421	88	95	42
BSE positive ⇔ BSE negative	LDA/QDA/ANN	843	~85 (ANN) 10 (LDA)	~8	260	96	92	29
BSE positive ⇔ BSE negative	PCA/LDA	481	36	13.4	160	82	93	30
	R-LDA	481	15	32.1	160	80	88	30
	ANN	481	19	25.3	160	93	93	30
	SVM	481	100	4.8	160	88	99	30

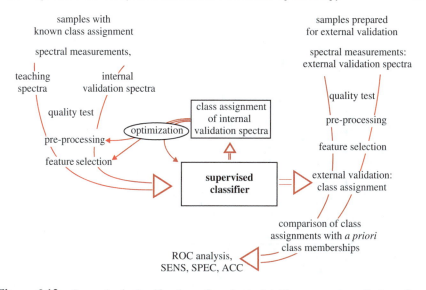

Figure 6.12 Supervised classification: flowchart detailing sequence of steps from spectral data acquisition to model evaluation by receiver operating characteristic (ROC) analysis.

In Figure 6.13 the root mean square error of prediction (RMSEP) is plotted as a function of the ratio between the number of teaching samples N_{teach} and the number of latent variables N_{para} which were used to train a PLS regression. After training the PLS algorithm with various numbers of spectra within the teaching set (ranging from 10 to 148 spectra) the various algorithms are applied to an independent data set of 99 blinded serum samples. After unblinding, the root mean square error between the predicted concentrations of the independent validation data and the actual concentrations are calculated. It can clearly be observed that the RMSEP dramatically deteriorates if the number of teaching samples is too low, *i.e.* if the aforementioned ratio falls to below approximately 5. At the same time, a root mean square error of calibration (which represents the accuracy of the algorithm re-evaluated within the teaching set itself) diminishes with the decreasing ratio! This is caused by the fact that, for a small number of samples or large numbers of "fit parameters" (*i.e.* latent variables) the PLS algorithm nicely adjusts to any fluctuation of spectral signatures, even if these are not at all related to the analyte. If the evaluation of this algorithm is performed with the identical data set to that used for calibration, it will of course reveal superior results as the algorithm was adjusted to ideally fit just that same data set. Thus, it is easy to obtain seemingly good results by overfitting the data if only the teaching data are used to assess the performance of the method.

Another indication for overfitting can be seen in Figure 6.7: at very high noise levels, *i.e.* at very low SNR, the algorithm tends to fit the noise, which gives superb results if evaluated with the same data (open circles) but which

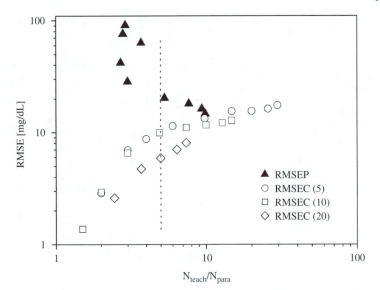

Figure 6.13   The root mean square error (RMSE) as derived from the calibration within the teaching set (RMSEC) or from evaluating the independent validation set (RMSEP) as a function of the ratio between the number of samples used to teach the algorithm and the number of parameters (latent variables); 5, 10, or 20 samples out of the teaching set were used for teaching. The dashed line indicates that the ratio should be larger than 5 in order to obtain reasonable values for independent validation. For $N_{teach}/N_{para} < 5$ the RMSEC decreases due to overfitting, but the RMSEP reveals that in fact only random variations had been fitted, which were not found in the independent validation set.

fails to supply valuable information in an independent validation set (closed symbols).

Although we have, so far, not been able to mathematically justify this rule of thumb of the minimum ratio of $N_{teach}/N_{para} \sim 5$, it has been our experience, and that of others, that consistent and dependable data can hardly be achieved at significantly lower ratios.

6.8   Conclusions

The technique of mid-IR microspectroscopy and imaging has great potential for the rapid and reliable identification of tissue structures not only for scientific research purposes but also in a real clinical set-up. The standardisation of the data acquisition and the assessment of the quality of the spectra constitute key factors for the successful transfer to clinical application. Furthermore the problem of overfitting and the role of independent validation have been discussed. In this chapter we also exemplified the question of standardisation of the hyperspectral image acquisition protocol and demonstrated how the interplay of unsupervised and supervised classification techniques can be utilised for classification model development. In particular, the concept of hierarchical or modular network classification, in combination with feature selection methods, dramatically enhances the capabilities of IR based classification techniques. We believe that the strategies outlined here may be successfully used in a great variety of applications in biomedical spectroscopy, specifically in histopathology.

6.9 Acknowledgements

We are grateful to Max Diem (Northeastern University Boston, USA), Dieter Naumann (Robert-Koch-Institut, Berlin, Germany) and Wolfgang Haensch (Robert-Rössle-Klinik at Max-Delbrück-Centrum Berlin, Germany) for fruitful discussions and support. Furthermore we would like to thank Jürgen Schmitt and Mark S. Novozhilov from Synthon GmbH (Heidelberg, Germany) for the excellent collaboration in the area of ANN analysis and D. Rohleder (Dioptic GmbH, Weinheim, Germany) and G. Kocherscheidt (Sirona GmbH, Bensheim, Germany) for the contributions to the quantitative analysis.

Appendix A: Noise and Reproduction Error

In an attempt to quantify the random error ("noise") and the systematic error ("reproduction error") with comparable metrics the signal observed overall is represented as:

$$S(\kappa) = S_t(\kappa) + N(\kappa) + E(\kappa) \tag{A1}$$

where S is the observed absorbance signal, S_t is the "true" absorbance signal in the absence of noise N and reproduction errors E, and κ is a running index number representing the wavenumber. It is convenient to define a value $Y(\kappa) = S(\kappa) - S_t(\kappa)$ which represents the difference between the expected spectrum and the observed spectrum and which should be a flat line at $Y(\kappa) = 0$ in the absence of noise and reproduction error. It is now assumed that the noise $N(\kappa)$ at index κ is independent of the noise $N(\kappa + 1)$ at the neighbouring wavenumber index. For r repetitions of a conceptually identical measurement, noise can then be approximated by:

$$N_{r,b}(\kappa) = \frac{1}{r} \sum_{j=1}^{r} \sqrt{\frac{1}{2b+1} \sum_{i=\kappa-b}^{\kappa+b} \left(Y_j(i) - \overline{Y}_j \right)^2} \quad \overline{Y}_j = \frac{1}{2b+1} \sum_{i=\kappa-b}^{\kappa+b} Y_j(i) \tag{A2}$$

Here, b determines the interval over which the noise of each measurement is averaged ($b > 1$). Typically b should correspond to a wavenumber region, which is similar to the typical line widths obtained in the spectra. For example, at an observed line width of $10\,\mathrm{cm}^{-1}$ and with a discretisation of the spectra into

$2\,\mathrm{cm}^{-1}$ intervals, the choice of $b=2$ would lead to averaging over $2b+1=5$ wavenumber intervals, *i.e.* over $10\,\mathrm{cm}^{-1}$.

Similarly, the reproduction error is defined as:

$$E_{r,b}(\kappa) = \sqrt{\frac{1}{r}\sum_{j=1}^{r}\left(\overline{Y}_j - \tilde{Y}\right)^2} \quad \tilde{Y} = \frac{1}{r}\sum_{j=1}^{r}\overline{Y}_j \tag{A3}$$

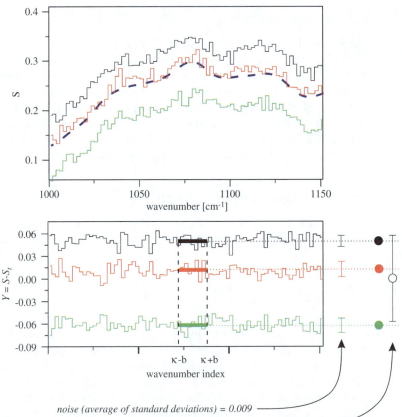

noise (average of standard deviations) = 0.009 ——————

reproduction error (standard deviation of averages) = 0.056 ——————

Figure 6.14 Deviation between the observed signal S and the expected signal S_t as a function of wavenumber index κ for a triplicate measurement of one sample. Three repetitions of a measurement of the identical sample, in general, deliver three slightly different spectra. The differences can be categorised into random contributions (noise) and systematic effects (reproducibility error). The error contributions are evaluated around a wavenumber index κ within a range from $\kappa - b$ to $\kappa + b$. Noise $N_{b,r}(\kappa)$ is estimated as the average over the standard deviations within each measurement while reproduction error $E_{b,r}(\kappa)$ is estimated as the standard deviation among the mean values of each of the three spectra within the given wavenumber index interval.

in a series of r repetitions ($r > 1$) of the conceptually identical measurements. These definitions are illustrated in Figure 6.14. While N estimates the average fluctuation from wavenumber to wavenumber, the reproduction error E enumerates the variations among the individual average spectra.

Appendix B: Differentiation Indices

Differentiation indices, or D-values, are obtained on the basis of Pearson's product momentum correlation coefficient between two spectra y_1 and y_2:

$$r_{y1y2} = \frac{\left(\sum_{i=1}^{n} y_{1i} \cdot y_{2i} \right) - n \cdot \bar{y}_1 \cdot \bar{y}_2}{\sqrt{\left(\sum_{i=1}^{n} y_{1i_{1i}}^2 - n \cdot \bar{y}_1^2 \right) \cdot \left(\sum_{i=1}^{n} y_{2i}^2 - n \cdot \bar{y}_2^2 \right)}} \tag{B1}$$

In eqn (B1), n is the total number of data points the IR spectra and $\bar{y}_1$ and $\bar{y}_2$ denote the average absorbance values. Furthermore, y_{1i} and y_{2i} are the *i-th* absorbance values of spectrum 1 or 2, respectively. The correlation coefficients are transformed to D-values by eqn (B2):

$$D_{y1y2} = (1 - r_{y1y2}) \cdot 1000 \tag{B2}$$

with $D_{y1,y2}$ varying between 0 ($r = 1$: highly correlated data/identity), 1000 ($r = 0$: uncorrelated data) and 2000 ($r = -1$: anti-correlated spectra).

References

1. E. Gazi, J. Dwyer, N. P. Lockyer, J. Miyan, P. Gardner, C. Hart, M. Brown and N. W. Clarke, *Biopolymers*, 2005, **77**(1), 18–30.
2. Fourier Transform Infrared Spectrometry, Second Edition P. R. Griffiths and J. A. de Haseth, Wiley-Interscience, New York, 2007.
3. G. L. Carr, *Rev. Sci. Instrum.*, 2001, **72**(3), 1613–1619.
4. S. W. Bruun, A. Kohler, I. Adt, G. D. Sockalingum, M. Manfait and H. Martens, *Appl Spectrosc.*, 2006, **60**(9), 1029–39.
5. "Atmospheric Vapor Compensation on the Spectrum 100 FT-IR and 100N FT-NIR Spectrometers" Technical note from PerkinElmer: http://las. perkinelmer.com/content/TechnicalInfo/TCH_Spectrum100FTIRAtmos-VaporConc.pdf.
6. P. Lasch and D. Naumann, *Biochim Biophys Acta (BBA) - Biomembranes*, 2006, **1758**, 814–829.
7. J. Möcks, G. Kocherscheidt, W. Köhler, W. Petrich, In Biomedical Vibrational Spectroscopy and Biohazard Detection Systems, edited by A. Mahdevan-Jansen, M. G. Sowa, G. J. Puppels, Z. Gryczynski,

T. Vo-Dinh, J. R. Lakowicz. *Proc. SPIE-Int. Soc. Opt.* Eng. 5321, 2004, 117–123.

8. D. Rohleder, G. Kocherscheidt, K. Gerber, W. Kiefer, W. Koehler, J. Moecks and W. Petrich, *J. Biomed. Opt.*, 2005, **10**, 031108.

9. D. Nanmann, in: *Biomedical Optical Spectroscopy*, edited by Anita Mahadevan-Jansen, Wolfgang Petrich, Robert R. Alfano, Alvin Katz, *Proc. SPIE-Int. Soc. Opt.* Eng. 6853, 2008, 68350G.

10. P. Heraud, B. R. Wood, J. Beardall and D. McNaughton, *J. Chemometrics*, 2007, **20**, 193–197.

11. H. Martens and T. Naes, *Multivariate Calibration*, John Wiley & Sons, 1989.

12. J. P. Marques de Sa, *Pattern Recognition: Concepts, Methods and Applications*, Springer-Verlag, Berlin, Heidelberg, New York, 2001.

13. J. B. McQueen, in L. M. LeCam, J. Neymann (Eds), Proceedings of Fifth Berkeley Symposium on Mathematical Statistics and Probability. 281–297, 1967.

14. J. C. Bezdek, *Pattern recognition with fuzzy objective function algorithms*, Plenum Press, New York, 1981.

15. P. C. Mahalanobis, *Proceedings of the National Institute of Science of India*, 1936, **12**, 49–55.

16. D. Helm, H. Labischinski and D. Naumann, *J. Microbiol. Methods*, 1991, **14**(2), 127–142.

17. P. Lasch, W. Haensch, D. Naumann and M. Diem, *Biochim. Biophys. Acta (BBA) - Molecular Basis of Disease*, 2004, **1688**(2), 176–186.

18. CytoSpec Website. http://www.cytospec.com [01. May 2008].

19. J. H. Ward and J. Americ, *Stat Assoc.*, 1963, **58**, 236.

20. P. Lasch, A. Pacifico and M. Diem, *Biopolymers*, 2002, **67**(4–5), 335–338.

21. T. Kohonen, Self-Organizing Maps. Series in Information Sciences, Vol. 30. Springer, Heidelberg. Second ed. 1997.

22. N. M. Amiali, M. R. Mulvey, J. Sedman, A. E. Simor and A. A. Ismail, *J. Microbiol. Methods*, 2007, **69**(1), 146–53.

23. H. Yang, I. R. Lewis and P. R. Griffiths, *Spectrochimica Acta Part A: Molec. Biomolec. Spectrosc.*, 1999, **45**(14), 2783–2791.

24. Wikipedia Website. http://en.wikipedia.org/wiki/Supervised_learning [01. May 2008].

25. A. Oust, T. Møretrø, C. Kirschner, J. A. Narvhus and A. Kohler, *J. Microbiol. Methods*, 2004, **59**(2), 149–62.

26. R. Goodacre, E. M. Timmins, P. J. Rooney, J. J. Rowland and D. B. Kell, *FEMS Microbiol. Lett.*, 1996, **140**(2–3), 233–239.

27. C. A. Rebuffo, J. Schmitt, M. Wenning, F. von Stetten and S. Scherer, *Appl. Environ. Microbiol.*, 2006, **72**(2), 994–1000.

28. J. Schmitt, M. Beekes, A. Brauer, T. Udelhoven, P. Lasch P and D. Naumann, *Anal. Chem.*, 2002, **74**, 3865–3868.

29. P. Lasch, J. Schmitt, M. Beekes, T. Udelhoven, M. Eiden, H. Fabian H, W. Petrich and D. Naumann, *Anal. Chem*, 2003, **75**(23), 6673–6678.

30. T. C. Martin, J. Moecks, A. Belooussov, S. Cawthraw, B. Dolenko, M. Eiden, J. von Frese, W. Köhler, J. Schmitt, R. Somorjai, T. Udelhoven, S. Verzakov and W. Petrich, *Analyst*, 2004, **129**, 897–901.
31. M. Romeo, F. Burden, M. Quinn, B. R. Wood and D. McNaughton, *Cell Mol. Biol. (Noisy-le-grand)*, 1998, **44**(1), 179–187.
32. P. Lasch and D. Naumann, *Cell Mol. Biol.*, 1998, **44**(1), 189–202.
33. P. Lasch, W. Haensch, L. Kidder, E. N. Lewis and D. Naumann, *Appl. Spectrosc.*, 2002, **56**(1), 1–9.
34. P. Lasch, M. Diem, W. Haensch and D. Naumann, *J. Chemometrics*, 2007, **20**, 209–220.
35. T. Udelhoven, M. Novozhilov and J. Schmitt, *Chemom. Intell. Lab. Syst.*, 2003, **66**, 219–226.
36. Synthon Website. http://www.synthon-analytics.de [01. May 2010].
37. A. Gasser and M. Kamel, *J. Intell. Rob. Syst.*, 1998, **21**, 117–129.
38. T. Udelhoven, D. Naumann and J. Schmitt, *Appl. Spectrosc.*, 2000, **54**(10), 1471–1479.
39. A. Staib, B. Dolenko, D. J. Fink, J. Frueh, A. E. Nikulin, M. Otto, M. S. Pessin-Minsely, O. Quarder, R. Somorjai, U. Thienel, G. Werner and W. Petrich, *Clin. Chim. Acta*, 2001, **308**, 79–89.
40. W. Petrich, B. Dolenko, J. Frueh, M. Ganz, H. Greger, S. Jacob, F. Keller, A. E. Nikulin, M. Otto, O. Quarder, R. L. Somorjai, A. Staib, G. Werner and H. Wielinger, *Appl. Opt.*, 200, **39**, 3372–3379.
41. J. Früh, S. Jacob, B. Dolenko, H.-U. Häring, R. Mischler, O. Quarder, W. Renn, R. Somorjai, A. Staib, G. Werner, W. Petrich. in Biomedical Vibrational Spectroscopy II, edited by A. Mahadevan-Jansen, H. H. Mantsch, G. J. Puppels. *Proc. SPIE-Int. Soc. OPt. Eng.* 4614, 2002, 63–69.
42. W. Petrich, K. B. Lewandrowski, J. B. Muhlestein, M. E. H. Hammond, J. L. Januzzi, *et al.*, Infrared spectroscopy can aid the triage of patients with acute chest pain, *Analyst*, 2009, **134**(6), 1092–1098.

CHAPTER 7

Synchrotron Radiation as a Source for Infrared Microspectroscopic Imaging with 2D Multi-Element Detection

G. L. CARR,[a] L. M. MILLER[a,b] AND P. DUMAS[c]

[a] National Synchrotron Light Source, Brookhaven National Laboratory, Upton, NY, USA; [b] Department of Biomedical Engineering, Stony Brook University, Stony Brook, NY, USA; [c] Synchrotron Soleil, St. Aubin, Gif sur Yvette, France

7.1 Introduction

In this chapter we consider developments in the use of infrared (IR) synchrotron radiation for microspectroscopy based on instruments having two-dimensional (2D) array detectors. As with microspectroscopy using single-element detectors (SEDs), we anticipate that the IR synchrotron radiation source is likely to play a significant role in the field of biological and biomedical spectroscopy. That role is not as a replacement for the standard thermal source, but as a complement where the synchrotron source is used to address problems on a local scale in combination with larger area surveys conducted using the standard IR source.

For a broad range of applications, the principal motivation for synchrotron-based IR microspectroscopy is to achieve significantly greater lateral resolution, typically at the diffraction limit, in combination with superior signal-to-noise without resorting to prohibitively long acquisition times. Though the brightness advantage of the source[1] is its most important quality – especially for

RSC Analytical Spectroscopy Monographs No. 11
Biomedical Applications of Synchrotron Infrared Microspectroscopy
Edited by David Moss
Published by the Royal Society of Chemistry, www.rsc.org

microscopy – it has other unique and useful features. For example, it is a continuum source with spectral coverage from the very far-IR up through X-rays, it is pulsed (on the picosecond time scale) and it possesses well defined polarization states. The source can be used in pump-probe studies of dynamics,[2,3] and it is particularly well suited to experiments where multiple and widely varying photon energies or wavelengths are required. An example is combined X-ray (fluorescence, absorption or diffraction) and IR microscopic analysis on the same sample,[4–6] which represents an activity of growing interest in the synchrotron radiation community. The broad spectral coverage and high brightness of synchrotron radiation reach well into the far-IR to below 1 THz.[7]

For microspectroscopy, the synchrotron serves as a high brightness source having diffraction-limited dimensions. When focused using a microscope, the illuminated spot is typically on the order of 10 μm, but the intensity in that region is about 100 times greater than for a thermal blackbody source such as the globar™. Therefore, confocal aperturing with $3\times3\,\mu m^2$ effective aperture size delivers good signal-to-noise and is routinely used with the synchrotron source. This confocal aperturing is an important ingredient for achieving a high level of spatial resolution and contrast.

With few exceptions,[8–11] synchrotron-based IR microscopy studies have been performed using instruments equipped with a single-element detector (SED). Univariate (and multivariate) images are produced by scanning the sample in a known and predetermined manner relative to the aperture, collecting complete spectra at each scan location. The quality and resolution of IR chemical images produced with the high brilliance synchrotron source have had an impact on many scientific disciplines, especially in the life sciences. In this regard, the study of individual single cells has probably benefited the most from the use of the synchrotron source.[12,13] In principle, a point-to-point examination of contiguous areas of any size can be conducted. However, the speed of data acquisition in point mapping protocols generally precludes the examination of large sample regions at high spatial and spectral resolution. This is especially true for *in vivo* studies of cells and other time-sensitive systems.

In recent years, tremendous progress in data acquisition time has been made by the use of large format array detectors.[14] These consist of numerous, small detector elements arranged in a 2D grid, often referred to as focal plane arrays (FPAs). These array detectors are akin to the ubiquitous modern digital camera, except that they are designed to sense light in the mid-IR, the spectral range best known for vibrational chemical spectroscopy. Accordingly, FPA-based microscopes are increasingly being introduced into synchrotron facilities at IR beamlines. The types of multi-element detectors that are (or have recently become) available for commercial microscope and spectrometer systems are briefly described, focusing on the characteristics relevant to the IR synchrotron radiation source. As with the SED microspectrometer, we can anticipate that an FPA microspectrometer will be capable of a spatial resolution limited only by optical effects when used optimally with the synchrotron source. To address this, we review the qualities of the IR synchrotron radiation source and the issues affecting the optical performance of the microspectroscopy instrument, in particular those that limit the quality of IR images. We then describe how the synchrotron source can be used with an FPA microscope to achieve both spatial oversampling and the signal-to-noise necessary for applying image deconvolution algorithms for enhancing resolution and contrast, and we report some early results. Though there have been only a few studies reported where an FPA equipped microspectrometer has used a synchrotron source, the results are encouraging, and significant developments can be expected over the next few years. Still, the synchrotron source is not expected to be competitive with laboratory-based thermal sources when larger areas are to be imaged and the highest spatial resolution is not required. While there have been major advances in near-field techniques for reaching beyond the diffraction limit, in some cases using free-electron laser sources, these have not involved the use of large area imaging methods and therefore are outside the scope of this review.

Lastly, we mention some other ways in which a 2D multi-element array detector could be used with synchrotron radiation to enhance measurements, such as for time-resolved chemical imaging on a sub-millisecond time scale.

7.2 Optical Issues for Infrared Microspectroscopy

7.2.1 The Standard Infrared Microspectrometer

In its typical configuration, the IR microspectrometer is a Fourier spectrometer (interferometer) combined with a microscope and single-element IR detector such as photoconductive $Hg_{1-x}Cd_xTe$ (MCT) designed to reach down to a frequency of $\sim 600\,cm^{-1}$. The optical path through the microscope can be configured to perform either transmission or reflection type measurements. These microscopes typically employ Schwarzschild-type optics both as a condenser to illuminate the sample and as an objective to collect the IR radiation from specific locations of the sample and deliver it to the detector. These instruments also function as optical microscopes to allow the user to visualize the sample prior to an IR measurement. The optical path for viewing may actually be reversed from the IR path, in which case the roles of the objective and condenser are swapped. As a result, it is typical for the objective and condenser to be labelled according to a conventional optical microscopy layout with the objective above the sample and the condenser below. An optical schematic for a typical microspectrometer using a single-element detector in a transmission configuration is shown in Figure 7.1. For convenience we show Schwarzschild optics having a rear focus at a finite distance (*i.e.* finite conjugate).

In the microspectrometer schematic of Figure 7.1, the two objectives share a common focus at the specimen, *i.e.* they are confocal. However, we add an additional constraint that the two apertures (labelled upper and lower in the figure) are both present and matched to define the same sample region when using the term confocal. Whether used with a single aperture, or in a confocal setup, the instrument spectroscopically samples just one location at a time.

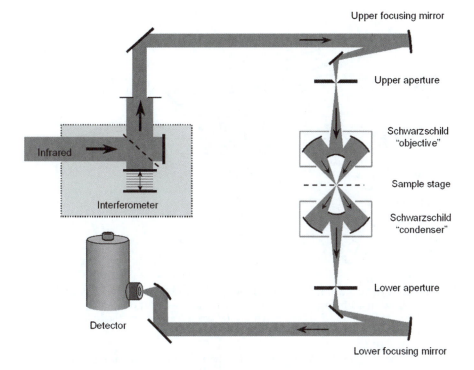

Figure 7.1 Schematic for a scanning infrared microspectrometer system using a single-element detector and the possibility for confocal operation where aperturing is used both before and after the sample.

Accordingly, an image can be built up by scanning the specimen through the focused beam in a raster-style fashion, stopping at each location to collect a full IR spectrum. The time spent at each location can range from seconds to minutes depending on the requirements for spectral resolution and signal-to-noise. Thus, the time to collect an image (often referred to as a spectroscopic map) is usually measured in hours, with collection times over 10 hours not uncommon.

When equipped with a conventional thermal source, high quality spectra can be collected from individual areas down to about 25 µm, which is above the limits imposed by diffraction. The latter depends on wavelength and the particular microscope optical configuration, but is typically below 10 µm.

7.2.2 The Schwarschild Microscope Objective

The Schwarzschild objectives are a catoptric design based on two spherical (or approximately spherical) mirrors centered on a common optic axis. Typical magnifications range from 6× up to 74×, with 15× and 32× (or 36×) used most frequently. Numerical apertures (NA) vary from about 0.3 up to nearly 0.7,

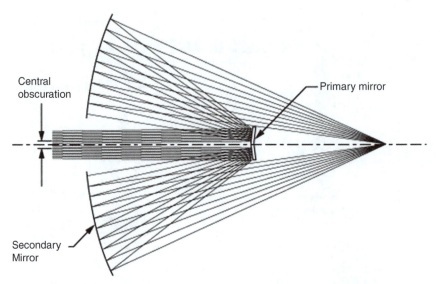

Figure 7.2   Schematic of a Schwarzschild-type objective having finite conjugates. The
central obscuration results in a reduction ($\sim 20\%$) of the throughput and
redistribution of energy within the beam waist at the focus.

with a value of ~ 0.6 typical (except when a longer working distance is
necessary). Schwarzschilds are available with both finite back conjugates and
"infinity-corrected" designs, in which the objective's back conjugate is designed
for a collimated beam, allowing error-free incorporation of polarizers, filters
and/or beamsplitters. A schematic of a Schwarzschild-type objective is displayed in Figure 7.2.

In the mid-IR, these objectives deliver diffraction-limited performance over
an area extending 100–200 µm from the optic axis. Having no refractive components, they are also free from chromatic aberrations. However, the design
causes the central portion of the objective's aperture to be obscured, losing up
to 25% of the aperture area. As we will show later, this obscuration leads to
significant diffraction effects when compared to a conventional microscope
objective.

7.2.3   The FPA Infrared Microspectrometer

Fortunately, such a microscope can now be fitted with a multi-element detector
to measure multiple areas simultaneously. One multi-element detector
approach is to construct quasi-1D arrays from the same $Hg_{1-x}Cd_xTe$ photoconductive material as for a single-element detector, providing essentially the
same spectral range and performance. Each element can be individually biased
and read-out by an AC coupled amplifier having bandwidth compatible with
FT-IR scanner velocities of 1 cm/s or above. Larger area, 2D arrays are usually
made with photovoltaic elements (*i.e.* photodiodes) to avoid the current biasing

requirements for photoconductors. One consequence of replacing the single-element detector with a multi-element array is that the light source illuminating the sample can no longer be constrained using an aperture, *i.e.* the optical system cannot be operated with two confocal apertures for improved resolution and contrast. At the same time, the large number of pixels in the FPA system presents the opportunity for point spread function (PSF) image deconvolution and the ability to improve both the spatial resolution and the contrast. We will return to this topic in a subsequent section.

Material quality and performance issues tend to limit the spectral response for $Hg_{1-x}Cd_xTe$ photodiode arrays to above $850\,cm^{-1}$. In a standard FPA used for spectroscopic imaging, the signal from each and every photodiode element (*i.e.* "pixel") is collected simultaneously and summed into a set of "wells" over a time period known as the integration time. The signal for all of the wells, referred to as a "frame", is then transferred to read-out electronics and the frame is output using one or several multiplexer circuits. During the read-out period, all of the pixels can be collecting and integrating the next frame such that the FPA can be sensing almost continuously if the integration time is set to match the frame read-out time. The read-out time depends on the dimensions of the array, but is significantly slower (by more than an order of magnitude) than for the single or few-element detector. An additional limitation is that the time-integrated signal at each element cannot exceed a maximum value (count) determined by the "well depth" in the read-out electronics. Even in the absence of the spectrometer's source, each array pixel senses the IR from the $T \sim 295\,K$ thermal background, adding to the signal count (and noise). For slow read-out rates and maximum integration time, the thermal background can cause the wells to become full without any IR from the spectrometer. In such situations, the array must be rendered blind for a portion of the time, resulting in an integration time less than ideal (*i.e.* "dead time"). Ideally, one wants the modulated IR light from the spectrometer to dominate over the thermal background, and the read-out rate to be sufficiently fast that the signal at each pixel can be integrated during the entire time interval between read-out events (*i.e.* no dead time). A fast read-out rate also allows for faster Fourier transform IR (FTIR) scanner velocities, which can be beneficial for avoiding a variety of signal noise sources often encountered at frequencies below a few hundred hertz.

The process of collecting a chemical image using an FPA is somewhat different. The same process takes place with an FPA detector, except that the spectra are collected in parallel (Figure 7.3).

The schematic in Figure 7.3 shows a simplified layout for an FTIR microspectrometer based on a staring-type FPA detection system. The FTIR spectrometer is similar (if not identical) to that used in the scanning instrument, except that it may require step-scan to allow time for reading out all pixels of the FPA at each optical retardation (scanning mirror position) of the interferometer. The system also uses Schwarzschild objectives, but the apertures for constraining the microscope's sensitive location are left open, *i.e.* they do not provide any spatial discrimination. Thus, the microscope's first objective

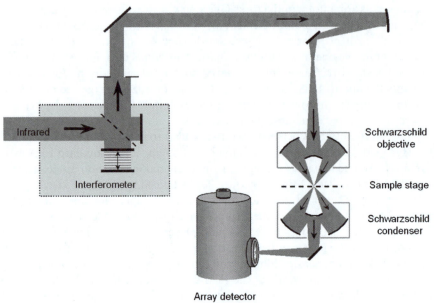

Figure 7.3   Schematic for an imaging infrared microspectrometer system using a focal
plane array (FPA) detection system. Note that the "upper" and "lower"
apertures must be left open or removed to allow light to fall onto the entire
area to be imaged.

illuminates a rather large area, and this illuminated region is then imaged onto
the FPA detector by the second Schwarzschild objective. Spatial discrimination
is provided by the individual pixels of the detector, each one serving as its own
"aperture". Because there is no matching aperture for the illumination objective, this system does not meet our definition of confocal.

Though the images shown in the figures and from FPA systems are of high
quality, there is not an exact representation of the specimen's properties.
Instead, the images are degraded due to limitations of the optical system and
other factors (*e.g.* detector fidelity, phase correction, apodization, *etc.*). However, the most significant limitation for high spatial resolution imaging is the
diffraction of light, which is discussed in the next section.

7.3   The Synchrotron Infrared Source

The characteristics of IR synchrotron radiation have been discussed by
numerous authors and in Chapter 3 of this volume. Rather than cover all of the
details of such sources, our intent is to describe primarily the features relevant
to multi-element imaging systems and techniques. The reader interested in a

broader discussion of sources and research activities at various facilities is referred to a collection of articles on these topics.[15]

7.3.1 Basic Properties of the Synchrotron Infrared Source

(See also Chapter 3.) Synchrotron radiation is the light produced by a beam of relativistic charged particles, most often electrons, in synchrotron storage rings. These storage rings involve combinations of magnets that bend (and focus) the electron beam onto an approximately circular orbit. They can accelerate and store hundreds of milliamperes of current for hours at a time. The slow loss of electron beam, due to scattering, is managed by replenishing the beam on a regular basis. Two source types have been exploited for IR spectroscopy, both using the light generated as the electrons pass through the magnets that bend the beam around the machine's orbit: (1) the dipole bend source and (2) the edge radiation source. The former represents the light generated as the electrons are transversely accelerated along the curved orbit inside the body of the magnets while the edge source stems from light generated as the electrons transition between a straight and a curved trajectory at the entrance (or exit) of a bending magnet's field. The source type in use at a particular synchrotron facility depends, in large part, on the physical design and construction of the storage ring itself rather than an intrinsic performance advantage of one type over the other.

Relativistic effects cause the emitted radiation to span a large spectral range, often reaching into the hard X-ray spectral range. In contrast to free electron lasers (FELs), synchrotron radiation is a spectrally continuous (*i.e.* "white") source of light. Thus the source can be used with the same conventional FTIR spectrometer designed for use with a thermal-type source. The other important characteristic relevant to microspectroscopy is the very high brightness of the source, the result of the intrinsically small source size and narrow emission angle for synchrotron radiation. The apparent source size is usually not determined by the physical electron beam dimensions, typically 100 µm or less, but rather by the limits imposed by diffraction. As a result, an IR microspectrometer using a synchrotron source can probe diffraction-limited sample regions with little loss in intensity and signal at the detector. When diffraction-limited, the root mean square (RMS) source dimensions are:

$$\sigma_{\text{bend}} \sim 1.6 \left(\rho \lambda^2 \right)^{1/3} \quad \text{[dipole bend source]} \tag{7.1}$$

$$\sigma_{\text{edge}} \sim \gamma \lambda \quad \text{[edge source]} \tag{7.2}$$

where λ is the wavelength of interest, ρ is the bend radius in the dipole magnet and $\gamma = 1957.E$ (where E is the electron energy in GeV) is the relativistic mass enhancement factor. As an example, a typical storage ring source has an electron beam energy of 3 GeV ($\gamma = 5871$) and a bending radius on the order of 10 m. The dipole bend RMS source size at $\lambda = 10$ µm is about 0.6 mm while the edge source is on the order of 6 cm. These source dimensions pertain to optical extractions limited to the natural opening angle of the emitted light, typically measured in milliradians (mrad). The RMS half-angles for emission are:

$$\theta_{\text{bend}} \sim 0.6(\lambda/\rho)^{1/3} \quad \text{[dipole bend source]} \tag{7.3}$$

$$\theta_{\text{edge}} \sim 1/\gamma \quad \text{[edge source]} \tag{7.4}$$

Note that the dipole bend emission angle at $\lambda = 10\ \mu\text{m}$ is 11 mrad, while for edge radiation the emission is below 0.1 mrad. Either of these sources can be suitably de-magnified in a microscope to yield RMS spotsizes on the order of the wavelength λ. It should be noted that the $1/\gamma$ emission angle for edge radiation assumes that the full source dimension is available for extraction. In practice, the source is often truncated by the storage ring's metal vacuum chamber, resulting in larger opening angles that also depend on wavelength. A horizontal collection angle greater than these values can result in additional horizontal source segments, turning the point-like, diffraction-limited source into an extended source. One could imagine collecting and extracting the entire horizontal swath of light from a bending magnet source, but the mechanical design of the dipole magnets and the electron beam vacuum chamber set a practical limit of $\sim 30°$ or less. Note that this only pertains to the dipole bend source because the edge source is, by definition, limited to the fringe field region of the magnet (typically less than 1° into the dipole). Though IR edge radiation by itself nearly always behaves as a point-like source, it evolves into bending magnet radiation that can be part of the extracted flux.

7.3.2 Infrared Microspectroscopy using the Synchrotron Source

The first description was by Ugawa *et al.* using the UVSOR facility in Okazaki (Japan).[16,17] This work clearly showed the potential for the synchrotron IR source in microspectroscopy, but the practical performance was limited by motion of the storage ring's electron beam. This was followed by a custom microscope design for work with high-pressure diamond anvil cells at the NSLS in Upton, NY (USA)[13] and by successful tests (also at the NSLS) with a commercial microscope system.[18,19] The success was due, in large part, to significant improvements to the stability of the electron beam in the NSLS VUV synchrotron storage ring, resulting in a very useful signal-to-noise when measuring sample dimensions at the diffraction limit. This, plus successful measurements on scientifically interesting problems,[20–24] led to a rapid development of IR microspectroscopy beamlines at facilities around the world. To this day, synchrotron IR microspectroscopy is performed mainly using a single-source optical extraction combined with an IR microspectrometer and single-element detector.

Though the number of synchrotron IR beamlines around the world is not small, examples of actual performance results with a multi-element FPA detector microscope have been limited to just a few facilities. The first demonstration of synchrotron-based IR microspectroscopy using a 2D multi-element FPA detection system was by Moss *et al.* and used the dipole edge source beamline at the ANKA synchrotron facility in Karlsruhe (Germany).[10] Only a portion of the array was illuminated. Still, the intensity at this region was much higher than for the standard thermal source and indicated the potential advantage of using synchrotron radiation. The resolution improvement, relative to the thermal source, was somewhat inconclusive and the demagnification onto the array was not sufficient to test deconvolution feasibility. Still, this was an important first step in the development of FPA microspectrometers for synchrotron sources.

It is important to understand and discuss the imaging performance of IR microspectrometer systems in order to compare various designs and measurement methods. We start by considering how images are produced and define metrics for the quality of these images. For a complete understanding, it is essential also to review previous work demonstrating how synchrotron radiation (SR)-based microspectroscopy delivers images at the theoretical limits of performance, and that this limit is at least a factor of 2 better than for a typical FPA system.

7.4 Imaging at the Diffraction Limit

In general, the goal for synchrotron-based FTIR microspectroscopy is to deliver images at the diffraction limit. This usually means setting one (or both) of the microscope's apertures to define a region somewhat smaller than the diffraction limit for the respective objective. For convenience, we use $d = \lambda/NA$ to define this diffraction-limited dimension. This is consistent with the calculated Full Width at Half Maximum (FWHM) of the Schwarzschild diffraction pattern and is also confirmed by experimental resolution studies on test specimens (to be shown later in this section).

7.4.1 Imaging and the Point Spread Function

In general, diffraction and other image degrading effects such as optical distortion are quantitatively given by the point spread function (PSF), which describes how the intensity from each point on the object is distributed in the image plane. A detailed discussion of the PSF and how it affects image formation can be found in most optics text books, so we present only a brief discussion here. Although the PSF can vary with position in the image plane (*e.g.* when optical aberrations are present, or from vignetting), we will assume that it is a function only of the wavelength of light, λ. This assumption will be valid for a properly designed and aligned Schwarzschild objective at locations within about 100 μm of the central (optical) axis at the specimen location. *This is certainly the case for an image produced by raster scanning of the specimen through the focused beam.* With this, we can write an expression relating the observable image to the true image as:

$$O_\lambda(x, y) = T_\lambda(x, y) \otimes P_\lambda(x, y) \tag{7.5}$$

where O is the observable image intensity, T is the true image intensity, P is the optical point spread function and the $\otimes$ symbol represents the convolution operation. Note that each is a function of position x, y in the image for a particular wavelength λ, so the convolution is two-dimensional (over both x and y coordinates). The actual image could be degraded even further depending on the spatial sampling, noise, and optical misalignment. We will assume that the system is properly aligned and the pixel sampling density is sufficiently high not to limit the image resolution (although this is not necessarily the case for some FPA systems).

This PSF (or diffraction pattern) for a model Schwarzschild objective is shown in Figure 7.4, along with a standard Airy pattern for comparison.

Although the Schwarzschild has a somewhat narrower central maximum, the first order diffraction maximum (ring surrounding the central peak) is much larger than for the Airy pattern. The significance of this can be understood by integrating the intensity (imaging sensitivity) as a function of radial distance from the optic axis, shown in Figure 7.5. Only slightly more than half of the intensity is located in the central peak, with the balance in the first and higher order diffraction rings. The effect is apparent when scanning across a sharp test specimen, as shown in the absorption profiles of Figure 7.6. Rather than a steep transition at the edge, the absorption signal increases gradually over a distance of more than 10 µm. Using a 10% to 90% criterion, the transition width is about twice the wavelength (13 µm wide for $\lambda \sim 6$ µm). The expected profile can be predicted, based on eqn (7.5), by convolving the Schwarschild diffraction pattern with an abrupt edge. This is also shown in Figure 7.6 as a solid line and clearly demonstrates that the spatial resolution is controlled by diffraction specific to the Schwarzschild objective.

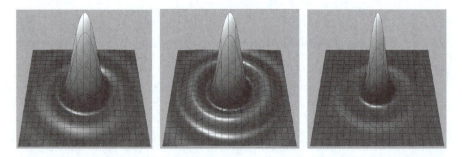

Figure 7.4   Point spread function (PSF) for three optical arrangements. *Left panel:* the Airy function for a standard, full-aperture objective. *Middle panel:* a Schwarzschild objective having a central obscuration of 25%. *Right panel:* two matched Schwarzschild objectives used in a confocal arrangement.

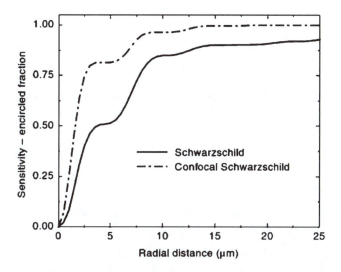

Figure 7.5   Radial integration of the optical sensitivity for a model Schwarschild objective when used in a standard (single-aperture) configuration (solid curve for the encircle intensity) and for the dual-aperture confocal configuration (dashed curve). A wavelength of 6 μm and Schwarschild optic having NA = 0.65 were assumed.

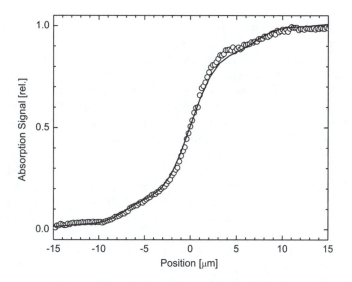

Figure 7.6   Absorption line scan across a sharp edge in a polymer photoresist using the λ = 6 μm amide vibration spectral feature. Raw absorption signal as a function of distance (open circles) and the expected profile (solid line) assuming a perfectly sharp edge convolved with a calculated point spread function.

The PSF for the Schwarzschild objective results in lower contrast than for a comparable glass-type objective,[25] but is remedied significantly when a second aperture is employed at the detector to make the optical system confocal. An advantage of such a single-element microscope is that it can be fitted with a variety of IR detectors for operation over a very broad spectral range. A disadvantage of the microspectrometer using a single-element detector is that imaging is only achieved by raster scanning the sample through the microscope's focus spot, a time-consuming process that limits some measurements to perhaps 20×20 pixels.

7.4.2 Performance with the Synchrotron Source and a Single-Element Detector

As noted above, microspectroscopy imaging with a single element detector is performed by moving the specimen through a micro-focused IR beam in a raster-like fashion. In particular, a complete spectrum is collected individually at each point of the specimen to be imaged, with the spotsize controlled by apertures and the stepsize usually set to have the sampled spots overlap somewhat. The material content of each location is then determined by spectral analysis, and the chemical content (or other piece of spectral information) is used to define the colour or intensity of a given pixel, yielding an image. Since the time spent at each spot (a pixel in the final image) is usually at least 15 seconds, the time required to build up a 64×64 pixel image approaches 18 hours. Desirable characteristics for the resulting image are: 1) high signal-to-noise at each pixel, 2) good lateral spatial resolution, and 3) good contrast fidelity (*i.e.* the observed change in a particular spectral intensity from point-to-point accurately matches the specimen's actual composition). High spatial resolution and image contrast are enhanced by reducing the effective aperture size for a raster-scan type IR microspectrometer system. When the aperture defines a region comparable to or smaller than the wavelength of interest, the resolution will be diffraction limited. This assumes ample signal-to-noise to render an acceptable image, and the thermal source typically does not offer sufficient performance. This is particularly true when using confocal aperturing, and illustrates the benefit of the synchrotron source's high brightness. Figure 7.7 shows a spectrum recorded at the MIRAGE beamline at LURE (France) for a $5\,\mu m$ thick section of brain tissue, deposited onto a gold-coated substrate (reflection mode), recorded with an aperture of $3 \times 3\,\mu m^2$.[26] The spectrum was collected as 64 co-added interferometer scans, each at $8\,cm^{-1}$ resolution, for a total collection time of $30\,s$. One should note that the "extra noise" below $\sim 1200\,cm^{-1}$ is a consequence of severe signal loss due to diffraction and the long wavelength cutoff behaviour of such a small aperture (less than half the wavelength). This confocal configuration allows one to reach a resolution of $\lambda/2$.[25] Figure 7.8 shows an image collected using the $3 \times 3\,\mu m^2$ confocal aperturing and synchrotron radiation. The sample consists of gold

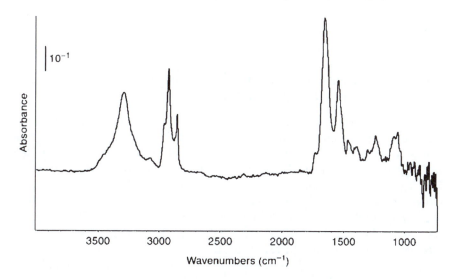

Figure 7.7 Synchrotron IR spectrum of a brain section (5 μm thickness), recorded in reflection mode, using a dual aperture of $3 \times 3 \, \mu m^2$, 64 scans at $8 \, cm^{-1}$ resolution.

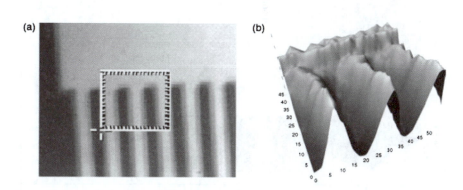

Figure 7.8 (a) Optical image of gold wires which have been coated on a silicon wafer. The gold layers are themselves covered with a thin polymer film. (b) Chemical image of the polymer overlayer (peak height of the ν_{as} CH$_2$, at $\sim 2920 \, cm^{-1}$), recorded with a dual aperture of $3 \times 3 \, \mu m^2$, 64 scans at $4 \, cm^{-1}$ resolution.

wires deposited onto a silicon wafer with a protective polymer overlayer. The high contrast image shows the absorption strength due to the –C–H$_2$ stretch modes in the 3 μm wavelength range. The specimen was spatially oversampled by raster-scanning through the beam in 1 μm steps. The spectrum for each pixel is an average of 64 FTIR scans at a resolution of $4\,\text{cm}^{-1}$. The total measurement time was 157 minutes.

7.4.3 Comparing Synchrotron IR Imaging with Internal Source-based FPA Imaging

We next look at some simple imaging examples where results using the single-element detector microspectrometer, often operating in a confocal mode, are compared against image results for a multiple element linear array detector microspectrometer. The single-element detector system was a modified Spectra-Tech Irμs microspectrometer set for $3\,\mu\text{m}\times3\,\mu\text{m}$ effective confocal apertures and using the high-brightness synchrotron source at NSLS beamline U4IR. The linear array (16 staggered elements) microspectrometer was a Perkin-Elmer Spotlight with a thermal IR source and magnification to yield a $6.25\,\mu\text{m}$ effective pixel dimension. The results for the Spotlight[7] are shown in Figure 7.9 along with results for the Irμs in Figure 7.10 for comparison. We should note

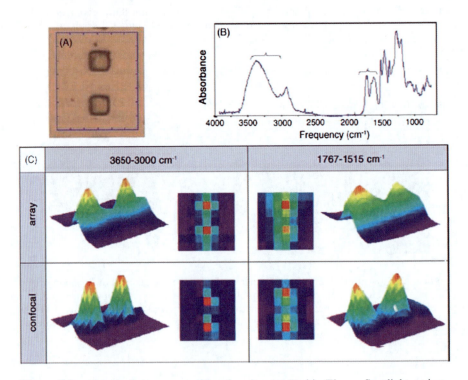

Figure 7.9 Chemical images produced using a Perkin-Elmer Spotlight micro-spectrometer system with $6.25\,\mu\text{m}$ effective pixel spacing. The specimen was a patterned layer of photoresist squares $12\,\mu\text{m}$ on a side – see visible image in panel (A). The photoresist absorption spectrum is shown in panel (B). Images in panel (C) were obtained based on absorption features around $3\,\mu\text{m}$ wavelength (left panel) and $6\,\mu\text{m}$ wavelength (right panel).

(a)

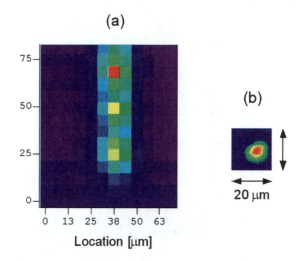

Location [μm]

Figure 7.10 Another set of chemical images for the same photoresist specimen as Figure 7.9 over the same 6 μm wavelength range. Left panel (a): three photoresist squares measured using the Perkin-Elmer Spotlight with linear array detector and conventional thermal source. Right panel (b): one of the same photoresist squares measured using a Spectra-Tech Irμs with confocal optics, single-element detector, and synchrotron infrared source. Stepsize was set to 0.5 μm.

that the results from the Spotlight are limited by the effective pixel dimension and diffraction effects, and are not a result of any technical deficiency. As expected, the confocal arrangement using a smaller effective aperture and an order of magnitude smaller sampling interval yielded much better resolution and contrast.[26] However, while the images produced using the Spotlight required just a few minutes of measurement time, the raster scanning with the Irμs required several hours to complete. This clearly shows the limitations of the raster scanning method for achieving high resolution chemical images for areas more than a few tens of microns on a side.

The next example shows a more practical comparison between two similar instruments, again using an internal source for the linear array instrument and a synchrotron-based single detector microscope, operating in confocal mode. A 6 μm thick porcine skin section has been imaged using two types of microscope: a Nicolet/Spectra-Tech Continuum microscope with the synchrotron source, and a Perkin Elmer (Spotlight) microscope with the thermal IR source. The optical image of the section analysed is displayed in Figure 7.11a, while Table 7.1 provides the characteristics of each microscopes used. One can note the presence of the three main layers of the skin section: the stratum corneum (external layer) of about 6 to 10 μm thickness, the epidermis (100 to 150 μm) and the dermis. The chemical images of lipids, recorded by integrating the frequency region 2800–3000 cm^{-1}, are displayed in Figure 7.11b and c. It is clearly evident that the image contrast is higher in the case of the microscope and synchrotron source combination.

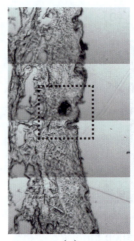

(a)

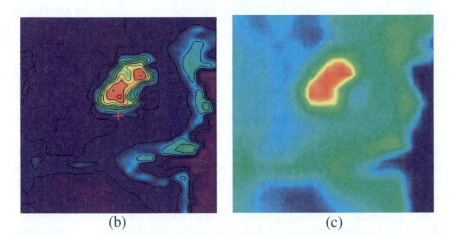

(b)   (c)

Figure 7.11   (a) Optical image of a 6 μm thick porcine skin section. The dotted black square is the region selected for analysis. (b, c) Chemical images of the lipid distribution across the same skin section. The area is 160 μm wide and 200 μm high. (b) Recorded with the synchrotron IR microscope; (c) recorded with a 2D array detector-based microscope.

A linear scan across the stratum corneum (SC) provides more evidence of the loss of information, while operating at the diffraction limit, with a non-confocal microscope. Figure 7.12 shows the successive spectra recorded across the SC.

It is important to realize that the detailed analysis of the biomolecular composition and protein secondary structure of the SC has important implications for understanding the barrier function of the outermost skin layer.[27] On average, human SC is 10 to 20 μm in thickness, and consists of flattened, anucleated and protein-rich cells embedded in a multi-lamellar lipid matrix composed mainly of ceramides.[27] It is believed that the penetration pathway is

Table 7.1 Data acquisition parameters for the two instruments used in the study shown in Figure 7.11.

	Continuum microscope (Thermo Fischer) Source = synchrotron	*Spotlight microscope (Perkin Elmer) Source = globar*
Detector type	One single MCT[a] detector	16 element array MCT[a] detector
Individual detector size	50 μm	25 μm
Spectral resolution	4 cm^{-1}	4 cm^{-1}
Number of scans	32 scans per point	32 scans per pixel
Area size	Projected aperture size = 6×6 μm^2	Projected point size = 6.2×6.2 μm^2
Total acquisition time	3.3 hours	22 minutes

[a]MCT, mercury cadmium telluride

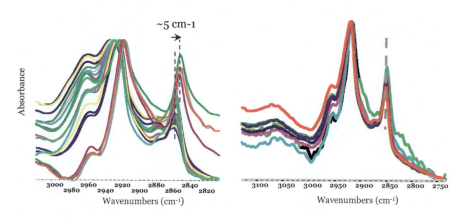

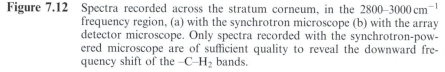

Figure 7.12 Spectra recorded across the stratum corneum, in the 2800–3000 cm^{-1} frequency region, (a) with the synchrotron microscope (b) with the array detector microscope. Only spectra recorded with the synchrotron-powered microscope are of sufficient quality to reveal the downward frequency shift of the –C–H$_2$ bands.

controlled by the intercellular lipid domain.[27] For pharmaceutical purposes, understanding the biochemical and biophysical properties of SC lipids is of paramount importance for developing strategies to enhance the transdermal penetration. The chemical composition of the SC is quite unusual for a biological membrane, since the lipids do not hydrate and form solid phases, with an orthorhombic symmetry.[27] An order–disorder transition occurs between 40 and 70 °C, which has been studied by IR spectroscopy in order to model the lipid composition of the SC.[28–30] The ordered phase is characterized by a lower frequency value of for both the symmetric and antisymmetric stretching modes of the –C–H$_2$ molecular sub-structure. Such analysis, in the study shown here, can only be made with the synchrotron-based IR microscope.

To summarize, in this section we have shown how single-point micro-spectroscopy using the synchrotron IR source and small, confocal aperturing can deliver improved resolution and contrast for biological imaging when compared with various array detector microscopes using a thermal source, but these improvements come with a serious handicap – extremely long measurement times.

7.4.4 Diffraction Effects and Issues for PSF Deconvolution

One method for improving the measured edge sharpness and contrast is to operate the microscope using a confocal optical arrangement. According to our definition, the confocal IR microscope uses two apertures, one to spatially constrain the illumination and the other to constrain the region being detected. If the two objectives are matched, and apertures are set to the same effective size, the same PSF applies for both illumination and detection and the combined (or total) sensitivity function is the $(PSF)^2$. This leads to significant improvements in the imaging quality. For example, the first order diffraction ring (plus higher orders) is immediately reduced, and the central peak narrows substantially. Figure 7.5 above shows that more than 80% of the sensitivity lies in the central maximum, and essentially all of the sensitivity is captured within 2λ of the optical axis.

Therefore, we expect that a confocal microscope will deliver images having much higher fidelity (both improved resolution and contrast). This is illustrated in Figure 7.13, comparing the non-confocal absorption profile of Figure 7.6

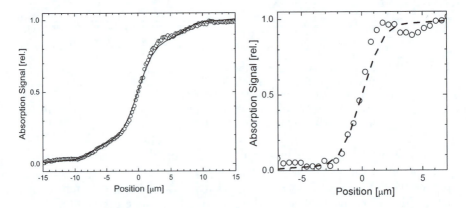

Figure 7.13   Absorption profiles for a line scan across the sharp edge of a thin polymethacrylate (PMMA) PMMA film, providing a measure of spatial resolution and image contrast. Open circles are experimental results and solid or dashed lines are the calculated profiles based on the diffraction PSF. Left: absorption profile recorded in nonconfocal mode with a single aperture (same data as Figure 7.6, shown again here for comparison). Right: absorption scan across the same photoresist edge, but with the microspectrometer set for confocal apertures.

with the confocal case. The edge sharpness improves by nearly a factor of 3, although additional noise is present due to the lower throughput of the confocal system.

Given that the measured profile is a convolution of the true profile with the optical PSF (which can be calculated), we have an opportunity to improve the fidelity of the profile by deconvolution. If $P_\lambda(x,y)$ is known or can be determined, then the operation can be inverted, a process known as deconvolution. This is accomplished with help from the convolution theorem

$$\text{FT}\{O_\lambda(x,y)\} = \text{FT}\{T_\lambda(x,y)\} \times \text{FT}\{P_\lambda(x,y)\} \qquad (7.6)$$

where $FT\{\}$ denotes a Fourier transform (2D our case) and "$\times$" is simple multiplication. The true image is then:

$$T_\lambda(x,y) = FT^{-1}[FT\{O_\lambda(x,y)\}/FT\{P_\lambda(x,y)\}] \qquad (7.7)$$

with $FT^{-1}[]$ representing the inverse Fourier transform. In practice, the observable image, *i.e.* the image intensity distribution following the optical system, cannot be precisely measured. For example, the image can never be sampled as a truly continuous function, so we have the usual issues associated with discrete Fourier transforms and sampling theorems. Since the sampling density of the deconvolved image is the same as the "input" (observed) image, the observable image must be sampled at a rather high spatial density. There are also errors due to noise introduced by fluctuations in the illumination intensity and noise in the detector. This problem is compounded by the deconvolution process since $FT\{P_\lambda(x,y)\}$ approaches zero for short length scales (high spatial frequencies), resulting in large fluctuations in the calculated image. The remedy is similar to that employed in FTIR spectroscopy, where an apodization function is applied prior to Fourier transforming in order to reduce the spectral resolution and eliminate "ringing" around narrow spectral features. In our case, however, the apodization process is applied to the spatial data and serves to limit the degree of resolution improvement. There is one more analogy with FTIR spectroscopy: higher signal-to-noise is needed in the raw data to achieve better spectral resolution while maintaining the same signal-to-noise in the resulting spectra. Thus, the degree by which the spatial resolution of an image can be improved is strongly dependent on the signal-to-noise of the raw image.

We can illustrate the potential for PSF deconvolution using the measurement data shown in Figure. 7.13 (left panel). This is an absorption profile for a line scan across the sharp edge of a photoresist pattern, so the deconvolution is only one-dimensional for this illustration. The required PSF is calculated for a model 32× Schwarzschild objective and a simple "boxcar" apodization serves to truncate the higher spatial frequencies and limit the resolution improvement, in one case set to 5 μm (medium) and the other to 1.6 μm (strong). The results are shown in Figure 7.14, along with the original (uncorrected) absorption profile for comparison. The edge sharpness improves by nearly a factor of 3 to a value of 4.4 μm and features associated with the first order diffraction ring are

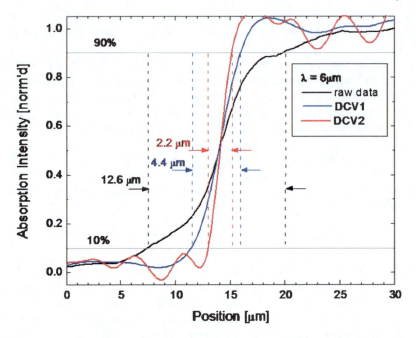

Figure 7.14   Deconvolution of the absorption line scan profile from Figure 7.13 (left panel). Black line: raw absorption signal as a function of distance. Blue line: deconvolution of the original data with the calculated PSF, with high spatial frequencies heavily filtered (suppressed). Red line: deconvolution of the original data, but with less filtering of high spatial frequencies.

reduced for the medium case. Preserving more high frequencies in the deconvolution reduces the edge width down to 2.2 µm, but oscillations in the absorption strength with 5% amplitude have now appeared. Therefore we conclude that a factor of 2 to 4 resolution improvement could be achieved by PSF deconvolution, assuming the S/N of Figure 7.13 (left panel) is available.

It is worth noting that a comparable degree of edge sharpness and contrast is achieved when the microspectrometer is run with diffraction-limited confocal apertures, and without any deconvolution. This is illustrated by comparing Figure 7.13 (right panel) with Figure 7.14. Unfortunately, the signal levels are reduced considerably when compared to the single-aperture case such that only the strongest absorption features can be accurately measured during a tolerable scan time.

We can illustrate a subtle diffraction artifact associated with diffraction when working with a single aperture. As can be seen in the single-aperture Schwarzschild PSF (Figure 7.4, middle panel), there is significant intensity in the first order diffraction ring around the PSF central maximum. When measuring a sample having a circular structure (for example, the lipid-rich wall of a well-formed red blood cell) with a diameter somewhat larger than this ring, the absorption intensity will reach a peak when the PSF is centered on the circular

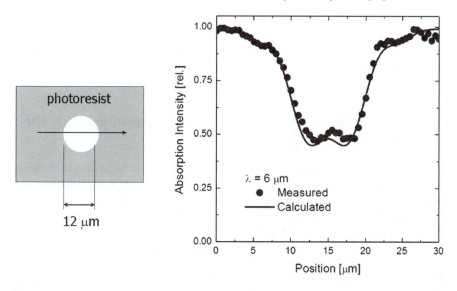

Figure 7.15 Absorption line scan, at the amide I vibration frequency of $\sim 1650\,\text{cm}^{-1}$, across a $12\,\mu\text{m}$ diameter hole in a $2\,\mu\text{m}$ thick layer of photoresist. Note the false peak near the centre of the hole (position $= 15\,\mu\text{m}$).

structure. This can give the (false) indication of a feature at the centre. In Figure 7.15 we show such a result for a circular hole in a thin layer of photoresist. At the amide I vibrational wavelength of about $6\,\mu\text{m}$, the first order diffraction ring just reaches to the edge when the beam is centered on the hole. Though the hole is completely empty, the microspectrometer results indicate an apparent absorption "bump" at the center. Also note the rather poor contrast in that the apparent absorption inside the hole does not drop below 40%.

The effects of diffraction are clearly evident in synchrotron-based IR microspectroscopy using the single-element detector (SED). Assuming high signal-to-noise, the spatially over-sampled images might be successfully deconvolved to yield a spatial resolution surpassing the conventional limits of diffraction. However, such an approach is not practical owing to the limited number of pixels that can be measured over an acceptable measurement period when raster-scanning. Indeed, having a small number of pixels requires an accurate measure of the true PSF itself to yield a successful deconvolution. This represents yet another difficult hurdle, given that the PSF depends on the detailed illumination of the Schwarzschild objective's entrance pupil, including the polarization. An alternative would be to adopt a more advanced deconvolution algorithm capable of determining both the deconvolved image *and* the relevant PSF. Such deconvolution methods are well known in visible light microscopy and astronomy, but require a reasonable degree of spatial over-sampling, a sufficiently large dataset (image size) and excellent signal-to-noise. Commonly used algorithms include Lucy–Richardson and "blind" deconvolution, the latter relaxing the requirement to have a precise knowledge of the instrument point spread function. This is potentially advantageous as the nature of the synchrotron source can make highly homogeneous illumination of the objective's entrance pupil challenging. This will be the subject of a later section of this review.

7.5 Focal Plane Array IR Microspectroscopy with the Synchrotron Source

When used with a thermal source, the 2D FPA for an IR microspectrometer is reasonably well illuminated. One can therefore ask whether the technique can benefit from the high brightness synchrotron source, especially given its typically point-like characteristics. Given the significantly greater brightness of the synchrotron source, it should be possible to sacrifice some (but not all) of this advantage in order to illuminate a significant sample area for imaging by the detector array. This was successfully demonstrated by Moss *et al.*,[10] where an IR microspectrometer equipped with a true 2D array detector was first operated at a synchrotron beamline (ANKA, Karlsruhe, Germany). The potential for an improved signal-to-noise using the synchrotron source was demonstrated (Figure 7.16).

As a next step, Carr *et al.* proposed using a 74× high-magnification objective such that each pixel of the FPA detector sampled a region significantly smaller than the diffraction limit, as would be needed to successfully achieve spatial

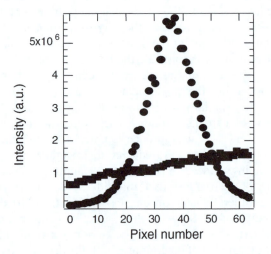

Figure 7.16   Intensity profiles across the central pixel row of a 64×64 pixel focal plane array detector, recorded through a 36× objective and integrated over the 1000–4000 cm^{-1} spectral range, comparing synchrotron (circles) and Globar (squares) as sources. The data show that the synchrotron beam can be spread out to provide a strong signal over a limited portion of the array. (From Moss *et al.* 10.)

deconvolution for both resolution and contrast improvement.[31] In comparison to a 15× objective having the same numerical aperture, the intensity at each pixel would be 25 times smaller. Thus, having a source 100× brighter would easily offset this reduction. Ideally, one also desires to homogeneously illuminate a significant portion of the array and utilize as many pixels as feasible, as can be done with the inherently extended (*i.e.* source size much greater than the diffraction limit) thermal source.

7.5.1   Matching the Dipole Bend Source to the FPA Microspectrometer

At this point, we return to the intrinsic characteristics of the dipole-type synchrotron radiation source. Recall that the radiation from a single electron is emitted into a opening angle of $\theta_v(\lambda) \sim (\tilde{\lambda}/\rho)^3$ where ρ is the bending radius of the electron beam. This result is basically a statement that the physical dimensions of the source (the transverse size of the electron orbit) and the opening angle meet the conditions for a diffraction-limited source. For a storage ring with a 2 m bend radius and a wavelength of 7 μm ($\sim 1400\,\mathrm{cm}^{-1}$), the full opening angle is 20 milliradians, or about 1°. This applies to both the horizontal and vertical emission angles. Of course, the source orbit spans a full 360°, although broken into segments by the individual dipole magnets. Therefore, from the perspective of an observer walking around the ring's perimeter, each time the observer's angle changes by 1°, a new diffraction-limited portion of the orbit, distinguishable from the previous segment, is viewed. The span of a dipole varies with the storage ring design, but most provide at least 10°. The National Synchrotron Light Source (NSLS) VUV/IR ring uses 45° dipoles, although only about 30° are physically available owing to obstructions from downstream accelerator components. Even if only 16° were collected, each pixel of a 16 element linear array could be illuminated with synchrotron radiation having ideal brightness.

In follow-on work at the NSLS VUV/OR ring, Carr *et al.* demonstrated the use of dipole bending magnet radiation as an extended source to fill a significant portion of a rectangular array without the need for significant defocusing.[32] The source was collected in four individual segments that were separately collimated and brought to the instrument and FPA microscope, each segment illuminating a separate portion of the array. In this manner, a 64×64 pixel area was made usable. A schematic of this multiple-source optical extraction is shown in Figure 7.17, and an absorption map for a collection of polystyrene spheres measured at a wavelength of approximately 3 μm is shown in Figure 7.18.[33] The same optical extraction was also used to image the protein distribution in oral mucosa cells.[34] This basic idea was optimized and fully implemented by Hirschmugl and co-workers at the IRENI beamline of the SRC synchrotron facility (Stoughton, WI, USA), using a 12-segment beam geometry.[11,35] Not only is a 34 μm×34 μm (64×64 pixels) region nicely illuminated, the imaging performance has been found to be excellent.

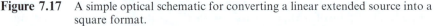

Figure 7.17   A simple optical schematic for converting a linear extended source into a
square format.

7.5.2   Initial Results using the Synchrotron Source and FPA

As shown in Figure 7.3, the FPA microscope does not have explicit apertures to
define the spatial sensitivity. Instead, each detector pixel serves this purpose.
With only one aperture, the system is not confocal according to our definition,
and the PSF will have about half of its intensity in the first and high order
diffraction rings. Edge sharpness (10% to 90% transition) will be approximately 2λ at best. We also illustrated some diffraction effects in section 7.4.4
(and Figure 7.15 in particular) associated with circular objects. The same type
of sample can be investigated using the FPA microspectrometer, revealing
similar behaviour (see Figure 7.19).

At present, a common FPA for use with an FPA imaging microspectrometer
is the Santa Barbara Focal Plane SB-161, 128×128 HgCdTe (MCT) photodiode array. While all 128×128 pixels are available, smaller regions can be read
out more rapidly and provide more flexibility in matching the read-out rate and
integration time to the FTIR scanner and source intensity. Indeed, some rectangular geometries can be read out more quickly and provide a simpler match
to the extended dipole bending magnet source. Some example results from the
NSLS/Brookhaven group are shown in Figure 7.20. Note the more distinct

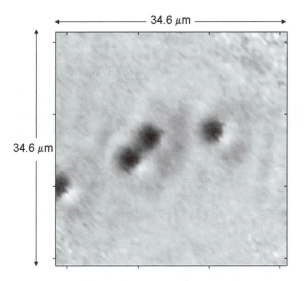

Figure 7.18 Absorption map at the –C–H$_2$ stretch spectral range around 2900 cm^{-1} for a collection of 2.5 μm diameter polystyrene spheres, measured using the multiple-beam synchrotron source at NSLS with beamline U10 and a 64 × 64 pixel MCT FPA microspectrometer. Darker regions correspond to greater absorption. (Source: R.L. Jackson *et al.* 33.)

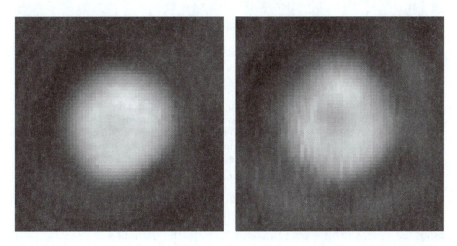

Figure 7.19 Absorption signal for a ∼2 μm thick photoresist in the vicinity of a ∼14 μm diameter hole using a 74× objective to yield 0.54 μm per detector pixel. Each field represents a 34.6 μm×34.6 μm region. Darker regions indicate larger absorption. *Left panel*: absorption at 6 μm wavelength. *Right panel*: same region as for the left panel, except for a wavelength of 7 μm. Note the appearance of a false absorption signal at the centre of the hole, plus the 3-fold symmetry resulting from the spoked assembly that holds the Schwarzschild secondary mirror in position.

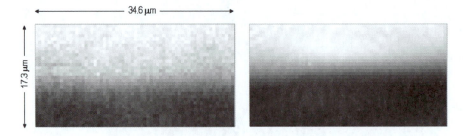

Figure 7.20   Absorption signal at a sharp photoresist edge using a 74× objective to
yield 0.54 μm per detector pixel. *Left panel*: conventional thermal spectrometer source. *Right panel*: Same photoresist edge, but measured using
the U10 synchrotron source. Same total measurement time for both
images. Note the significantly improved signal-to-noise for the synchrotron source.

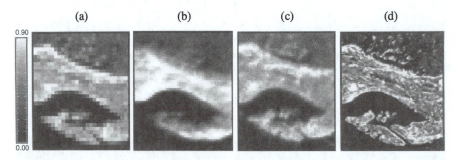

Figure 7.21   Infrared images of a prostate gland section using four different microspectrometers and sources. (*a*) A Perkin-Elmer Spotlight microspectrometer using point mapping and a geometric aperture set to 10 μm
by 10 μm. (*b*) The same Spotlight but using its quasi-linear array with an
effective pixel dimension of 6.25 μm by 6.25 μm. (*c*) A Varian FTIR with
focal plane array and effective pixel dimension of 5.5 μm by 5.5 μm. For
(*a*) through (*c*), a conventional thermal source was used. (*d*) A Bruker
FTIR with Hyperion FPA microspectrometer connected to the IRENI
synchrotron beamline and a 0.54 μm by 0.54 μm effective pixel dimension. (From Reference 34.)

edges, improved contrast, and much reduced pixel noise for the image collected
using the synchrotron source in comparison to the thermal source.

   We also show early results from the IRENI imaging beamline at the Wisconsin SRC, comparing different instrumentation and sources on a biological
tissue section (Figure 7.21). The spectral range was not indicated, so we assume
these images are for the full range of the instrument and detector (up to at least
$4000\,\mathrm{cm}^{-1}$, or 2.5 μm). In this comparison, each image was produced at a pixel
dimension considered to be "standard" for the particular combination of
instrument, detector, and source. Note the sharper image produced by the very

small effective pixel dimensions enabled using the high-brightness synchrotron source. The enhanced resolution and contrast are a direct result of the higher signal-to-noise at each pixel, including the dominance of the modulated synchrotron IR over the unmodulated T = 295 K blackbody radiation that lands on the FPA.

7.5.3   Basic PSF Deconvolution with FPA Microspectrometers

The FPA imaging microspectrometer system offers some opportunity to enhance image quality through PSF deconvolution. The method is commonly employed in visible light microscope images. The method may not be directly transferrable to the IR system due to the much broader spectral range of interest for molecular spectroscopy, the Schwarzschild diffraction pattern (as opposed to an Airy pattern) and the limited image field for which diffraction-limited performance is achieved. The distortion that occurs is only significant for image locations rather far from the instrument's optic axis, and for the shorter wavelengths. This is illustrated in Figure 7.22, showing the calculated image for a circular disc when imaged at $\lambda = 3\,\mu$m. The object's appearance is dominated by diffraction when centred on the optic axis (left) and also when it lies 100 μm from the axis (middle). However, when moved to 283 μm (equivalent to the corner position of a 400 μm × 400 μm image), distortion becomes significant. Such variations in the PSF would be difficult to deconvolve without some prior knowledge of the distortion. Including the Schwarzschild PSF in a deconvolution algorithm presents no particular obstacles, especially if the algorithm is of the "maximum likelihood" type that automatically converges to a suitable PSF given appropriate starting conditions. The real challenge lies in the broad spectral coverage – more than a factor of 4 in wavelength – sometimes needed for molecular identification.

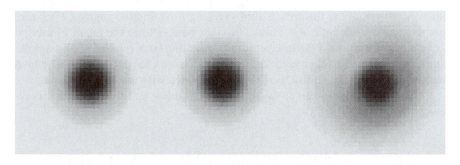

Figure 7.22   Calculated images for a 3 μm diameter circular object at a wavelength of 3 μm and at three different locations within the spectrometer's potential viewing area, illustrating diffraction blurring and optical distortion. Left: centred on the instrument's optical axis; middle: object located 100 μm from the optical axis; right: object located 283 μm from the optical axis.

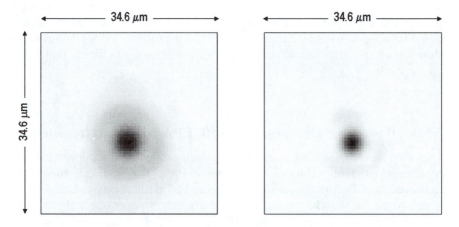

Figure 7.23 Example maximum likelihood deconvolution using the measured intensity at $\lambda = 4.3\,\mu m$ through a small pinhole aperture. *Left panel*: raw
intensity, showing the blurring effect of diffraction. Darker regions correspond to higher intensity. *Right panel*: deconvolved image. Note both
the narrower pattern and the significantly reduced halo, the latter being
highly beneficial for image contrast. (Source: Reference 36.)

We should point out that PSF deconvolution of a 2D image with a large
number of pixels will be less dependent on detailed prior knowledge of the
optical system PSF. A commonly used PSF deconvolution approach is the
Lucy–Richardson algorithm that assumes a Poisson noise distribution in
the image and determines the most likely (or probable) deconvolution result.
This type of algorithm can also be invoked to determine a most likely PSF as
well (often referred to as a "blind" deconvolution), providing the opportunity
to refine the PSF when it is not accurately known. Images with more than
32×32 pixels are typically needed for this process to be successful. As an
example, in the left panel of Figure 7.23 we show a 64×64 pixel image of the
intensity distribution through a sub-wavelength (estimated diameter $< 3\,\mu m$)
pinhole at a wavelength of $4.3\,\mu m$. The right panel shows the result of a blind
2D deconvolution on that intensity map, resulting in a smaller spot diameter
and much reduced background intensity away from the central spot. The final
image still shows some diffraction effects associated with the 3-fold symmetry of
the Schwarzschild objective, suggesting that additional resolution improvements may be feasible.

7.5.4 Opportunities for Advanced 2D Image Deconvolution

As noted above, a key motivation for using an extended IR synchrotron source
is to allow for a high degree of spatial oversampling and enable image
deconvolution to improve the spatial resolution. For visible light deconvolution, the spectral range is limited to $0.4\,\mu m$ to $0.7\,\mu m$, *i.e.* less than a factor of 2,

and is usually divided into just three colours (red, green and blue) with no physical constraints on their relative proportions in the final image. On the other hand, IR chemical imaging spans the wavelength range from $\sim 2.5\,\mu m$ to beyond $12\,\mu m$, a factor of 5 or more in wavelength. More importantly, the resulting spectral images must be useful for chemical identification, based on the information contained in the ~ 1000 data point spectrum at each pixel. Initial results from the Wisconsin SRC group suggest that the image for each and every spectral point can be spatially deconvolved and reassembled to produce physically reasonable chemical spectra.[11,35]

7.6 Conclusions

In our concluding section we point out several other measurement opportunities that might be pursued using FPA detector technology. To date, the majority of IR chemical imaging has been performed using HgCdTe photoconductors and photodiode arrays, with the latter covering the 2 to $12\,\mu m$ wavelength range. These MCT photodiode arrays would not be readily available were it not for their application in thermal sensing and imaging. Of course there is chemical information at longer wavelengths, motivating the development of other array technologies. These include arrays based on the blocked impurity band (BIB) detector,[37] or the quantum well IR photodetector (QWIP),[38] commonly used by the astronomical research community. We can hope that such longer wavelength array detector technology will eventually become economically feasible for use with the FTIR microscope.

Other opportunities may present themselves for combining FPAs with the synchrotron source. For example, the multiplex advantage of FTIR spectrometers is negated when compared with a dispersive spectrometer and linear array detector. This suggests that one could design a microspectrometer using an FPA detection system, with one dimension of the array being used for spatial information and the other for spectral information.[39] The system would produce 2D spectral images by the "push broom" method. An extended line source would match such an instrument particularly well. However, a useful comparison of this technique with other spectroscopic imaging methods will require a detailed design and performance analysis and is beyond the scope of this chapter.

The current generation of focal plane readout electronics allows small to moderate size sections of the array to be read out at rates approaching $10\,kHz$. Thus, each pixel can sense at 100 microsecond intervals. This time scale is of interest for biological systems, including *in vivo* processes at the cellular level. We can foresee the use of such FPA equipped microscopes, combined with narrow-band IR filters, to capture such processes. Assuming such processes cannot be repetitively triggered, signal-averaging at each pixel is not easily achieved. This places additional demands on the signal-to-noise available at each pixel and motivates the use of the synchrotron IR source.

Finally, we note that there are many examples of biomedical imaging involving three-dimensional (3D) datasets, in some cases similar to those obtained from computerized tomographic scanning, but also for confocal fluorescence microscopy.[40,41] These datasets are amenable to more advanced 3D deconvolution methods,[42,43] and may point to future IR FPA imaging methods involving the synchrotron source. Such deconvolution methods may also prove useful in analysing IR spectroscopic images where two dimensions are spatial (the x and y pixel coordinates) and the third is spectral (the frequency coordinate).[44] Whereas the microscope's PSF defines the blurring in the spatial dimension, the FTIR's apodized resolution function, or a Gaussian (inhomogeneous broadening) or Lorentzian (lifetime broadening) line shape might define the other. In this manner the spectral integrity at a given pixel might be maintained or controlled. This will be an important topic for future work.

In summary, there have been significant developments in the use of imaging IR microspectrometers with the high-brightness synchrotron source. The quality of the images collected with sub-wavelength effective pixel dimensions has enabled the use of image deconvolution methods for improving the effective resolution and contrast. The limits for such methods are not yet known, but initial results are promising.

Acknowledgements

Portions of the work were supported by the US Department of Energy under contract DE-AC02- 98CH10886, and by the National Institute of Health under RR023782 and AR052778. We are grateful to G. Nintzel, M. Caruso and D. Carlson for technical support, and to R.J. Smith, G.D. Smith and A. Acerbo for assisting with some of the measurements and for allowing early access to some of the measurement results presented here.

References

1. W. D. Duncan and G. P. Williams, Infrared Synchrotron Radiation, *Appl. Opt.*, 1983, **22**, 2914.
2. G. L. Carr, High-resolution microspectroscopy and sub-nanosecond time-resolved spectroscopy with the synchrotron infrared source, *Vib. Spectrosc.*, 1999, **19**, 53–60.
3. G. L. Carr, R. Lobo, J. LaVeigne and D. Reitze, Exploring the dynamics of superdonductors by time-resolved far-infrared spectroscopy, *Phys. Rev. Lett.*, 2000, **85**, 3001–3004.
4. P. Dumas and L. Miller, Biological and biomedical applications of synchrotron infrared microspectroscopy, *J. Biol. Phys.*, 2003, **29**(2–3), 201–218.
5. P. Dumas and L. M. Miller, The use of synchrotron infrared microspectroscopy in biological and biomedical investigations, *Vib. Spectrosc.*, 2003, **32**, 3–21.
6. L. M. Miller, Q. Wang, R. J. Smith, H. Zhong, D. Elliott and J. Warren, A New Sample Substrate for Imaging and Correlating Organic and Trace-

Metal Composition in Biological Cells and Tissues, *Anal. Bioanal. Chem.*, 2007, **387**(5), 1705–1715.

7. R. P. S. M. Lobo, J. D. LaVeigne, D. H. Reitze, D. B. Tanner and G. L. Carr, Performance of new infrared beamline U12IR at the National Synchrotron Light Source, *Rev. Sci. Instrum.*, 1999, **70**, 2899–2904.

8. L. Miller and R. J. Smith, Synchrotrons versus globars, point-detectors versus focal plane arrays: Selecting the best source and detector for specific infrared microspectroscopy and imaging applications, *Vib. Spectrosc.*, 2005, **38**(1–2), 237–240.

9. G. L. Carr, and G. D. Smith, unpublished.

10. D. Moss, B. Gasharova and Y.-L. Mathis, Practical tests of a focal plane array detector microscope at the ANKA-IR beamline, *Infrared Phys. Technol.*, 2006, **49**, 53–56.

11. P. Heraud, S. Caine, G. Sanson, R. Gleadow, B. R. Wood and D. McNaughton, Focal plane array infrared imaging: a new way to analyse leaf tissue, *New Phytol.*, 2007, **173**, 216–225.

12. J. N. Nasse, R. Reininger, T. Kubala, S. Janowski and C. J. Hirschmugl, Synchrotron infrared microspectroscopy imaging using a multi-element detector (IRMSI-MED) for diffraction-limited chemical imaging, *Nucl. Instrum. Meth. Phys. Res. A*, 2007, **582**, 107–110.

13. P. Dumas, G. D. Sockalingum and J. Sulé-Suso, Adding synchrotron radiation to infrared microspectroscopy: what's new in biomedical applications?, *Trends Biotechnol.*, 2007, **25**(1), 40–44.

14. L. Miller, M. J. Tobin, S. Srichan, P. Dumas, About the use of synchrotron radiation in infrared microscopy for biomedical applications, In: *Biomedical Applications of FTIR spectroscopy*, P. Harris, Editor, 2008.

15. R. Bhargava and I. W. Levin, *Spectrochemical Analysis using Infrared Multichannel Detectors*, 2005, Oxford, Blackwell Publishing.

16. See, for example "*Infrared Synchrotron Radiation*", P. Roy and Y.-L. Mathis, eds. il Nuovo Cimento (1999).

17. A. Ugawa, H. Ishii, K. Yakushi, H. Okamoto, T. Mitani, M. Watanabe, K. Sakai, K. Suzui and S. Kato, Design of an instrument for far-infrared microspectroscopy using a synchrotron radiation source, *Rev. Sci. Instrum.*, 1992, **63**, 1551–1554.

18. R. J. Hemley, H. K. Mao, A. F. Goncharov and M. V. V. S. Hanfland, Synchrotron infrared spectroscopy to 0.15 eV of H_2 and D_2 at megabar pressures, *Phys. Rev. Lett.*, 1996, **76**, 1667.

19. J. Reffner, G. L. Carr, S. Sutton, R. J. Hemley and G. P. Williams, Infrared microspectroscopy at the NSLS, *Synch. Rad. News*, 1994, **7**, 30–37.

20. G. L. Carr, J. A. Reffner and G. P. Williams, Performance of an infrared microspectrometer at the NSLS, *Rev. Sci. Instrum.*, 1995, **66**, 1490–1492.

21. A. F. Goncharov, E. Gregoryanz, H. K. Mao, Z. Liu and R. J. Hemley, Optical evidence for a nonmolecular phase of nitrogen above 150 GPa, *Phys. Rev. Lett.*, 2000, **85**, 1262–1265.

22. N. Jamin, P. Dumas, J. Moncuit, W. H. Fridman, J. L. Teillaud, G. L. Carr and G. P. Williams, Highly resolved chemical imaging of living cells by using synchrotron infrared microspectrometry, *Proc. Natl. Acad. Sci. USA*, 1998, **95**, 4837–4840.

23. N. Guilhaumou, P. Dumas, G. L. Carr and G. P. Williams, Synchrotron infrared microspectrometry applied to petrography in micrometer-scale range: Fluid chemical analysis and mapping, *Appl. Spec.*, 1998, **52**, 1029–1034.

24. J. L. Bantignies, G. L. Carr, P. Dumas, L. M. Miller and G. P. Williams, Applications of Infrared Microspectroscopy to Geology, Biology, and Cosmetics, *Synch. Rad. News*, 1998, **11**, 31–36.

25. F. Polack, R. Mercier, L. Nahon, C. Armellin, J. P. Marx, M. Tanguy, M. E. Couprie and P. Dumas, Optical design and performance of the IR microscope beamline at SuperACO-France. *SPIE* (eds G. L. Carr, P. Dumas), 1999. **3575**: 13.

26. G. L. Carr, Resolution limits for infrared microspectroscopy explored with synchrotron radiation, *Rev. Sci. Instrum.*, 2001, **72**(3), 1613–1619.

27. V. Schreiner, G. S. Gooris, S. Pfeiffer, G. Lanzendörfer, H. Wenck, W. Diembeck, E. Proksch and J. Bouwstra, Barrier Characteristics of Different Human Skin Types Investigated with X-Ray Diffraction, Lipid Analysis, and Electron Microscopy Imaging, *J. Invest. Dermatol.*, 2000, **114**, 654–660.

28. B. Ongpipattanakul, M. L. Francoeur and R. O. Potts, Polymorphism in stratum corneum lipids, *Biophys. Acta*, 1994, **1190**, 115–122.

29. M. Lafleur, *Can. J. Chem.*, 1998, **76**, 1501–1511.

30. D. J. Moore, M. E. Rerek and R. Mendelsohn, *Biochem. Biophys. Res. Comm.*, 1997, **231**, 797–801.

31. G. L. Carr, O. Chubar, and P. Dumas, *Multichannel Detection with a Synchrotron Light Source: Design and Potential, in Spectrochemical Analysis Using Infrared Detectors*, I. W. L. R. Bahrgava, Editor. 2006, Blackwell Publishing.

32. G. L. Carr, *et al.*, presented at the 2005 Workshop on Infrared Spectroscopy and Microscopy with Accelerator-Based Sources [Rathen, Germany].

33. R. L. Jackson, and G. L. Carr, in preparation. 2010.

34. L. M. Miller and P. Dumas, Chemical imaging of biological tissue with synchrotron infrared light, *Biochim. Biophys. Acta.*, 2006, **1758**(7), 846–57.

35. M. J. Walsh, A. Hammiche, T. G. Fellous, J. M. Nicholson, M. Cotte, J. Susini, N. J. Fullwood, P. L. Martin-Hirsch, M. R. Alison and F. L. Martin, Synchrotron FTIR Imaging for the Identification of Cell Types within Human Tissues, in "WIRMS 2009: Proceedings of the 5th Int'l Workshop on Infrared Microscopy and Spectroscopy with Accelerator Based Sources", A. Predoi-Cross and B. E. Billinghurst, eds. (AIP Conf. Proc. #1214, p. 105).

36. A. Acerbo, in preparation. 2010.

37. J. Leotin, Far infrared photoconductivity studies in silicon blocked impurity band structures, *Infrared Physics and Technology*, 1999, **40**,

p. 153; see also N. Lum, *et al. Low- noise, low-temperature, 256x256 Si:As IBC staring FPA*. Proc. SPIE-Int. Soc. Opt. Eng., v. **1946**, pages 100 (1993).

38. B. F. Levine, Quantum Well Infrared Photodetectors, *J. Appl. Phys.*, 1993, **74**, see also S. D. Gunapala, S. V. Bandara *Quantum Well Infrared Photodetector (QWIP) Focal Plane Arrays* in "Semiconductors and Semimetals" (2000), v. **62** Chapter 4 (Academic Press).

39. U. Schade, private communication.

40. D. J. Stephens and V. J. Allan, Light Microscopy Techniques for Live Cell Imaging, *Science*, 2003, **82**, 300.

41. B. K. Ford, C. E. Volin, S. M. Murphy, R. M. Lynch and M. R. Descour, Computed Tomography-Based Spectral Imaging For Fluorescence Microscopy, *Biophys J. v.*, 2001, **80**, 986–99.

42. J. A. Conchello and E. W. Hansen, Enhanced 3-D reconstruction from confocal scanning microscope images. I. Deterministic and maximum likelihood reconstructions, *Appl. Opt. v.*, 1990, **29**, 3795–3804; see also J. Markham and J.-A. Conchello, *Fast maximum- likelihood image-restoration algorithms for three dimensional fluorescence microscopy. J. Opt. Soc. Am. A, v.* **18**, (2001) p. 1062.

43. P. J. Shaw, *Comparison of wide-field/deconvolution and confocal microscopy for 3D imaging*, in *Handbook of Biological Confocal Microscopy*, 2nd ed., J. B. Pawley, ed., Plenum, New York, 1995, pp. 373–387.

44. P. Lasch and D. Naumann, Spatial resolution in infrared microspectroscopic imaging of tissues, *Biochim. Biophys. Acta*, 2006, **1758**, 814–829.

CHAPTER 8

Scattering in Biomedical Infrared Spectroscopy

PAUL BASSAN AND PETER GARDNER*

School of Chemical Engineering and Analytical Science, Manchester Interdisciplinary Biocentre (MIB), University of Manchester, 131 Princess Street, Manchester M1 7DN, UK

8.1 Introduction to Scattering in Infrared Spectroscopy

In an infrared (IR) spectroscopy experiment, the usual aim is to measure the absorbance spectrum of a sample relative to a background spectrum of zero absorbance. The absorbance is calculated by taking the negative logarithm (usually of base 10) of the transmittance spectrum. An ideal sample is one which produces an absorbance spectrum that represents light loss due to absorption by the sample, as Figure 8.1(a) shows. In the case of samples that do not have a flat surface, such as most biomedical samples (single cells and tissue sections), light is lost owing to scattering. Scattering results in light path deviation such that it does not reach the detector as illustrated in Figure 8.1(b); this loss of light is evident in the absorbance spectra of single cells, which have recently become of interest.[1–12]

In the diagnostic application of synchrotron radiation Fourier transform IR (SR-FTIR) we are interested in measuring single cell spectra, or high resolution images of cells/tissue. The measured spectra are often distorted by scattering effects such that the biochemical information is rendered unreliable. This

RSC Analytical Spectroscopy Monographs No. 11
Biomedical Applications of Synchrotron Infrared Microspectroscopy
Edited by David Moss
© Royal Society of Chemistry 2011
Published by the Royal Society of Chemistry, www.rsc.org

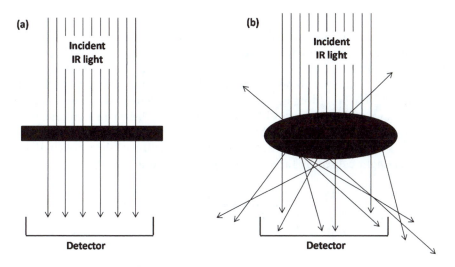

Figure 8.1 (a) Incident light absorbed by sample. (b) Incident light is absorbed by sample and scattered such that it does not reach the detector.

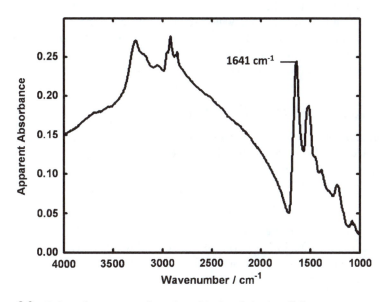

Figure 8.2 Infrared spectrum of a cultured isolated single cell from a prostate cancer (PC-3) cell line as measured with a synchrotron source.

unreliability takes the form of broad oscillating baselines, shifted peak positions and distorted peak shapes.[12,13] This is evident in Figure 8.2, which shows the IR spectrum of a cultured single isolated prostate cancer (PC-3) cell. The predominant cause of these distortions is Mie scattering.[12–15]

8.2 Mie Scattering

In 1908 Gustav Mie formulated his theory on light scattering by small particles.[16] In mid-IR spectroscopy the wavelengths of light are typically between 1 and 10 μm, which is of a similar magnitude to a cell and its components such as the nucleus.[13] This inevitably results in Mie scattering where a significant amount of light is lost (does not reach the detector) owing to the morphological characteristics of the sample. In 1957, van de Hulst published an approximation equation for the Mie scattering efficiency, Q, stated in eqns (8.1) and (8.2).[17] The term Q essentially describes the loss of light as a function of the wavelength (λ) of incident light caused by the scattering particle, and is given by:

$$Q = 2 - \frac{4}{\rho}\sin \rho + \frac{4}{\rho^2}(1 - \cos \rho) \tag{8.1}$$

where

$$\rho = \frac{2\pi d(n - 1)}{\lambda} \tag{8.2}$$

where n and d denote the ratio of the real refractive indices of particle to surrounding medium, and the diameter of the scattering particle, respectively. In this case, the medium is air, for which the real refractive index is essentially 1, hence n simplifies to the real refractive index of the scattering particle. To understand the characteristics of resonant Mie scattering, the complex refractive index requires explanation and is the subject of the next section.

8.3 Complex Refractive Index

The complex refractive index, η, comprises a real and an imaginary part which are denoted n and k respectively. The relation between these terms is concisely given by

$$\eta = n - ik \tag{8.3}$$

where $i =$ the imaginary unit, $\sqrt{-1}$.

The real and imaginary parts of the complex refractive index govern important characteristics of a material's interactions with light and are discussed in more detail in the next sections.

8.3.1 The Imaginary Refractive Index, k

The absorbance spectrum, A, of a sample with an effective optical path length through the sample, l, and the absorption index, α, are related by

$$A(\tilde{\nu}) = \alpha(\tilde{\nu})l \tag{8.4}$$

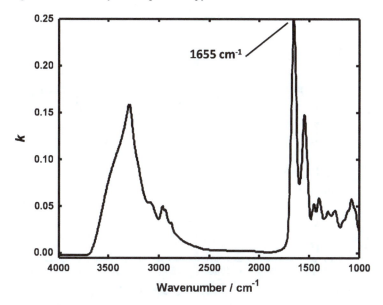

Figure 8.3 Example imaginary refractive index, k, spectrum of Matrigel (an extra-cellular matrix).

where

$$\alpha(\tilde{\nu}) = 4\pi\tilde{\nu}k(\tilde{\nu}) \tag{8.5}$$

where k = imaginary part of the complex refractive index.

The imaginary part of the complex refractive index, k, when plotted against wavenumber looks very similar to the original absorbance spectrum, except it has a different y-axis scaling. The absorbance spectrum and the k spectrum can be approximated as directly proportional for any calculations involving Mie scattering. In Figure 8.3 an example k spectrum of Matrigel (an artificial extracellular matrix, composed of cellular components such as protein and lipids) is shown and will be used throughout this chapter to illustrate various concepts.

8.3.2 The Real Refractive Index, n

The real and imaginary parts of the complex refractive index are related by a mathematical expression called the Kramers–Kronig transform and one can be calculated from the other.[18,19] The Kramers–Kronig transform for calculating the real refractive index from the imaginary part is defined as:

$$n(\tilde{\nu}) = \langle n \rangle + \frac{2}{\pi} P \int_0^\infty \frac{s \times k(\tilde{\nu})}{s^2 - \tilde{\nu}^2} ds \tag{8.6}$$

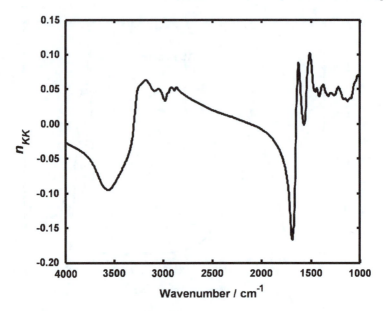

Figure 8.4   The n_{KK} spectrum of the Matrigel spectrum in Figure 8.3.

where $\langle n \rangle$ is the average refractive index, and P denotes the Cauchy principal of improper integrals, needed in this case when $s = \tilde{\nu}$ as a division by zero occurs, creating a singularity; s is the subject of the integration.

The output from the Kramers–Kronig transform is the real refractive index spectrum minus the average real refractive index, hereafter referred to as n_{KK}.

$$n_{KK} = n(\tilde{\nu}) - \langle n \rangle = \frac{2}{\pi} \mathrm{P} \int\limits_{0}^{\infty} \frac{s \times k(\tilde{\nu})}{s^2 - \tilde{\nu}^2} ds \qquad (8.7)$$

Performing the Kramers–Kronig transform on a k spectrum such as Figure 8.3, the n_{KK} spectrum can be calculated and is shown in Figure 8.4. As can be seen at each position where an absorption band appears in Figure 8.3, a derivative-like bandshape is observed in the n_{KK} spectrum in Figure 8.4.

8.4   Resonant Mie Scattering (RMieS)

The van de Hulst approximation equation for the Mie scattering efficiency, Q, was used to approximate non-absorbing spherical particle scattering characteristics.[17] Highly scattering biomedical medical samples such as single cells are absorbing materials,[15] and hence this equation could not fully describe all the scattering phenomena observed, namely an effect previously known at the "dispersion artefact".[20,21] In 2009 Bassan *et al.* published a modification to the van de Hulst equation to describe absorbing particles, and compared theoretical simulations to measured data of isolated single poly(methyl methacrylate) (PMMA) microspheres.[14] The match between theory and observed data was good, showing that this modification was a satisfactory approximation for resonant Mie scattering (RMieS).

The modification to the van de Hulst approximation was to use a real refractive index spectrum of the PMMA instead of using a constant value such as 1.4. To illustrate this concept here, we have simulated a resonant Mie scattering efficiency curve of a theoretical spherical particle of Matrigel. The n spectrum used is Figure 8.4, with a value of 1.3 added to it to act as an average real refractive index, this value is considered to be typical of biomedical samples.[13] Using a particle radius of 4 μm, theoretical Q curves can be computed and are shown in Figure 8.5.

Figure 8.5 shows two theoretical Q curves for the same sized particle, however the two curves look very different and this is due to the refractive index used to compute them. Figure 8.5(a) shows a non-resonant Mie scattering curve computed using a constant refractive index of $n = 1.3$, and Figure 8.5(b) shows the equivalent resonant Mie scattering curve computed using the n_{KK} spectrum shown in Figure 8.4. The effect of having a changing refractive index is that the Q curve changes from a smooth sinusoidal-like function to a more complicated version which exhibits spectral features. The two curves share the same broad oscillation; however the resonant Mie scattering curve shows a sharp decrease in intensity on the high wavenumber side of the amide I band and a shift of peak maximum position.

The band shape observed in a resonant Mie scattering curve is dependent upon the gradient of the broad oscillating baseline at that position. In Figure 8.5(a) a section is highlighted where the gradient of the Q curve is positive (it is important to note that the *x*-axis is reversed). Absorption bands at the position of a positive gradient result in a sharp decrease in intensity on the high wavenumber side of the band and, conversely, bands present at a negative gradient position exhibit a sharp increase in intensity on the high wavenumber side of the band.[14] For the vast majority of biomedical cases, such as single cells, the Q curve has a negative gradient at the amide I position and this results in the derivative-like line shape often observed. In Figure 8.5(b) the peak-like feature of the amide I band is shifted to a lower wavenumber, from 1655 to 1626 cm^{-1}, which is a shift of 19 cm^{-1}. This shift is significant as it may entirely change the interpretation of the biochemistry of a sample when the absorbance spectrum is analysed. In almost all IR measurements of single cells and tissues the amide I band is shifted; the only scenario in which it is not shifted is when the sample surface is flat.

This resonant Mie scattering effect is the cause of the phenomenon previously referred to as the "dispersion artefact", which was an inaccurate description because the effect was not solely due to dispersion and it is not an artefact because it can be explained fully. With an understanding of the physics behind resonant Mie scattering, preliminary algorithms based on the extended multiplicative signal correction (EMSC) have been developed.[22] The EMSC

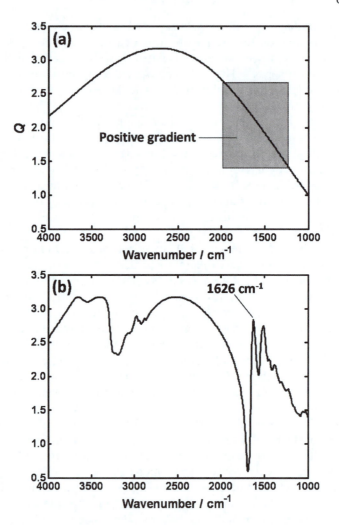

Figure 8.5   (a) A non-resonant Mie scattering curve for a particle of 4 µm radius with
$n = 1.3$. (b) Resonant Mie scattering curve computed for a theoretical
4 µm radius particle of Matrigel using the refractive index shown in
Figure 8.4 with an added value of 1.3.

algorithm requires explanation before the RMieS-EMSC can be explained, this
is covered in the next section.

8.5   Extended Multiplicative Signal Correction (EMSC)

The EMSC is a model-based multivariate data pre-processing method and is
based on linear statistical regression modelling.[23,24] It can be extended to

handle non-linear effects such as in this case Mie scattering, which would otherwise require complicated non-linear mathematical modelling. This is implemented by representing the non-linear mathematical modelling by a low-rank multivariate bilinear model.[20]

The EMSC is best explained with the use of a simple example, such as a spectrum which is affected by constant offset and sloping baselines simultaneously, as shown in Figure 8.6.

Figure 8.6 shows a spectrum of Matrigel that is affected by contributions consisting of an offset and a sloping baseline, which are represented by the dashed lines. This situation can be represented by a simple model in which the spectrum is simply the superposition of the pure absorbance spectrum (of interest to us) and a constant and sloping baselines (which we want to remove). This model can be concisely written as:

$$\vec{Z}_{Raw} = c + m\vec{\nu} + h\vec{Z}_{Ref} + \vec{E} \qquad (8.8)$$

where Z_{Raw} = the raw spectrum to be corrected, c = value of constant offset baseline, m = gradient of sloping baseline, Z_{Ref} = a reference spectrum and E = un-modelled residual variance. Symbols with arrows above them denote vectors, *i.e.* a row or column of numbers such as the absorbance values which make up a spectrum.

A reference spectrum, Z_{Ref}, is required to give the model stability during the parameter estimation stage. Ideally Z_{Ref} should be the pure absorbance

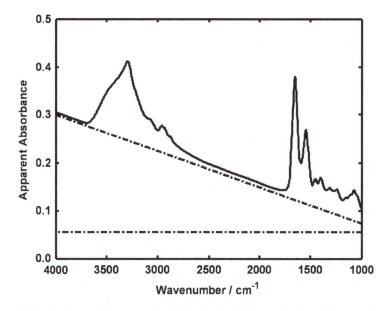

Figure 8.6   A theoretical spectrum of Matrigel (solid trace) affected by an offset and sloping baselines (dashed traces).

spectrum of the sample to be corrected, however this is rarely known, so a reasonable approximation has to be used. The spectra of biomedical samples share a similar profile which enables a spectrum such as that of Matrigel to be used as a reference spectrum.

Once the parameters c, m, and h are estimated for eqn (8.8) the spectrum can be corrected using the following formula:

$$\vec{Z}_{\text{Corr}} = \frac{\vec{Z}_{\text{Raw}} - \left(c + m\vec{v}\right)}{h} \tag{8.9}$$

The parameter estimation is done using a least squares fitting algorithm, which for the above example is a trivial problem. The EMSC is a powerful and versatile tool for correcting spectral problems, even non-linear effects such as Mie scattering. In 2008 Kohler *et al.* pioneered the first subspace based model to correct for non-resonant Mie scattering, which was very successful at removing the broad oscillating baselines.[20] In 2009 a collaboration between the labs of Achim Kohler (Nofima Mat. Norway), Hugh Byrne (DIT, Ireland) and Peter Gardner (University of Manchester, UK) produced the first resonant Mie scattering correction (RMieS-EMSC). This algorithm is described in the next section.

8.6   Resonant Mie Scattering Correction using the Extended Multiplicative Signal Correction (RMieS-EMSC)

A spectrum affected by resonant Mie scattering is currently modelled as the superposition of the *pure* absorbance spectrum of the sample and a scattering curve:[22]

$$\vec{Z}_{\text{Raw}} = \vec{Z}_{\text{Pure}} + \vec{Z}_{\text{Scatter}} \tag{8.10}$$

where Z_{Pure} and Z_{Scatter} represent the pure absorbance and RMieS spectrum; again symbols with arrows above denote vectors. Single cells and tissue samples are rarely the single homogeneous spherical particles that Mie theory describes; instead there may be multiple scattering particles in which case Z_{Scatter} will be comprised of multiple scattering curves from each particle:

$$\vec{Z}_{\text{Scatter}} = \vec{Z}_{\text{Sc}_1} + \vec{Z}_{\text{Sc}_2} + \dots \vec{Z}_{\text{Sc}_n} = \sum_n \vec{Z}_{\text{Sc}_n} \tag{8.11}$$

Each measured sample comprises multiple scattering curves of which the particle refractive index and diameter are unknown. This information is not easily available and not feasible to acquire when correcting large numbers of spectra and hence a method is required to deal with this situation. The solution

used in the RMieS-EMSC was to produce a database comprising many resonant Mie scattering efficiency, Q, curves for different combinations of particle
diameter and real refractive index properties.[22]

8.6.1 Construction of Mie Scattering Efficiency Database

To construct a database of Mie scattering efficiency, Q, curves, suitable ranges
of values for the scattering particle diameter, d, and real refractive index, n,
need to be chosen. The particle diameter range chosen was 4 to 10 µm as a
starting point; the algorithm allows the user full control over parameter ranges.
The real refractive index needs to be dealt with in a slightly different way to
cover the many possible characteristics.[22]

To cover the possibilities for n, the Kramers–Kronig transform is performed
on the reference spectrum, Z_{Ref}. This gives an n_{KK} spectrum which is normalised such that its minimum value is (-1), the reason for this is discussed later.
At this point it is useful to introduce two new terms, a and b, which represent
the average real refractive index and the "resonance parameter", respectively,
and together form the overall n spectrum:

$$n = a + b.n_{KK} \qquad (8.12)$$

The "resonance parameter" b enables full control over the absorptivity of the
sample, *i.e.* if the sample is non-absorbing then a value of $b = 0$ describes the
situation. The values of a represent the average real refractive index, which
have been chosen to range between 1.1 and 1.5. The range of values that b can
take is limited by the value of a, as the n spectrum cannot have values below 1.
For example if $a = 1.3$, then the n_{KK} spectrum (which has a minimum value of
-1) can be multiplied by a range of b values from 0 to 0.3. This ensures that n
never goes below 1 because this is not possible in the case of biomedical samples. Therefore, in general b can range between a value of 0 and $(a–1)$. To create
our database of Q curves, 10 equidistant values of a, b, and d are used in the
following ranges:

d: 4 to 10 µm
a: 1.05 to 1.50
b: 0 to $(a-1)$

Using 10 equidistant values for the three parameters results in 1000 Q curves.
These Q curves could in theory be directly used in an EMSC model such as:

$$\vec{Z}_{Raw} = c + m\vec{\nu} + h\vec{Z}_{Ref} + \sum_{m=1}^{1000} w_m \vec{Q}_m + \vec{E} \qquad (8.13)$$

where w_m is the weight applied to Q_m.

This model however would result in a very large and complex linear
regression problem during which a least squares fitting algorithm would most

likely fail because there is too much co-linearity between the Q curves. This co-linearity would cause problems as the least squares fitting algorithm would have difficulties deciding how to place the various weights w as many of the curves seem to be similar, this is often called rank deficiency. An elegant solution to solve this issue of rank deficiency was implemented by Kohler *et al.* in 2008, and was to decompose the Q database into a lower number of spectra (typically fewer than 10) which describe all the variance in the database.[20] These spectra are obtained from decomposing the Q database using principal component analysis (PCA), which is described in the next section.

8.6.2   Decomposition of the Resonant Mie Scattering Efficiency Database, Q

The Mie scattering efficiency database can be reorganised into a matrix (a block of numbers) in which each row is an individual Q curve. This matrix hereafter is denoted as $\mathbf{Q}$, where the bold font indicates a matrix according to standard linear algebra notation. If our raw spectrum to be corrected comprises 2000 absorbance values, then the 1000 Q curve matrix will have 1000 rows and 2000 columns (a 1000×2000 matrix).

Using principal component analysis (PCA) this $\mathbf{Q}$ matrix can be summarised as a small number (such as 10) of principal components (PCs). The $\mathbf{Q}$ matrix can be decomposed into a scores matrix, $\mathbf{T}$, and a loadings matrix, $\mathbf{P}$, and the relationship between them is:

$$\mathbf{Q} = \mathbf{TP'} \tag{8.14}$$

where the dash next to $\mathbf{P}$ represents a matrix transpose (an operation where the matrix is reorganised such that the rows are arranged into columns or vice versa). The multiplication here is a matrix multiplication, details of which can be found in any linear algebra text book.

Of use here is the loadings matrix, $\mathbf{P}$, as this essentially gives a summary of the $\mathbf{Q}$ matrix in the form of about 10 curves. The number of curves to choose from is much higher than 10, however the first few curves describe the majority of the information in the database; the remaining curves describe negligible information and noise. These curves multiplied by the correct factor and added correctly can reconstruct any of the original 1000 Q curves with negligible loss of information. Another important property of these loadings is that they form an orthonormal (orthogonal and normalised) set of vectors, which is a useful property to have during the least squares fitting part of the RMieS-EMSC algorithm.[22]

These loading vectors can be incorporated into an EMSC model to give:

$$\vec{Z}_{\text{Raw}} = c + m\vec{\nu} + h\vec{Z}_{\text{Ref}} + \sum_{i=1}^{10} \vec{g}_i p_i + \vec{E} \tag{8.15}$$

where g_i = weight applied to loading vector p_i. The sigma term describes the scattering contributions in the spectrum, which can describe any combination of the curves in the **Q** matrix. Rearrangement of this equation can yield a corrected spectrum:[22]

$$\vec{Z}_{\text{Corr}} = \frac{\vec{Z}_{\text{Raw}} - \left(c + m\vec{\nu} + \sum_{i=1}^{10} \vec{g}_i p_i \right)}{h} \qquad (8.16)$$

To calculate the corrected spectrum, 13 parameters need to be estimated, 2 for the constant and sloping baselines, 1 for the multiplicative factor and 10 for the loadings which describe the scattering curve. These parameters are estimated using a least squares fitting algorithm as before.

8.7 Evaluation of the RMieS-EMSC Algorithm

As with any correction algorithm it is important to test it on a data set where the correct answer is already known. In a study by Bassan *et al.* a simulated data set was used.[22] The algorithm was tested to see how well it could recover two different "cell populations" and how well it could recover the correct position of the amide I band. Hence, to validate the algorithm, a simulated data set where all constituent effects were known was created. The set was simulated to form two groups (clusters) of data, each with 25 spectra. These spectra were created by adding together a number of Gaussian curves with various peak position heights and widths. The spectra were created so that they visually appeared to be similar to the fingerprint region of typical biomedical IR spectra, however no two spectra were identical. A spectrum of a thin layer of Matrigel was used as a "template" to acquire peak parameters. A random number generator was used to vary the positions ($\pm 1\,\text{cm}^{-1}$), heights ($\pm 20\%$) and widths ($\pm 2.5\%$) of peaks within each spectrum. The second data set was subject to the same random variation as the first but was intentionally given higher intensities (by 0.1 absorbance) at the $1300\,\text{cm}^{-1}$ and $1740\,\text{cm}^{-1}$ peaks so that the two groups of data would appear different when analysed using PCA.[22] These simulated data and the corresponding score plot for the first two principal components are shown in Figure 8.7(a) and (b).

As expected, two distinct clusters of spectra can be observed. This two-component model accounted for 81.9% of the variance in these "ideal" data. Figure 8.7(c) shows the result of scattering on the perfect spectra and the fact that they can no longer be separated by PCA. Using the RMieS-EMSC correction algorithm it was shown that this data set could be corrected to recover two separate groups under PCA (Figure 8.8).[22]

More importantly the authors also showed that the correct position of the amide I band could be recovered to within $1\,\text{cm}^{-1}$ as shown in Figure 8.9.[22] Figure 8.9(a) shows the position of the amide I band influenced by scattering. The original position of the peak was set to $1655 \pm 1\,\text{cm}^{-1}$, as indicated by the left-most vertical line. The actual peak positions of the simulated scattering

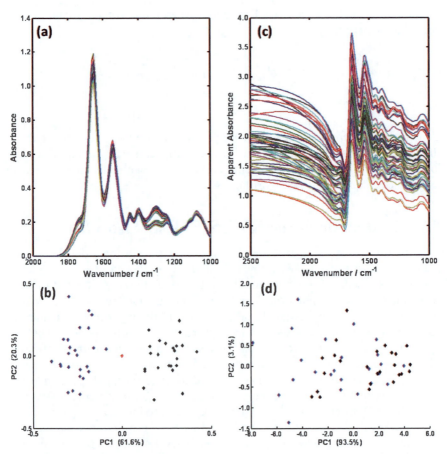

Figure 8.7   (a) Simulated data and (b) corresponding PCA scores plot. (c) Artificially
distorted spectra from (a) using RMieS curves, and (d) corresponding
PCA scores plot.

data range from 1635.9 to 1647.0 cm^{-1} with a mean of 1642.3 cm^{-1}.
Figure 8.9(b) shows that the amide I bands are closely grouped with positions
ranging from 1653.0 to 1656.4 cm^{-1} and a mean of 1654.5 cm^{-1}. This is
important since the position of the amide I band can be used to determine the
secondary structure of proteins.[22]

8.8   Correction of Real Spectra

Having established that the correction algorithm works on a simulated set of
data,[22] it is important to demonstrate that it can be used to correct real spectra
of single biological cells. Figure 8.10 shows the algorithm applied to the highly
scattering prostate (PC-3) cell spectrum in Figure 8.2. The application of the

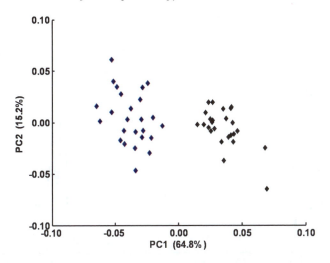

Figure 8.8 Principal component analysis (PCA) scores plot of the simulated scattered spectra in Figure 8.7(c) after application of the RMieS-EMSC algorithm.

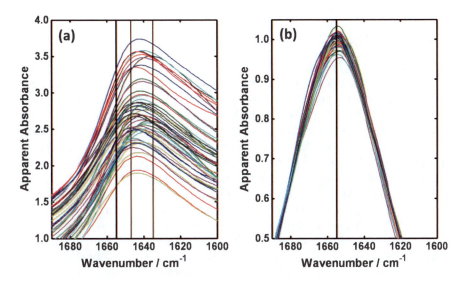

Figure 8.9 Amide I band peak positions of: (a) artificially scattered spectra and (b) corrected spectra using the RMieS-EMSC.

RMieS-EMSC correction has completely removed the oscillating baseline and the derivative-line shape that was present at the amide I band. As a result, the position of the amide I band has shifted from 1641 to 1654 cm^{-1}, which is likely to be very close to the true position of this absorption band.[22]

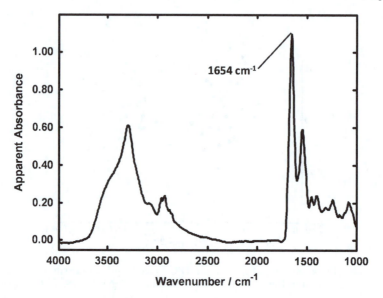

Figure 8.10   Corrected spectrum of a single isolated prostate cancer (PC-3) cell from Figure 8.2 using the RMieS-EMSC.

8.9   Conclusions

We have shown in this chapter that IR spectra of biological samples, particularly single isolated eukaryotic cells, are difficult to measure without the presence of distortion due to scattering. This distortion gives rise to oscillating baselines and derivative-like band shapes, making any kind of biological interpretation of the raw data unreliable. Until recently the origin of these distortions were not well understood and so the correction algorithms used to recover a pure absorption spectrum often failed. Thus the problem of scattering became the biggest single barrier to making real progress in the biological interpretation of SR single cell data and was specifically targeted as a problem to be solved by the DASIM single cell spectroscopy group. In a collaboration involving the Gardner group (Manchester, UK), the Kohler group (Ås, Norway), the Byrne group (Dublin, Ireland) and the Dumas group at Synchrotron SOLEIL (France), experiments on a simple model system of PMMA microspheres enabled an understanding of the underlying phenomenon to be developed.[14] Once the phenomenon was understood a new correction algorithm was developed that could be used to correct spectra of highly scattering samples.[22] It was rigorously tested and shown to be robust on simulated test data and is now being used by some twenty groups world-wide on real data sets from biological cells.

This important step represents a watershed in the development of SR-FTIR microscopy. Now that pure absorption spectra can be extracted from highly scattering data sets, experiments can be performed in which the subtle biological differences between similar sets of cells, or cells treated with and without cytotoxic and other chemotherapy agents, can finally be elucidated.

References

1. N. Jamin, P. Dumas, J. Moncuit, W. H. Fridman, J. L. Teillaud, L. G. Carr and G. P. Williams, *Proc. Natl. Acad. Sci. USA*, 1998, **95**, 4837–4840.
2. J. Lee, E. Gazi, J. Dwyer, M. D. Brown, N. W. Clarke and P. Gardner, *Analyst*, 2007, **132**, 750–755.
3. P. Lasch, A. Pacifico and M. Diem, *Biopolymers*, 2002, **67**, 335–338.
4. P. Lasch, M. Boese, A. Pacifico and M. Diem, *Vib. Spectrosc.*, 2002, **28**, 147–157.
5. P. Dumas and L. Miller, *Vib. Spectrosc*, 2003, **32**, 3–21.
6. P. Dumas, N. Jamin, J. L. Teillaud, L. M. Miller and B. Beccard, *Faraday Discuss.*, 2004, **126**, 289–302.
7. E. Gazi, J. Dwyer, N. P. Lockyer, P. Gardner, J. Miyan, C. A. Hart, M. D. Brown, J. H. Shanks and N. W. Clarke, *Biopolymers*, 2005, **77**, 18–30.
8. E. Gazi, J. Dwyer, N. P. Lockyer, J. Miyan, P. Gardner, C. A. Hart, M. D. Brown and N. W. Clarke, *Vib. Spectrosc.*, 2005, **38**, 193–201.
9. E. Gazi, P. Gardner, N. P. Lockyer, C. A. Hart, N. W. Clarke and M. D. Brown, *J. Lipid Res.*, 2007, **48**, 1846–1856.
10. D. A. Moss, M. Keese and R. Pepperkok, *Vib. Spectrosc.*, 2005, **38**, 185–191.
11. M. J. German, A. Hammiche, N. Ragavan, M. J. Tobin, L. J. Cooper, N. J. Fullwood, S. S. Matenhelia, A. C. Hindley, C. M. Nicholson, N. J. Full-wood, H. M. Pollock and F. L. Martin, *Biophys. J.*, 2006, **90**, 3783–3795.
12. J. Sule-Suso, D. Skingsley, G. D. Sockalingum, A. Kohler, G. Kegelaer, M. Manfait and A. El Haj, *Vib. Spectrosc.*, 2005, **38**, 179–184.
13. B. Mohlenhoff, M. Romeo, M. Diem and B. R. Wood, *Biophys. J.*, 2005, **88**, 3635–3640.
14. P. Bassan, H. J. Byrne, F. Bonnier, J. Lee, P. Dumas and P. Gardner, *Analyst*, 2009, **134**, 1586–1593.
15. P. Bassan, H. J. Byrne, J. Lee, F. Bonnier, C. Clarke, P. Dumas, E. Gazi, M. D. Brown, N. W. Clarke and P. Gardner, *Analyst*, 2009, **134**, 1171–1175.
16. G. Mie, "Contributions to the optical properties of turbid media, in particular of colloidal suspensions of metals" Leipzig, *Ann. Phys.*, 1908, **330**, 377–445.
17. H. C. van de Hulst, *Light scattering by small particles*, Dover, New York, 1981.
18. R. de and L. Kronig, *J. Opt. Soc. Am.*, 1926, **12**, 547–557.
19. H. A. Kramer, *Atti. Congr. Intern. Fisica Como.*, 1927, **2**, 545–557.
20. A. Kohler, J. Sule-Suso, G. D. Sockalingum, M. Tobin, F. Bahrami, Y. Yang, J. Pijanka, P. Dumas, M. Cotte, D. G. van Pettius, G. Parkes and H. Martens, *Appl. Spectrosc.*, 2008, **62**, 259–266.

21. J. K. Pijanka, A. Kohler, Y. Yang, P. Dumas, S. Chio-Srichan, M. Manfait, G. D. Sockalingum and J. Sulé-Suso, *Analyst*, 2009, **134**, 1176.
22. P. Bassan, A. Kohler, H. Martens, J. Lee, H. J. Byrne, P. Dumas, E. Gazi, M. D. Brown, N. W. Clarke and P. Gardner, *Analyst*, 2010, **135**, 268–277.
23. H. Martens, J. P. Nielsen and S. B. Engelsen, *Anal. Chem.*, 2003, **75**, 394.
24. H. Martens and E. Stark, *Anal. Chem.*, 2003, **9**(8), 625–35.

Section 3
Case Studies

CHAPTER 9

Synchrotron Based FTIR Spectroscopy in Lung Cancer. Is there a Niche?

JOSEP SULE-SUSO

Associate Specialist and Senior Lecturer in Oncology, Cancer Centre, University Hospital of North Staffordshire and Keele University, Stoke-on-Trent, UK

9.1 Introduction

Lung cancer is one of the main types of cancer and is associated with a huge socio-economic impact worldwide. It is the most common cancer in terms of both incidence and mortality, with 1.35 million new cases per year and 1.18 million deaths; the highest rates are in Europe and North America.[1] The incidence of lung cancer is currently lower in developing countries.[2] However, the incidence is expected to increase in the next few years, notably in China[3] and India,[4] because of the increased smoking pattern in these countries. In fact, about 90% of lung cancer cases are thought to be caused by tobacco.[5] Furthermore, there is a clear dose–response relationship between lung cancer risk and the number of cigarettes smoked per day, degree of inhalation, and age at initiation of smoking. Someone that has smoked all his or her life has a lung cancer risk 20–30 times greater than a non-smoker. However, the risk of lung cancer decreases with time since smoking cessation.[5]

It is widely accepted that lung cancer is associated with a bad prognosis. The overall 5-year survival rates vary from 8.9% in developing countries to 15% in the United States.[1] There are several reasons linked to this. However, two

RSC Analytical Spectroscopy Monographs No. 11
Biomedical Applications of Synchrotron Infrared Microspectroscopy
Edited by David Moss
© Royal Society of Chemistry 2011
Published by the Royal Society of Chemistry, www.rsc.org

important issues associated with the poor prognosis are the diagnosis and treatment of lung cancer. The most significant factor for survival in lung cancer is the stage of disease at diagnosis. This is entirely dependent upon how early in its development the tumour is diagnosed. Lung cancer tends not to cause symptoms in its early stages, whilst in patients with more advanced disease the symptoms are non-specific (cough, breathlessness, pain, *etc.*).[6] Therefore, it is difficult to diagnose cancer at an early stage when surgical excision of the tumour is still feasible. In fact, more than 50% of patients with a lung tumour have advanced disease at the time of diagnosis. In most of these cases, surgery is not possible and patients need to undergo a course of chemotherapy and/or radiotherapy. On the other hand, it is not uncommon to be unable to identify the subtypes of lung cancer from biopsies (squamous, adenocarcinoma, large cell). With the development of new drugs and drug combinations specific for certain subtypes of lung cancer, this means that some patients may not receive the most appropriate treatment. The second important issue associated with the poor prognosis of lung cancer is the fact that this disease is usually resistant to both chemotherapy and radiotherapy (RT).

It is therefore obvious that further research is required in order to improve the diagnosis and treatment of lung cancer. In fact, it would be ideal to have a technique that could characterize biomarkers for early diagnosis and for tumour response to treatment. More important, it would be desirable that these biomarkers could be detected at single cell and subcellular levels. Synchrotron based FTIR (S-FTIR) spectroscopy could be such a technique.[7] The possibility to study cells at both single cell and subcellular levels[8] could be useful to assess the presence of spectral biomarkers to be used for screening, diagnosis, and to assess the effects of chemotherapy drugs and/or RT on cancer cells. While there is a huge amount of literature showing that it might be possible to use FTIR spectroscopy in cancer diagnosis and treatment assessment, a clinical application has not been developed yet. Therefore, the aim of this chapter is to take the reader through: first, how to implement S-FTIR spectroscopy in the management of lung cancer; second, where the possible difficulties might arise; third, how to translate this into a clinical application. The three main areas to be discussed here are screening, diagnosis and treatment.

9.2 Lung Cancer Screening

One of the problems in the diagnosis of lung cancer is that when the disease presents with symptoms it is usually too late to treat radically. This is even more important in those tumours presenting in the periphery of the lung. Figures 9.1a and 9.1b show the example of a lung tumour in the left lung. The images have been taken with a chest X Ray and a computed tomography (CT) scan, respectively. In this case, by the time the patient presented with symptoms (cough, breathlessness) the tumour was greater than 5 cm in its largest dimension. While surgery could still be an option depending on how fit the patient is, this is not the case for most patients. Therefore, it is important to

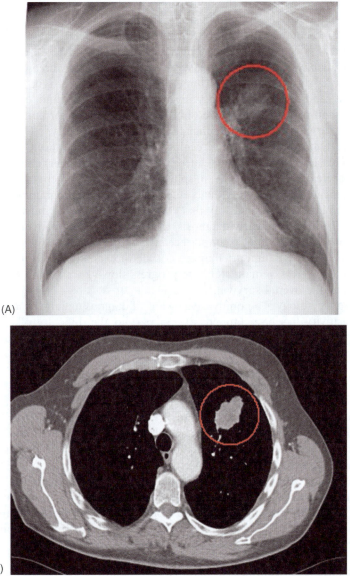

Figure 9.1 Radiological images of a tumour of the left lung taken with a chest X ray (a) and a computed tomography (CT) Scan (b).

diagnose lung cancer when it is still small enough to undergo radical treatment. One possible way to achieve this is by screening the population at risk. The population segment most likely to develop lung cancer is people over 50 years of age who have a history of smoking.

The aim of screening is to detect disease at a stage when cure or control is possible.[9] An ideal screening tool should satisfy the following conditions.

(i) The disease in question should be a major public health problem with a clear understanding of the natural history. (ii) There should be a suitable, safe, sensitive, acceptable, and cost-effective method to detect latent disease (*i.e.*, disease that is causing no symptoms). (iii) There should be an effective treatment available for the early disease so detected.[10] These three conditions are met, up to certain extent, by lung cancer, making this type of cancer an excellent candidate for screening. Several studies using sputum cytology, chest X-rays, and/or CT scans have shown that screening for lung cancer can improve survival but does not change lung cancer mortality.[11] However, there is currently no recommended screening strategy for lung cancer.[12] The reasons why increased survival in lung cancer (patients living longer) is not associated with a decreased mortality, *i.e.*, in spite of living longer patients still die of lung cancer, include: (i) The individuals entering a screening programme differ from the general population in a way that could affect the overall mortality for lung cancer. (ii) In spite of early diagnosis, the time of death remains unchanged so the patient survives longer simply because the disease was diagnosed earlier. (iii) Less aggressive tumours that are not life threatening (including neoplasms that would regress, remain stable or progress slowly) are diagnosed, therefore increasing the overall survival.[13]

Sputum cytology is a noninvasive means of diagnosing lung cancer although the sensitivity has been limited, due in part to the difficulty of consistently obtaining high-quality sputum specimens. However, sputum induction is a well-tolerated, safe procedure,[14] and new techniques are being developed for acquiring high-quality sputum specimens for diagnostic purposes.[15] In addition, identifying malignant cells in sputum requires a high degree of expertise by experienced personnel.[15] However, the recent development of automated quantitative cytometry to study the presence of cancer cells by identifying abnormal nuclear size and shape, nucleoli,[16] abnormal DNA,[17] and molecular biomarkers has renewed the interest in sputum collection for lung cancer screening.[12] Abnormal cells may be recognized in sputum up to 20 months before a lung cancer diagnosis is made,[6] and furthermore a link between abnormal sputum cytology and lung cancer development has been described.[6] At present, with sputum samples, up to 80% of central tumours can be diagnosed. However, sensitivity is low for the detection of small peripheral lung cancers.[17]

It is therefore obvious that there is no method well established yet to screen lung cancer using sputum samples. The possibility of using S-FTIR spectroscopy to study single cells in sputum samples could represent a new way to screen for lung cancer. Preliminary work should be aimed, first, at assessing whether S-FTIR spectroscopy could differentiate between normal lung cells, lung cancer cells, and non-malignant abnormal cells. While this has already been shown in cancer of the cervix,[18] further work is required to assess whether this is possible in lung cancer. Second, pilot studies including at least 50 patients should be carried out. Should such studies yield positive results, then a multicenter study including a larger number of patients should be established in order to assess, amongst others, the sensitivity and specificity of S-FTIR spectroscopy to diagnose dysplastic and/or tumour cells in sputum samples, and whether using S-FTIR spectroscopy for screening purposes would improve not only lung cancer survival but also reduce the number of deaths caused by lung cancer. In other words, not only patients would be living longer but also fewer numbers of patients would die from their lung cancer.

9.3 Lung Cancer Diagnosis

Patients who present with a suspected lung cancer must undergo a biopsy to obtain tissue to be analysed by a pathologist in order to confirm or rule out the presence of lung cancer. One of the problems in the diagnosis of lung cancer is the difficulty sometimes experienced in obtaining good biopsy samples. Usually, bronchoscopy (introducing a tube through the windpipe down to the lung) is carried out to obtain a tissue sample. However, it might not be easy to obtain a sample if the tumour is in the periphery of the lung and away from the main bronchi (see Figure 9.1b). In these cases, other diagnostic techniques such as CT scan guided biopsies or a surgical approach are required. Furthermore, in the case of the patient undergoing a bronchoscopy, it might be easy in some cases to obtain a good biopsy sample because the tumour is easily accessible (Figure 9.2a). However, in other cases only small amounts of cells can be obtained as the tumour can either not be seen or causes just slight changes in the bronchus (Figure 9.2b). In the latter case, samples may be obtained with a small number of cells that might be abnormal, with features suggesting malignancy, but not enough to reach the diagnosis of cancer. At this point, patients will undergo further biopsies with all the discomfort and side effects that these entail, and delaying their treatment. It is sometimes difficult for a

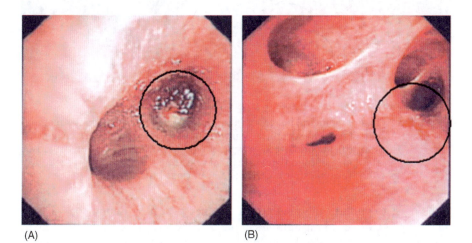

(A)   (B)

Figure 9.2   Bronchoscopy images of a lung tumour appearing as a white mass (a) and of a lung tumour showing just redness in the mucosa of the bronchus (b). The tumours have been ringed in black for clarity.

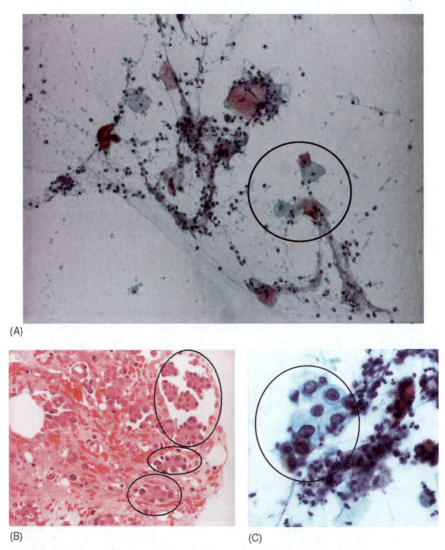

Figure 9.3 Histopathological images of (a) normal basal/intermediate lung squamous cells (blue) and superficial lung squamous cells (pink) (100× magnification); (b) abnormal cells suspicious of lung cancer but not diagnostic (100× magnification); (c) malignant cells from a squamous cell lung cancer (200× magnification). Cells have been ringed for clarity.

pathologist to confirm the presence or absence of a lung tumour based on a small number of cells in a biopsy sample as Figures 9.3a–c show. Malignant cells are characterized by larger nuclei, irregular chromatin, and irregular outline. However these features are sometimes not clear enough to reach a diagnosis. In these cases, S-FTIR spectroscopy could become a tool that could better characterize these abnormal cells and reach the diagnosis of lung cancer

Table 9.1   Tumours of the lung. (Modified from Reference (34).)

– Epithelial tumours:
   – Squamous cell carcinoma
   – Small cell carcinoma
   – Adenocarcinoma
   – Large cell carcinoma
   – Adenosquamous
   – Carcinomas with pleomorphic, sarcomatoid or sarcomatous elements
   – Carcinoid tumours
   – Carcinomas of salivary gland type
– Mesothelial tumours:
   – Mesothelioma
– Miscellaneous tumours:
   – Clear cell tumour
   – Germ cell neoplasms
   – Thymona
   – Melanoma
   – Others
– Lymphoma
– Secondary tumours
– Unclassified tumours

without the need for further biopsies. Obviously, it is important to assess, as mentioned before, whether S-FTIR spectroscopy can differentiate among normal lung cells, dysplasia, and malignant lung cancer cells, as has been shown in other types of tumours.[18]

Once it has been confirmed that a patient has lung cancer, the next step is to identify which type of lung cancer he/she has. This is important as treatment for the different types of lung tumours might vary. Pathologists have several tools (staining, immunohistochemistry) that help them to better characterize the type of lung cancer. However, this is not always possible and, in some instances, tumours have to be described, for example, as "undifferentiated" or "non-small cell carcinoma" without further differentiation. While it could be hypothesized that S-FTIR spectroscopy could help in these cases, the situation is rather complex. Table 9.1 shows a modified list of the different types of tumours that can be found in the lung.

Studies including large numbers of patients should be initiated to assess whether S-FTIR spectroscopy could indeed differentiate among some or all types of lung cancer. However, there are several difficulties:

• There is a great variety of lung tumours, some of them very similar in appearance.
• Implementing S-FTIR spectroscopy in tissue diagnosis in lung cancer in clinical practice and, by extension, to other types of cancer, requires this technique to be time efficient when compared with histological methods.
• In the case of metastasis to the lung from other types of tumours, it might be difficult to assess from which organ the primary tumour originated. While it has been described previously that FTIR spectroscopy could

differentiate between the different types of metastases in brain tissue,[19] further work is required to confirm this in lung cancer.

Therefore, the initial application of S-FTIR spectroscopy in lung cancer diagnosis might be aimed at helping to diagnose the disease in samples containing just a few cells, rather than in the study of tissue samples (millimetres to centimetres in size) which could be very time consuming and computationally difficult. However, this could be improved in the future with further technological developments in this area of research. Furthermore, the recent work on S-FTIR spectroscopy of stained samples opens up a new avenue towards what could be termed "spectral pathology", as stained cells deemed suspicious by pathologists could be studied directly with S-FTIR spectroscopy.[20]

9.4   Treatment of Lung Cancer

One of the principal precepts in medicine is in the Latin phrase "*primum non nocere*" (first, do no harm). It reminds a physician that he or she must consider the possible harm that any intervention might do. This is an important issue in oncology where treatment for cancer can have severe and perhaps life-threatening toxicity. Chemotherapy is an excellent example: while the treatment could help to control or cure the disease, it is associated with severe side effects. In the case of lung cancer, chemotherapy, together with surgery and/or radiotherapy, is one of the main treatments. However, the use of chemotherapy drugs in the mangement of lung cancer is associated with two main problems. First, lung cancer is usually resistant to chemotherapy. Second, we do not know at present, which is the best chemotherapy drug combination for each individual patient with lung cancer.

In the management of patients with lung cancer with chemotherapy, it is widely accepted that, regarding tumour response, a combination of two drugs is better than a single drug, but that a combination of three drugs is not better than two, and that three drugs also cause more side effects. Furthermore, most chemotherapy regimes for lung cancer contain a platinum compound in combination with a second chemotherapy agent. However, it is not clear yet which is the best second drug in this combination. Therefore, it would be ideal to have a tool that could help clinicians to decide, *a priori*, which would be the best combination regime for each individual patient. In fact, prediction of drug sensitivity for each patient, and the cell kill kinetics of the drug, may improve the outcome of treatment.[21]

It has been widely documented that FTIR spectroscopy can detect changes in cancer cells following the addition of different chemotherapy drugs.[22-30] However, work has been carried out recently to better assess whether these changes are due to physical or chemical properties of cells.[31-33] With the capability of S-FTIR spectroscopy to analyse whole single cells and at the subcellular level, it could be hypothesized that by incubating lung cancer cells with the different drugs used in the management of lung cancer, tumour cell sensitivity to chemotherapy could be assessed. However, there are several problems:

(i) tumour cell variability: while some cells in a tumour may be sensitive to a drug or combination of drugs, other cells within the same tumour may be resistant. (ii) Drug delivery: in order for a chemotherapy agent to act on tumour cells, it has first to reach the target tumour and be active on cancer cells. This might be difficult in some cases because, amongst other factors, of the localization of the tumour, a decreased blood flow to the tumour, and/or increased drug metabolism by tumour cells. Sensitivity to a given chemotherapy drug *in vitro* does not guarantee that the tumour will respond to the treatment. Therefore, S-FTIR spectroscopy may need to be considered as a tool to assess tumour cell resistance to chemotherapy drugs rather than tumour cell sensitivity, so that drugs that are not active *in vitro* are not given to the patient.

Similarly, it is not possible at present to predict which tumours will respond to RT. Patients with lung cancer receiving RT will have a dose of RT based on, amongst other things, the size of the tumour, the patient's general health, and the localization of the tumour. Furthermore, RT is also associated with side effects. Therefore, S-FTIR spectroscopy could also be used to assess tumour cell resistance or sensitivity to RT. In this way, clinicians could deliver doses of RT based on S-FTIR spectroscopy data which would lead to an increase in the benefits of RT and a decrease in the side effects. This is known in clinical practice as the therapeutic index or ratio (the amount of a therapeutic agent that causes a therapeutic effect in relation to the amount that causes toxic effects).

9.5 Conclusions

The possibility of studying single cells with S-FTIR spectroscopy needs to be explored further in cancer screening, diagnosis, and treatment assessment. While several studies have indicated that it is possible to use this technique in the mangement of patients with cancer, there is, at present, no clear application for S-FTIR spectroscopy in lung cancer in routine clinical practice. In order to achieve this, further multicentre studies including a large enough number of patients to produce robust, statistically significant data are required. With such data, a clear-cut application for S-FTIR spectroscopy in the management of cancer could be developed. At that point, synchrotrons could become important tools in clinical practice. Ultimately, benchtop spectrometers would have to produce robust data so that they can be used in clinical practice. However, until this occurs, IR beamlines at synchrotrons might need to be considered not only as research facilities, but also as service providers, which could then bring revenue to the synchrotron. Finally, it is therefore important that both physicians and spectroscopists continue to work together to develop the medical applications of S-FTIR spectroscopy, which would benefit both fields.

References

1. D. M. Parkin, F. Bray, J. Ferlay and Pisani Paola, Global Cancer Statistics, 2002, *CA Cancer J. Clin.*, 2005, **55**, 74–108.

2. World Health Organization. Gender in lung cancer and smoking research. 2004 (http://www.who.int/gender/documents/en/lungcancerlow.pdf).
3. B. Q. Liu, R. Peto, Z. M. Chen, J. Boreham, Y.-P. Wu, J.-Y. Li, T. C. Campbell and J.-S. Chen, Emerging tobacco hazards in China: 1. Retrospective proportional mortality study of one million deaths, *Br. Med. J.*, 1998, **317**, 1411–1422.
4. D. Behera and T. Balamugesh, Lung cancer in India, *Ind. J. Chest Dis. Allied Sci.*, 2004, **46**, 269–281.
5. J. E. Tyczynski, F. Bray and D. M. Parkin, Lung cancer in Europe in 2000: epidemiology, prevention, and early detection, *Lancet Oncol.*, 2003, **4**, 45–55.
6. C. Read, S. Janes, J. George and S. Spiro, Early Lung Cancer: screening and detection, *Primary Care Resp. J.*, 2006, **15**, 332–336.
7. P. Dumas, G. D. Sockalingum and J. Sulé-Suso, Adding synchrotron radiation to infrared microspectroscopy: what's new in biomedical applications? *Trends Biotechnol.*, 2007, **25**, 40–44.
8. J. Pijanka, A. Kohler, Y. Yang, P. Dumas, S. Chio-Srichan, M. Manfait, G. D. Sockalingum and J. Sulé-Suso, Spectroscopic Signatures of Single, Isolated Cancer Cell Nuclei using Synchrotron Infrared microscopy, *Analyst*, 2009, **134**, 1176–1181.
9. E. F. Patz, P. C. Goodman and G. Bepler, Screening for lung cancer, *New Engl. J. Med.*, 2000, **343**, 1627–1633.
10. A. K. Ganti and J. L. Mulshine, Lung cancer screening: panacea or pipe dream? *Ann. Oncol.*, 2005, **16** (Suppl 2), ii215–ii219.
11. J. M. Reich, Improved survival and higher mortality. The conundrum of lung cancer screening, *Chest*, 2002, **122**, 329–337.
12. I. Guessous, J. Cornuz and F. Paccaud, Lung cancer screening current situation and perspective, *Swiss Med. Wkly1*, 2007, **37**, 304–311.
13. A. Rossi, P. Maione, G. Colantuoni, F. D. Gaizo, C. Guerriero, D. Nicolella, C. Ferrara and C. Gridelli, Screening for lung cancer: New horizons? *Crit. Rev. Oncol./Hematol.*, 2005, **56**, 311–320.
14. C. E. Brightling, Clinical Applications of Induced Sputum, *Chest*, 2006, **129**, 1344–1348.
15. F. L. Johnson, B. Turic and R. Kemp, Improved Diagnostic Sensitivity for Lung Cancer Using an Automated Quantitative Cytology System and Uridine 5-Triphosphate-Induced Sputum Specimens, *Chest*, 2004, **125**, 157S–158S.
16. T. L. Petty, Sputum cytology for the detection of early lung cancer, *Curr. Opin. Pulm. Med.*, 2003, **9**, 309–312.
17. A. McWilliams, J. Mayo, S. MacDonald, J. C. leRiche, B. Palcic, E. Szabo and S. Lam, Lung Cancer Screening. A Different Paradigm, *Am. J. Respir. Crit. Care Med.*, 2003, **168**, 1167–1173.
18. B. R. Wood, L. Chiriboga, H. Yee, M. A. Quinn, D. McNaughton and M. Diem, Fourier Transform Infrared spectral mapping of the cervical transformation zone and dysplastic squamous epithelium, *Gynecol. Oncol.*, 2004, **93**, 59–68.

19. C. Krafft, L. Shapoval, S. B. Sobottka, G. Schackert and R. Salzer, Identification of primary tumours or brain metastases by infrared spectroscopic imaging and linear discriminant analysis, *Technol. Cancer Res. Treat.*, 2006, **5**, 291–298.

20. J. Pijanka, G. D. Sockalingum, A. Kohler, Y. Yang, F. Draux, P. Dumas, C. Sandt, G. Parkes, D. G. van Pittius, G. Douce, V. Untereiner and J. Sulé-Suso, Synchrotron based FTIR spectra of stained single cells. Towards a clinical application in pathology. *Lab Invest.* 2010, **90**, 797–807.

21. T. Mori, M. Ohnishi, M. Komiyama, A. Tsutsui, H. Yabushita and H. Okada, Prediction of cell kinetics of anticancer agents using the collagen gel droplet embedded-culture drug sensitivity test, *Oncol. Rep.*, 2002, **9**, 301–305.

22. J. Zhou, Z. Wang, S. Sun, M. Liu and H. Zhang, A rapid method for detecting conformational changes during differentiation and apoptosis of HL60 cell by Fourier Transform infrared spectroscopy, *Biotechnol. Appl. Biochem.*, 2001, **33**, 127–132.

23. K.-Z. Liu, S. M. Kelsey, A. C. Newland and H. H. Mantsch, Quantitative determination of apoptosis on leukaemia cells by infrared spectroscopy, *Apoptosis*, 2001, **6**, 269–278.

24. A. Gaigneaux, J.-M. Ruysschaert and E. Goormaghtigh, Infrared spectroscopy as a tool for discrimination between sensitive and multiresistant K562 cells, *Eur. J. Biochem.*, 2002, **269**, 1968–1973.

25. G. I. Dovbeshko, V. I. Chegel, N. Y. Gridina, O. P. Repnytska, Y. M. Shirshov, V. P. Tryndiak, I. M. Todor and G. I. Solyanik, Surface enhnaced IR absorption of nucleic acids from tumour cells: FTIR reflectance study, *Biopolymers*, 2002, **67**, 470–486.

26. J. Ramesh, M. Huleilel, J. Mordehai, A. Moser, V. Erukhimovich, C. Levi, J. Kapelushnik and S. Mordechai, Preliminary results of evaluation of progress in chemotherapy for childhood leukaemia patients employing Fourier transform infrared microspectroscopy and cluster analysis, *J. Lab. Clin. Med.*, 2003, **141**, 385–394.

27. F. Gasparri and M. Muzio, Monitoring of apoptosis of HL60 cells by Fourier Transform Infrared spectroscopy, *Biochem. J.*, 2003, **369**, 239–248.

28. J. Sulé-Suso, D. Skingsley, G. D. Sockalingum, A. Kohler, G. Kegelaer, M. Manfait and A. J. El Haj, FTIR microspectroscopy as a tool to assess lung cancer cells response to chemotherapy, *Vib. Spectrosc.*, 2005, **38**, 179–184.

29. R. K. Sahu, U. Zelig, M. Huleihel, N. Brosh, M. Talyshinsky, M. Ben-Harosh, S. Mordechai and J. Kapelushnik, Continuous monitoring of WBC (biochemistry) in an adult leukemia patient using advanced FTIR-spectroscopy, *Leuk. Res.*, 2006, **30**, 687–693.

30. F. Draux, P. Jeannesson, C. Gobinet, J. Sulé-Suso, J. Pijanka, C. Sandt, P. Dumas, M. Manfait and G. D. Sockalingum, IR spectroscopy evidences effect of non-cytotoxic (cytostatic) doses of antitumor drug on cancer cells, *Anal. Bioanal. Chem.*, 2009, **395**, 2293–2301.

31. A. Kohler, J. Sulé-Suso, G. D. Sockalingum, M. Tobin, F. Bahrami, Y. Yang, J. Pijanka, P. Dumas, M. Cotte, D. G. van Pittius, G. Parkes and H. Martens, Estimating and correcting Mie scattering in synchrotron-based microscopic FTIR spectra by extended multiplicative signal correction (EMSC), *Vib. Spectrosc.*, 2008, **62**, 259–266.
32. P. Bassan, H. J. Byrne, F. Bonnier, J. Lee, P. Dumas and P. Gardner, Resonant Mie scattering in infrared spectroscopy of biological materials – understanding the 'dispersion artifact', *Analyst*, 2009, **134**, 1586–1593.
33. P. Bassan, A. Kohler, H. Martens, J. Lee, H. J. Byrne, P. Dumas, E. Gazi, M. Brown, N. Clarke and P. Gardner, Resonant Mie Scattering (RMieS) correction of infrared spectra from highly scattering biological samples, *Analyst*, 2010, **135**, 268–277.
34. E. Brambilla, W. D. Travis, T. V. Colby, B. Corrinz and Y. Shimosato, The new World Health Organization classification of lung tumours, *Eur. Respir. J.*, 2001, **18**, 1059–1068.

CHAPTER 10

Head and Neck Cancer: Observations from Synchrotron-sourced Mid-infrared Spectroscopy Investigations[§]

MARK J. TOBIN,[a,†] JOHN M. CHALMERS,[b,‡,**] ANDREW T. HARRIS Bsc, MRCS[c] AND SHEILA E. FISHER[d,***] Msc, FDS, FRCS[d,e,*]

[a] Principal Scientist, Infrared Beamline, Australian Synchrotron, Clayton, Victoria 3168, Clayton, Australia; [b] School of Chemistry, University of Nottingham, University Park, Nottingham NG7 2RD, UK; [c] CR-UK Research Fellow. Faculty of Medicine and Health, University of Leeds, UK; [d] Senior Lecturer/Hon. Consultant in Maxillofacial Surgery, Faculty of Medicine and Health, University of Leeds, UK; [e] Faculty of Medicine and Health, Floor 6, Worsley Building, Clarendon Way, Leeds, West Yorkshire, UK, LS2 9LU

10.1 Introduction

In this chapter, we will look at the place of vibrational spectroscopy in head and neck (H&N) cancer diagnosis from the point of view of clinical need and

---

[§] Originally published as part of Chapter 5 in "Vibrational Spectroscopy for Medical Diagnosis", M. Diem, P.R. Griffiths and J.M. Chalmers (eds), John Wiley Ltd, Chichester (2008). Reproduced by kind permission of John Wiley & Sons Ltd.
[†] The author's address while the studies reported in this chapter were being undertaken was: Synchrotron Radiation Department, CCLRC Daresbury Laboratory, Warrington, Cheshire WA4 4AD, UK
[‡] Present address: VS Consulting, Stokesley, UK

RSC Analytical Spectroscopy Monographs No. 11
Biomedical Applications of Synchrotron Infrared Microspectroscopy
Edited by David Moss
© Royal Society of Chemistry 2011
Published by the Royal Society of Chemistry, www.rsc.org

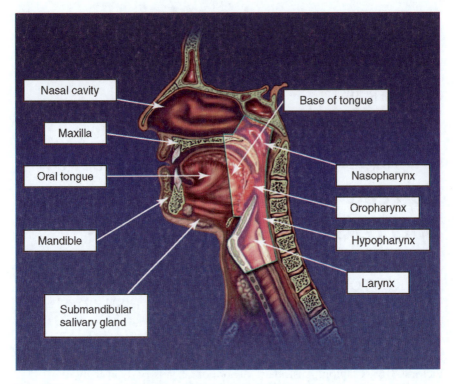

Figure 10.1   Anatomy of the upper aereodigestive tract. [Modified from illustration
designed by Sanofi-Aventis (with permission).]

biological diversity. Taking research in this field towards application requires
close collaboration and understanding between spectroscopists and biologists
and, as the technology moves closer to patient use, with doctors who have an
active interest in clinical research. Variability in both patient characteristics and
the disease itself has an impact on vibrational spectra taken from clinical settings and these aspects form an integral part of testing prior to trials in patients.

The term H&N cancer describes those cancers arising from the mucosal
lining of the upper aerodigestive tract and major specific structures within this
anatomical area, especially the four major salivary glands and the thyroid
gland. The area covered by this descriptive term is shown in Figure 10.1.

Head and neck cancer represents an ideal model for clinical study as the
disease is common globally, and particularly so in the emerging economies, it is
accessible for inspection and non-invasive or minimally invasive diagnostic
surveillance, it exhibits the full range of precancerous changes, carries a substantial risk of development of second primary tumours and has a relatively
short disease course, allowing robust outcome analysis. However, it is important for the basic scientist working in the field of biomedical diagnostic spectroscopy to understand the place of confounding factors that may have a
significant influence on results: smoking, ingested material, drugs (medical or

recreational) and even other disease states have the potential to have an impact on spectral features. Age, sex, ethnicity (especially skin or mucosal pigmentation), hormonal changes and mucosal thickness and structure may all have an impact. Some treatment effects, especially the fibrotic changes seen after radiotherapy or chemoradiotherapy, may alter tissue architecture and composition. These factors make it critically important for work towards diagnostic or therapeutic tools in cancer to be developed in strong basic science/clinical collaborative research teams, placing analysis of biomarkers in context in terms of diagnosis, confounding variables and outcome.

Work towards '*in-vivo*' and laboratory assessments has included other modalities as well as vibrational spectroscopy, especially fluorescence and elastic scattering spectroscopy (a real-time, *in vivo* optical technique that detects changes in the physical properties of cells). Such work falls outside the remit of this chapter but should be acknowledged, in that, for some applications, tools other than those based on vibrational spectroscopy may play a role and for others a multimodal approach, bringing different technologies together, may play a role.

Our work has involved both mid-infrared (mid-IR) and Raman spectroscopy, taking an approach by which a data set is evolved from single cell work in either cancer tissue or cell lines of known provenance towards evaluation of tissue samples, blood and potentially saliva and, if feasible, bone marrow.

10.2 Experimental Work

For the majority of our mid-IR studies, mostly undertaken prior to 2002, we used synchrotron-sourced mid-IR radiation coupled into a commercial Fourier transform (FT)-IR microscope to differentiate between cell types present in normal and malignant oral mucosa, producing tissue maps that allowed discrimination between normal and cancer tissue.[1,2] The FT-IR spectrometer was a Nicolet model 730 (Nicolet Instruments, Inc., Madison, WI, USA) interfaced to a Nic-Plan® IR microscope, fitted with a $32\times$ objective and equipped with a narrow-band (cut-off *ca.* $700\,\mathrm{cm}^{-1}$) mercury–cadmium–telluride (MCT) detector. The spectra were recorded at the Synchrotron Radiation Source (SRS), Daresbury Laboratory, UK. The high brightness (low divergence) of mid-IR radiation emanating from a synchrotron makes it eminently appropriate for high ($<10\,\mu\mathrm{m}$ diameter) lateral spatial single-point-mapping resolution studies using FT-IR microspectroscopy; this is exemplified in Figure 10.2. Since our studies, recent developments in both FT-IR instrumentation and synchrotron beamline design have significantly improved the speed of collection of high quality data at high spatial resolution.

Cryo-microtomed tissue sections, nominal thickness $5\,\mu\mathrm{m}$, were mounted on $0.5\,\mathrm{mm}$ thickness $\mathrm{BaF_2}$ windows and their spectra were recorded in transmission mode. Parallel sections were stained conventionally (with haematoxylin and eosin; H&E) to facilitate the identification of regions of particular interest.

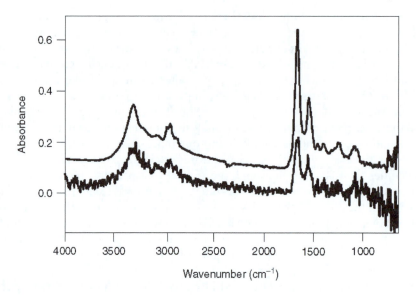

Figure 10.2 Comparison of synchrotron radiation-source (SRS) (upper) and Globar-source (lower) FT-IR spectra, $8\,cm^{-1}$ spectral resolution, 1024 scans, from a $10\,\mu m \times 10\,\mu m$ area on a tissue section. The upper spectrum has been offset for clarity.

Some of the sections used for IR examination were similarly stained after they had been studied spectroscopically. Cell culture samples were deposited on low-e glass slides (Kevley Technologies, USA) and examined in transflection mode: full details can be found in reference (1). The sampling area interrogated was 10 $\mu m \times 10$ μm for tissue section studies (unless stated otherwise) and $15\,\mu m \times 15\,\mu m$ for cultured cell samples, the latter allowing spectra to be recorded from single cells without interference from adjacent cells. Spectra were collected at $8\,cm^{-1}$ spectral resolution; 512 or 1024 scans were co-added for tissue section samples, 128 scans were co-added for cell culture samples. These gave adequate signal-to-noise ratio (SNR) for subsequent multivariate data analyses.

Because of the time requirements associated with single-point mapping measurements, it was only possible, compared with what can now be achieved using imaging mid-IR spectroscopy (see Chapter 7), to interrogate small data sets; nevertheless, studies such as these paved the way for highlighting the potential of mid-IR spectroscopy for medical diagnoses.

The potential of mid-IR spectroscopy to discriminate between oral squamous carcinoma cells (SCCa), normal gingival tissue and normal subgingival tissue, using key spectral bands, had previously been demonstrated, including studies concerned with the structure of nucleic acids and intracellular collagen.[3] Using the SRS beamline we were able to achieve spectral mapping at a lateral spatial resolution of 20 μm (or less), allowing single cell analysis.[1,2]

10.3 Mid-infrared Synchrotron Radiation FT-IR Studies of Oral Tissue Sections

Using fresh, unstained, air-dried sectioned tissue material mounted on a BaF_2 window, we identified regions of interest. A parallel H&E stained section was used for comparison, where necessary. Oral tissues containing normal and cancerous cells were used to test the capability of multivariate analysis techniques to discriminate between them. Pre-processing corrections were made to all tissue spectra for variations in sample thickness (normalization to the amide II band), baseline correction and water vapour absorption subtraction using routines within the Nicolet Omnic32™ software supplied with the FT-IR spectrometer. Sets of spectra between $1806 \, \text{cm}^{-1}$ and $938 \, \text{cm}^{-1}$ were then subjected to multivariate analysis using the Pirouette® version 3 software package (Infometrix, Inc., Woodinville, USA). This software included routines for hierachical cluster analysis (HCA), principal component analysis (PCA) and soft independent modeling of class analogies (SIMCA). Empirically, it was decided that all data sets would also be pre-processed, within the Pirouette® software package, as first-derivative (15-points) spectra with variance scaling; cross-validation (leave-one-out) was used for PCA evaluations.

Figure 10.3(a) shows a 4× magnification visual image from an H&E stained oral tissue section; the stroma and tumour regions are clearly discriminated by their light and dark purple stains, respectively. Figure 10.3(b) shows a 32× magnified visual image from a portion of a parallel, unstained section; the superimposed dashed white line separates the visually different morphologies. In sequence, five spectra were recorded from each of the three distinct regions using an aperture of $10 \, \mu\text{m} \times 10 \, \mu\text{m}$. The locations of these are marked by a + on Figure 10.3(b) and numbered 1–5 for the upper tumour region, 6–10 for the central stroma layer, and 11–15 for the lower tumour region. The 15 synchrotron-sourced FT-IR transmission spectra as recorded from these positions

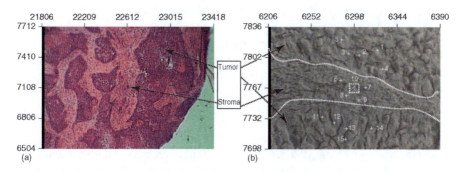

Figure 10.3 Visual images of an oral tissue section: (a) H&E stained, 4× magnification; (b) unstained, 32× magnification, showing the centres, marked with a +, of the $10 \, \mu\text{m} \times 10 \, \mu\text{m}$ areas from which FT-IR microscopy spectra were recorded. The white dotted lines on (b) have been superimposed to highlight the boundaries between the tumour and stroma regions. The white dashed square around point 10 delineates the aperture size.

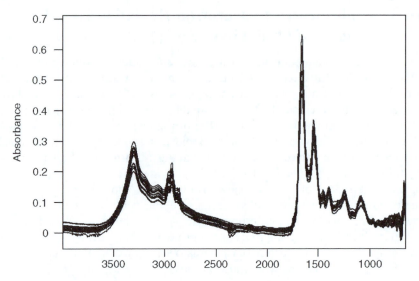

Figure 10.4 Overlaid SRS FT-IR absorbance spectra recorded from the 15
10 μm × 10 μm areas marked on Figure 10.3(b).

are shown as absorbance plots overlaid in Figure 10.4. Very clearly the inter-
spectral differences associated with path-length, baseline and water vapour
intrusion outweigh any inter-class absorbance variations. These spectra following pre-processing, and over the region selected for further interrogation by
multivariate analysis procedures, are shown in Figure 10.5 (the abscissa scale
inversion is a consequence of the way the data set was input into the Pirouette®
software package). The region contained 901 data points. The inter-class
spectral differences are still not obviously apparent over any intra-class variations. In Figure 10.6 an output dendrogram from an HCA, Euclidean distance
metric, group-average linkage, is shown. This simple example shows that the
unsupervised classification method clearly separates the set of spectra into two
distinct classes that represent stroma and tumour.

Overlays of the average of the normalized absorbance spectra from each of
three regions shown in Figure 10.3(b) are presented in Figure 10.7. The two
average spectra of the tumour regions (the arithmetic means of spectra 1–5 and
11–15) closely overlaid, but are readily distinguished from the average
spectrum of the stroma region, determined as the average of spectra numbers
6–11. From these normalized amide II band intensity data, some general
observations may be made. There is an apparent decrease in relative intensity of
the amide I band in connective tissue spectra compared with tumour. This is
accompanied by a general increase in relative absorbance over the range *ca.*
1500–1130 cm^{-1} in stroma over tumour, and a reduction in relative intensity of
the band at *ca.* 1090 cm^{-1} in stroma regions. However, it must be remembered
that these differences could have arisen as a consequence of the normalization
procedure employed.

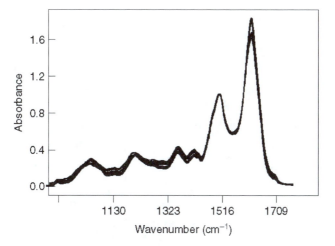

Figure 10.5 Pre-processed SRS FT-IR absorbance spectra over the range 1806 to 938 cm^{-1} recorded from the 15 10 μm×10 μm areas marked on Figure 10.3(b), and used as input data for multivariate analysis using Pirouette® software. (The abscissa scale reversal is a consequence of loading the data into the Pirouette software for multivariate analysis.)

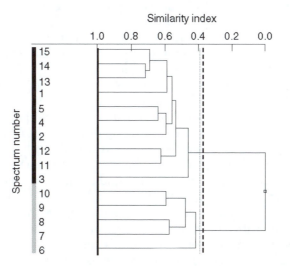

Figure 10.6 Dendrogram from an HCA of the 15 spectra shown in Figure 10.5.

Two other areas from the tissue section were similarly examined, and individually realized similar results. In total 44 single-point spectra were recorded from the three examinations of separate areas and combined into one pre-processed data set for subsequent HCA, the dendrogram from which is shown

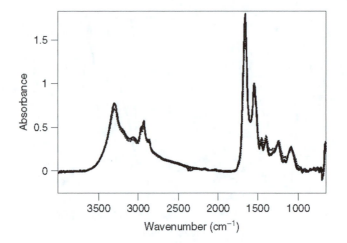

Figure 10.7   Overlaid normalized SRS FT-IR absorbance spectra of the average
                  spectrum recorded within each of the three regions shown in Figure
                  10.3(b). The two average spectra of the tumour regions, solid lines, (the
                  means of spectra 1–5 and spectra 11–15) are closely overlaid, but the
                  average spectrum (mean of spectra 6–10) from the stroma region, dashed
                  line, shows significant differences.

in Figure 10.8. This essentially classified the samples into two main groups
comprising those purporting to be from the tumour regions and those purporting to be from the stroma regions, excepting number 37, which was misclassified, according to our visual perception, as stroma. (Although by no
means clear, with hindsight closer visual inspection of the morphology of the
tissue section region concerned indicated some suggestion of 'mixing' of stroma
and tumour types in the regions around measurement point number 37, which
lay close to a boundary between two visually different morphologies.)

A white-light image of a fourth area of this tissue section is shown in
Figure 10.9. Superimposed on this picture is a grid highlighting the region
studied, and showing the individual area elements from which spectra were
recorded. A numbered grid is shown below, indicating the numbers allocated to
individual spectra. This area of the tissue section is more complex than the
previous three regions examined; a preliminary histopathological study showed
it to contain primarily three regions. These are indicated on the figure as:
tumour, stroma and an area of early keratinization. Keratin is the protein that
covers the surface of mucosa or skin. Keratinization is seen where a cancer
attempts to adopt the morphology of its parent tissue. As this represents a
favourable prognostic morphology the ability to identify its presence by IR
microspectroscopy is important.

Preliminary multivariate analysis very clearly showed that the spectrum
recorded from area element number 46 was atypical of the remainder in the set
and an outlier. When overlaid over the others in the set, this spectrum, even
after the pre-processing steps, clearly contained a more distorted baseline than

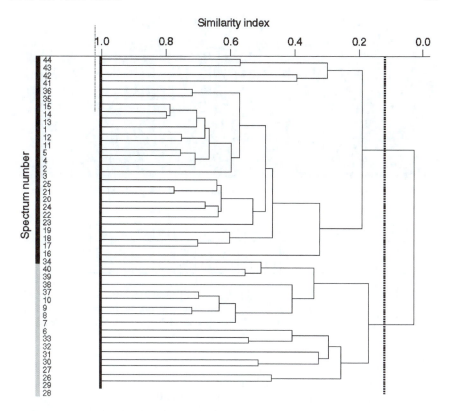

Figure 10.8 Dendrogram from an HCA of pre-processed mid-IR absorbance spectra for 44 spectra over the range 1810 to 900 cm^{-1} from the same tissue sample, see text for details.

the rest, and was therefore excluded from the data set and any further analyses. Preliminary PCA and SIMCA analyses of the data set, including a SIMCA prediction based on a model generated from the spectra numbered 1–44, highlighted that the spectra recorded from within the region marked as early keratinization were spectrally similar but very different from others in the data set. A simple HCA interrogation indicated the presence of three clusters. The more separated cluster contained all those spectra recorded from the region indicated as early keratinization, excluding number 88 but including number 55. Both spectra numbers 88 and 55 lie in the border between stroma and early keratinization, and close inspection of the white light image of Figure 10.9 clearly shows that the boundary, indicated by differing morphologies, meanders through area elements numbers 88, 72, 56 and 55. (The distinct grey-scale contrast between the left half and right half of the image of Figure 10.9 is artificial; it is a consequence of the image being a composite of two independent pictures corresponding to each half.) In a similar manner the boundary between the marked stroma and tumour regions does not follow a vertical line as

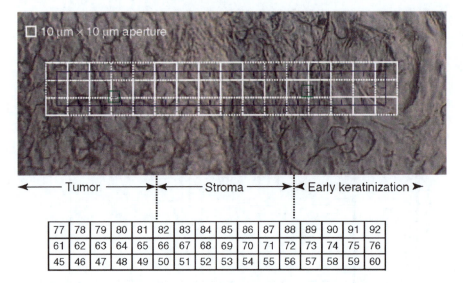

77	78	79	80	81	82	83	84	85	86	87	88	89	90	91	92
61	62	63	64	65	66	67	68	69	70	71	72	73	74	75	76
45	46	47	48	49	50	51	52	53	54	55	56	57	58	59	60

Figure 10.9   White light image of an area of oral tissue section, 32× magnification. The superimposed dashed white line grid on the image shows the areas (10 μm×10 μm) from which SRS FT-IR spectra were recorded. These were numbered as shown in the grid below the image.

indicated, but rather appears to meander somewhere through the area contained within the area elements 50–52, 65–67 and 80–82. Hence, there was some swap-over in the HCA of these boundary spectra from the cluster anticipated from that simply indicated by the preliminary histopathological survey. Closer histopathological inspection highlighted that there had been invasion of the stroma region by tumour within the vicinity of the boundary between the two layers. Since we were primarily concerned at this stage of our study with ascertaining spectral characteristics of essentially distinct classes of tissue cells, rather than gradation processes or mixed types, those within the two boundary regions, that is spectra corresponding to area elements 50, 51, 55, 56, 65, 66, 72, 81, 82 and 88, were excluded, along with the outlier 46, from the final multivariate analyses of the data.

Figure 10.10 shows similar factor 2 versus factor 3 projections from each of a PCA and SIMCA analysis of the residual set of 37 spectra. Those from the early keratinization region clearly form a distinct cluster well separated from the others; the spectra from the tumour are also well clustered and separated from those of the stroma region. The spectra from the central region of the stroma, although forming a cluster, are more dispersed, probably indicating that they are less 'pure', being possibly invaded from each side by characteristics of tumour and early keratinization. In particular, 71 and 87, which lie close to the stroma/early keratinization boundary seem to suggest perhaps some invasion of early keratinization characteristics. Based on the evidence above, three sub-groups of spectra were selected to generate the average spectra

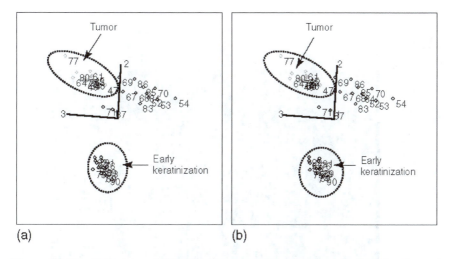

(a) (b)

Figure 10.10 PCA (a) and SIMCA (b) scores plot output projections for factor 2 versus factor 3 from the pre-processed absorbance spectra, excluding those recorded from the boundary regions, see text for details.

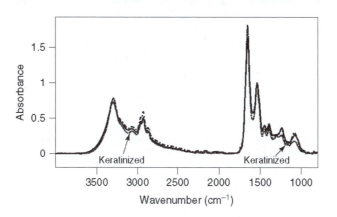

Figure 10.11 Overlaid, normalized SRS FT-IR absorbance spectra of the average spectrum recorded from within in each of the three regions shown in Figure 10.9, see text for details. The solid line (———) represents the average spectrum of the tumour region spectra; the dotted ($\cdots\cdots$) line is the average spectrum of the early keratinization region spectra; the average spectrum of the stroma region spectra is denoted by the dot–dash line (– – – – –).

characteristic of each region in Figure 10.11. The differences noted above in Figure 10.7 between the spectra from tumour and stroma regions is essentially repeated in the comparison in Figure 10.11, while the average spectrum characteristic of early keratinization is readily distinguishable in this comparison from the other two types.

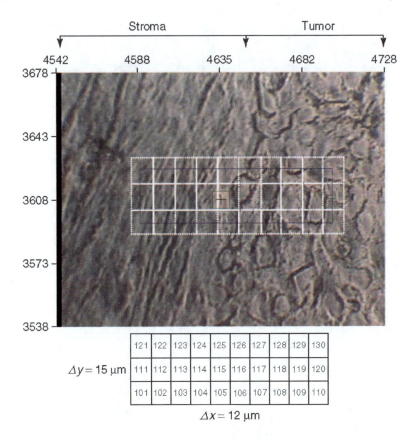

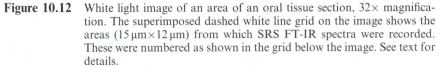

$\Delta x = 12\ \mu m$

Figure 10.12   White light image of an area of an oral tissue section, 32× magnification. The superimposed dashed white line grid on the image shows the
areas (15 μm × 12 μm) from which SRS FT-IR spectra were recorded.
These were numbered as shown in the grid below the image. See text for
details.

Figure 10.12 is a visual image of an oral tissue section from another patient.
The boundary between the two regions identified as stroma and tumour is
visibly distinct. An area of this section encompassing both regions and the
boundary was point-by-point mapped by synchrotron-sourced FT-IR microscopy, using a 12 μm × 15 μm aperture. In total 30 spectra were recorded, in
three rows of ten, as depicted by the grid superimposed on the visual image of
Figure 10.12. Below the tissue image shown in this figure is a duplicate grid in
which the numbers in the area elements correspond to the numbers allocated to
the recorded spectra. An HCA of these spectra did not resolve them into three
sets; spectra from the boundary region were mixed in with those from the
tumour classification. While both preliminary PCA and SIMCA three-dimensional (3D) plots of factors 1–3 could be oriented to indicate that the spectra
recorded from apertured areas overlaying the boundary (spectra numbers 105,

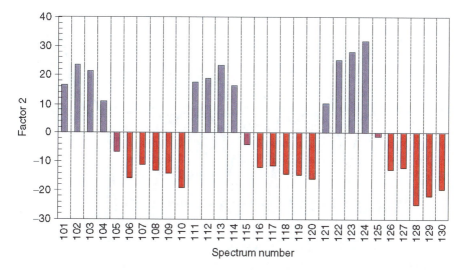

Figure 10.13 Bar-chart representation of the Class projection output for factor 2 of a SIMCA analysis of the pre-processed absorbance spectra numbered 101–130, see text for details.

155 and 125) were similar and had more bias towards tumour than other spectra within the tumour classification, such depictions were overly subjective. More objective pictures were obtained from the SIMCA outputs of 'Class Projection' and 'Class Distances'. The Class Projection provides a visual evaluation of the degree of class separation and is generated from a three-factor SIMCA of the spectra training set.

Figure 10.13 is a bar-chart representation of the "Class Projection" output for factor 2 from a SIMCA, in which the modulus of the class distinctions is 10 or greater for all spectra except numbers 105, 115 and 125. The values for these spectra are about 7, 4 and 1 respectively, suggesting much less distinction (particularly for spectrum number 125), as might be expected from spectra recorded along a boundary region. A class distance plot of for the two main categories is shown in Figure 10.14. The transition from stroma to tumour along a row of spectra is clearly evident in the regions of spectra numbers 105, 115 and 125; the other crossovers occur at the end of each row of spectra in the grid-map, *i.e.*, at spectra numbers 111 and 121.

Figure 10.15 is one plot projection of the three SIMCA factors, in which spectra 105, 115 and 125 lay nearer the tumour classification boundary spectra than others within the tumour cluster. As a consequence, we have categorized these three spectra separately from both tumour and stroma in our false-colour coding of the mapped area shown in Figure 10.16. This simple example of 'spectroscopic-staining' accords well with that from conventional histopathology H&E staining.

As a final example of some of our SRS FT-IR microscopy studies on oral tissue sections, Figure 10.17 shows a set of five white light images taken from an

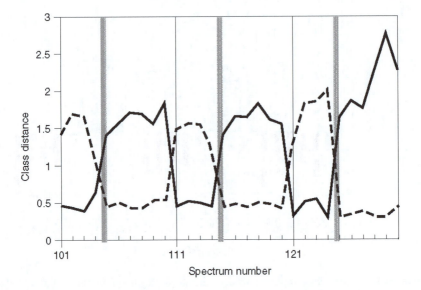

Figure 10.14   Class Distance plot output from a SIMCA analysis of the pre-processed absorbance spectra numbered 101–130, see text for details.

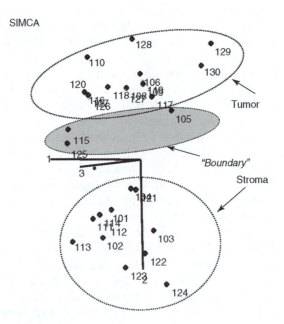

Figure 10.15   A SIMCA plot projection for factors 1, 2 and 3 from the analysis of the pre-processed absorbance spectra numbered 101–130, see text for details.

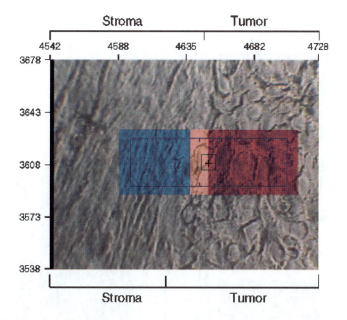

Figure 10.16 'Spectroscopic-staining' of the SRS FT-IR mapped area of the white light image shown in Figure 10.12. The tumour region is coloured in red; the stroma region is coloured blue; the boundary region is highlighted in purple. See text for details.

oral tissue section from a patient. Histopathological examination showed this to be a complex region containing stroma, tumour and necrotic tissue; the extent of each of these is indicated on Figure 10.17. A line profile of IR spectra, consisting of consecutive points, as depicted in Figure 10.17, was recorded across this region. The spectra recorded are numbered 201–256; there is some overlap between the consecutive images shown in Figure 10.17. Preliminary multivariate analyses clearly indicated that the spectrum number 246 was an outlier. Close inspection of the white light image shows that the position from which this spectrum was recorded lay close to a tear (parting) in the tissue section, so it is possible that this spectrum was affected by some stray-light intrusion. Spectrum number 255 was also considered an outlier, although the underlying reason for this was not so clear, except that it appears to lie on what is probably a ridge in the tissue section. Since borders between the stroma, tumour and necrotic regions were not distinct, four spectra encompassing each boundary region were excluded from further multivariate analyses. Thus in total 13 spectra were eliminated from the set of 58 spectra before further data processing. The spectra left out were numbered 211–214, 230–233, 244–247 and 225.

Figure 10.18 shows an HCA output from the remaining 45 spectra data set. The extreme spectra (numbers 252–258) from the necrotic region form a tight cluster, although those nearer the boundary (numbers 249–251) are clustered

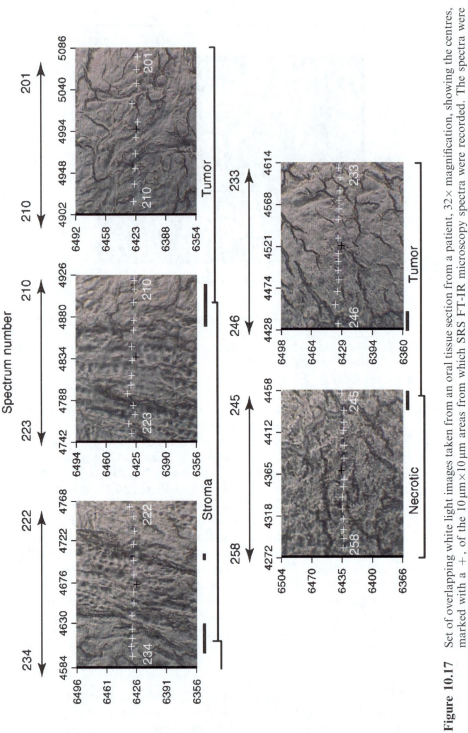

Figure 10.17 Set of overlapping white light images taken from an oral tissue section from a patient, 32× magnification, showing the centres, marked with a +, of the 10 μm × 10 μm areas from which SRS FT-IR microscopy spectra were recorded. The spectra were numbered 201–258. The histopathological diagnosis is indicated. Spectra excluded from the detailed multivariate analysis occur in the regions marked by the solid black rectangles beneath the images, see text for details.

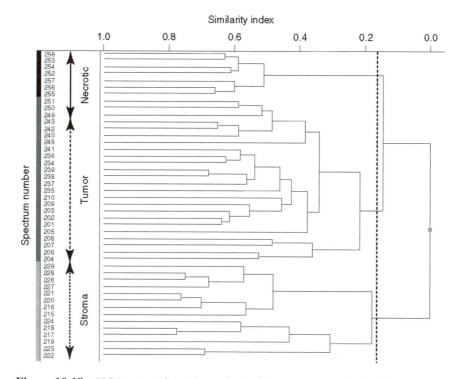

Figure 10.18 HCA output from the analysis of the pre-processed absorbance spectra numbered 201–210, 215–229 (excluding 225), 234–243 and 248–258, see text for details.

with the spectra associated with the tumour region. The spectra from the stroma region form a separate cluster, as depicted in Figure 10.18. Figure 10.19 is a PCA analysis of this 45 spectra data set, and yields a similar prediction to the HCA.

The examples described above serve to highlight the potential of mid-IR spectroscopy for the diagnosis of oral tissue sections, but emphasize its present reliance on the 'gold standard' of H&E histopathology staining. In addition, the difficulties that are associated with collecting spectra from complex samples with variable contour boundaries using a definite geometrically shaped aperture to delineate sample micro-regions for spectral analysis can be seen.

This work showed promise in the discrimination of cancer tissue.[1,2] On one occasion, the clinician's analysis of the section indicated non-cancer tissue, however the spectral profile was suggestive of cancer, a finding confirmed by subsequent inspection of the stained section. This work proved the feasibility of producing a robust data set for cancer/non-cancer evaluation at single cell level using the high lateral spatial resolution available through the use of synchrotron mid-IR radiation. Continuing improvements in bench-top spectrometers and sampling instrumentation now make translation of such initiatives to the clinic and clinical laboratory a feasible option.

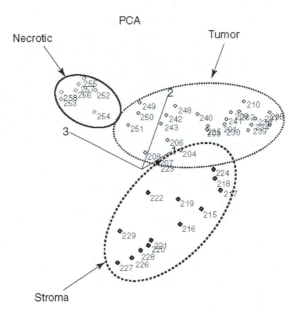

Figure 10.19 A PCA Scores plot output from the analysis of the pre-processed absorbance spectra numbered 201–210, 215–229 (excluding 225), 234–243 and 248–258, see text for details.

10.4 Mid-infrared Synchrotron Radiation FT-IR Studies of Cultured Cells

Mid-infrared spectroscopy has also been used to explore cell changes, as described by Tobin *et al.*,[1,2] to study cellular responses related to epidermal growth factor (EGF). Expression of the EGF receptor is known to be critically important in a number of epithelial cancers including those of the H&N and cervix.[4] These two pathologies share many similarities, and to explore the place of EGF further a carcinoma cell model (in this case derived from the cervix) was selected. These cultured carcinoma cell studies were undertaken in parallel with the oral tissue section studies discussed above with the goals of both gaining a better understanding of the inherent variability in the mid-IR spectra of such epithelial cells and further demonstrating the capability of synchrotron-sourced FT-IR microscopy to make informative measurements at the single cell level.[1] The cell culture samples were A431 cultured cervical cells, which were stimulated with the growth-stimulating hormone EGF [full details about the cells, their preparation, and sampling and storage can be found in reference (1)]. As stated above, the cells were examined using the transflection mode. Figure 10.20 shows an example of the mid-IR spectra of A431 cells cultured on a low-e glass slide after drying, in which can be seen cell nuclei, subcellular organelles, and fibres attached to the substrate. The white crosses overlaid on the figure indicate the positions (cells) from which spectra were recorded. The

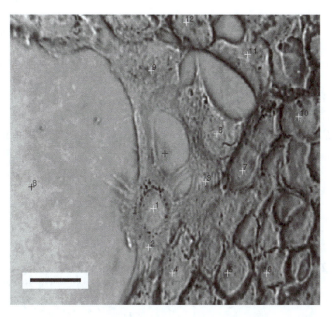

Figure 10.20 A431 cervical epithelial cells cultured on IR reflective glass slides. White crosses mark selected data collection points. Bar represents 30 μm. (Reproduced from reference [1] by permission of The Royal Society of Chemistry, © 2003.)

cell spectra, see example in Figure 10.21, were broadly similar in overall appearance to those recorded from the oral tissue sections.

In the cell study, prepared samples were exposed to EGF for periods of time between one minute and three hours by incubation in phosphate-buffered saline (PBS) solutions containing EGF at concentrations in the range 1×10^{-7} to 1×10^{-9} M. For each EGF concentration and each time interval, spectra were recorded from up to 30 individual cells from each sample. As a preliminary assessment of the spectral differences occurring as a consequence of the treatment, several spectral features were measured. These were: measurement of the amide I and amide II peak positions; the amide I and amide II absorption band areas (baseline corrected between $1800 \, cm^{-1}$ and $790 \, cm^{-1}$); the $\nu C{=}O$ ($1730 \, cm^{-1}$) peak area (baseline 1760–$1723 \, cm^{-1}$) ratioed (normalized) against the amide I band area; the $970 \, cm^{-1}$ band area (baseline 980–$940 \, cm^{-1}$) ratioed against the amide I area; the $1170/1155 \, cm^{-1}$ band area ratio (baseline 1182–$1142 \, cm^{-1}$). Although no changes in these spectral measurements individually correlated well with EGF concentration or time after stimulation with EGF, cross-correlation of some pairs did show EGF concentration and time dependences.[1] The most prominent of these was the one between the area of the $970 \, cm^{-1}$ absorption band area and the amide I peak maximum wavenumber position; see, for example, Figure 10.22. After an interval of one minute, similar correlations were observed for both cells treated with 1×10^{-8} M EGF

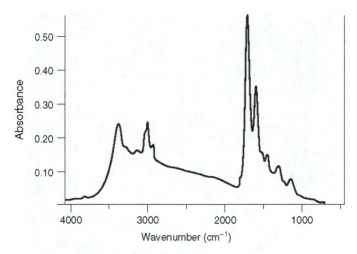

Figure 10.21   Infrared absorbance spectrum of single A431 cervical carcinoma cell. (Reproduced from reference [1] by permission of The Royal Society of Chemistry, © 2003.)

[Figure 10.22(a)] and untreated cells. After 10 minutes, half of the cells at this concentration showed both an increase in absorption area at $970\,cm^{-1}$ and a shift towards higher wavenumber of the amide I band maximum [Figure 10.22(b)]. A smaller fraction showed these effects at 45 minutes [Figure 10.22(c)]. This correlated change was still evident in some cells after three hours [Figure 10.22(d)]. At the lower concentration ($1\times10^{-9}\,M$ EGF), which is close to physiological levels of EGF, after only one minute [Figure 10.22(e)] the spectral correlation distribution was similar to that observed at 10 minutes for the $1\times10^{-8}\,M$ EGF solution [Figure 10.22(b)] and persisted to varying degrees after 10 minutes [Figure 10.22(f)] and 45, minutes [Figure 10.22(g)]. However, after three hours of exposure [Figure 10.22(h)] it had mostly returned to the unstimulated distribution. At $1\times10^{-7}\,M$ EGF concentration a slight change in the correlation distribution was observed after 45 minutes but the effect was weaker than than that shown for the two lower concentrations.

Tobin *et al.*[1] discussed some possible explanations for their empirically derived observed correlation. It is known that the amide I absorption band in such biological samples occurs primarily from the C=O stretching vibration of proteins within the sample. Changes in its peak maximum position are indicative of changes in the average secondary structure of proteins. The shift in this band maximum position in the mid-IR spectra recorded from stimulated cells from $1648\,cm^{-1}$ towards $1664\,cm^{-1}$ was considered characteristic of a shift in protein composition from mostly α-helical structures towards a higher level of turns and bends. This may be spectroscopic evidence of increased protein translation in response to the mitogenic signal of the EGF, resulting in elevated levels of incompletely folded protein within the cells.

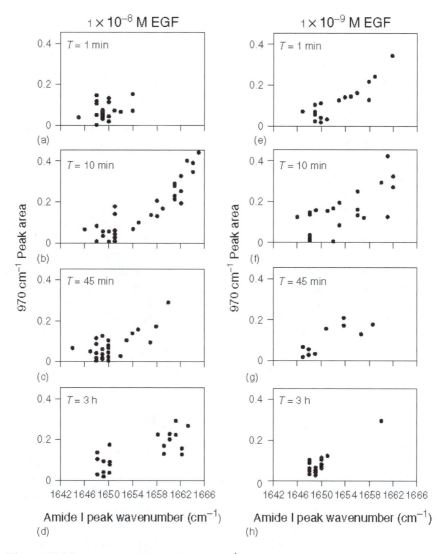

Figure 10.22 Correlation between 970 cm⁻¹ absorption peak area and Amide I peak wavenumber position for A431 cells at a time after addition of 1×10^{-8} M (left column) and 1×10^{-9} M EGF (right column). Each point represents one cell. (Adapted from reference 1 by permission of The Royal Society of Chemistry, © 2003.)

Such changes are investigated in molecular medicine as epigenetic phenomena, and their relationship to cancer behaviour is a matter of much interest. If such molecular changes can be explored by spectroscopy, this may well elucidate important changes in cell metabolism. Bruni *et al.* suggested changes perhaps related to high levels of DNA, collagen, carbonates and lipids in infiltrating cancers.[5] They emphasized the complexity of biological systems but pointed to the promise of perhaps predicting aspects of cancer behaviour based on single cell analysis. Holman et al.[6] provided evidence of a shift in amide I and amide II bands corresponding to changing protein morphologies, characteristic of dying cells, together with a significant increase in the intensity of an ester carbonyl C=O peak at 1743 cm^{-1}. Taken together, these papers indicate the promise of mid-IR spectroscopy in predicting biochemical events at the single-cell level.

One possible interpretation discussed for the simultaneous correlated increase in intensity of the weak absorption band at 970 cm^{-1} was that it also related to triggering of the protein phosphorylation cascade following EGF stimulation.[1] A similar increase in absorption was observed by Sanchez-Ruiz and Martinez-Carrion in the mid-IR spectra recorded from phosvitin and ovalbumin after the phosphorylation of the proteins at specific residues.[7] An alternative explanation, based on the work of Diem et al.[8] and Boydston-White et al.,[9] points to bulk changes in the DNA absorption properties being responsible for changes in relative intensity of the absorption band at 970 cm^{-1}. Whatever the reason(s), it is likely that high lateral spatial resolution spectroscopy at the sub-cellular level will be required to determine whether the observed spectral changes are the consequence of cytoplasmic events, such as protein phosphorylation and translation, and/or nuclear events, such as DNA replication.[1] The high SNR of mid-IR spectra attainable at diffraction-limited spatial resolution, ca. 4 μm at 1650 cm^{-1}, using synchrotron-sourced FT-IR microscopy will undoubtedly remain a key tool in these continuing types of study.

10.5   Raman Studies of H&N Samples

In view of its potential for 'in-vivo' probes, the feasibility of studying many cancer sites by Raman spectroscopy has been investigated. With H&N, Krishna et al. studied material from biopsies of five malignant and five 'normal' tissues.[10] They observed distinct differences through PCA, which was especially encouraging as the 'normal' tissue was taken from the same patients and, in the possible presence of field change, may not have been entirely so. In a study combining FT-IR and Raman microspectroscopy, this group has also recently described differences between cell lines that were and were not resistant to chemotherapeutic agents.[11] Prediction of response would be valuable in the clinical environment, ensuring targeting of cancers by an agent to which they are known to be responsive, whilst avoiding futile and expensive therapies which inevitably carry morbidity for the patient.

Much of the 'in-vivo' work in H&N as well as other sites has been led by Stone (a co-author of Chapter 4 in this book). His preliminary work, on laryngeal tissue, published in 2000,[12] used linear disciminant analysis to separate spectra with the most visible differences being seen at 850, 950, 1200 and 1350 cm^{-1}. He also concluded that the relative intensity of the nucleic acid peak increases with progression to malignancy. Not only do several publications indicate efficacy in the differentiation of non-malignant and malignant cell phenotypes but they also demonstrate an evolving consensus as to what they represent in terms of cell behaviour. It is our belief that attempts should be continued to relate spectral changes to known characterization and manipulation, and we are currently undertaking studies of thyroid and oral cell lines, using Raman spectroscopy and a neural network based mathematical model, with encouraging early results.

10.6   Conclusions

The primary aim of this chapter was to show the application of vibrational spectroscopy to H&N cancer and to indicate its promise for future clinical use. To ensure progression from the science laboratory to the clinic requires both a mutual understanding of spectroscopy and medicine and that the translational element be driven by clinical need. Work towards a clinical setting must take account of biological variation in both patients and cancers, and clear recommendations exist for the development of data sets for clinical trials taking into account patient variables. A good model to achieve this is the phased approach used in the pharmaceutical industries. Evaluation must commence with a Phase 1 trial, in which the safety of the protocols and the establishment of an appropriate candidate dosage schedule are established in a volunteer population. Any obvious confounding factors in a candidate data set are identified and removed in the Phase 1 trial. The study is continued through a Phase 2 trial, which is similar to Phase 1 but involving a small population of patients. Phase 2 trials usually require 30–40 patients. In the case of a proposed spectroscopic diagnostic tool or intervention, a trial of this size would allow any less than robust spectroscopic markers to be identified and removed and a final refined data set carried forward, and this methodology has been followed for colorectal adenocarcinoma with significant promise.[13] The Phase 2 trial also allows a sample calculation to be performed for a major (Phase 3) evaluation and pre-marketing trial, after which formal submission for regulatory approval is completed.

Vibrational spectroscopy carries much potential for cancer diagnosis and the development of predictive and prognostic models. Much of this consideration of H&N cancers can be generalized to other sites. The key message is that a basic understanding of the disease and its therapy facilitates scientific advances. Recent developments in technology make laboratory-based and even hand-held probes a feasible development within the next 5–10 years. If this promise is confirmed, vibrational spectroscopy will make a substantial contribution to clinical medicine. For this reason, careful collaboration between clinicians and spectroscopists forms the ideal base for robust research leading to results that are likely to form the basis for a dependable and reproducible dataset and to result in a direct clinical impact.

Acknowledgements

The spectroscopic studies that are reported in this chapter were undertaken primarily at the Synchrotron Radiation Source (SRS), Daresbury Laboratory,

UK, under a UK Engineering and Physical Sciences Research Council (EPSRC) funded project (GR/M62778) as part of its Physics for Healthcare programme. The collaborative project brought together researchers from the University of Nottingham, Leeds University, Derby Royal Infirmary, and Daresbury Laboratory. The authors are indebted to many members of these institutions that formed part of the research team of spectroscopists, clinicians and histopathogists, in particular: Professor Michael A. Chesters and Dr Frank J.M. Rutten from the School of Chemistry, Nottingham University; Mr Ian M. Symonds from the Department of Obstetrics and Gynaecology, University of Nottingham and Derby City General Hospital; Dr Richard Allibone from Queen's Medical Centre, University of Nottingham; and Dr Andy Hitchcock from the Derbyshire Royal Infirmary, Derby.

Our recent work, in Leeds, has been funded by Cancer Research UK, who fund the research undertaken by Andrew Harris through a Clinical Research Training Fellowship.

References

1. M. J. Tobin, M. A. Chesters, J. M. Chalmers, F. J. M. Rutten, S. E. Fisher, I. M. Symonds, A. Hitchcock, R. Allibone and S. Dias-Gunasekara, *Faraday Discuss.*, 2004, **126**, 27–39.
2. M. J. Tobin, F. Rutten, M. Chesters, J. Chalmers, I. Symonds, S. Fisher, R. Allibone and A. Hitchcock, *Eur. Clin. Lab.*, 2002, **21**, 20–22.
3. Y. Fukuyama, S. Yoshida, S. Yanagisawa and M. Shimizu, *Biospectroscopy*, 1999, **5**, 117–126.
4. K. K. Ang, B. A. Berkey, X. Tu, H.-Z. Zhang, R. Katz, E. H. Hammond, K. K. Fu and L. Milas, *Cancer Res.*, 2002, **62**, 7350–7356.
5. P. Bruni, C. Conti, E. Giorgini, M. Pisani, C. Rubini and G. Tosi, *Faraday Discuss.*, 2004, **126**, 19–26.
6. H. Y. Holman, M. C. Martin, E. A. Blakely, K. Bjornstad and W. R. McKinney, *Biopolymers (Biospectroscopy)*, 2000, **57**, 329–335.
7. J. M. Sanchez-Ruiz and M. Martinez-Carrion, *Biochemistry*, 1988, **27**, 3338.
8. M. Diem, S. Boydston-White and L. Chiriboga, *Appl. Spectrosc.*, 1999, **53**, 148A–161A.
9. S. Boydston-White, T. Gopen, S. Houser, J. Bargonetti and M. Diem, *Biospectroscopy*, 1999, **5**, 219–227.
10. M. C. Krishna, G. D. Sockalingum, J. Kurian, L. Rao, L. Venteo, M. Pluot, M. Manfait and V. B. Kartha, *Appl. Spectrosc.*, 2004, **58**, 1128–1135.
11. M. C. Krishna, G. Kegelaur, I. Adt, S. Rubin, V. B. Karftha, M. Manfait and G. D. Sockalingum, *Biopolymers*, 2006, **82**, 462–470.
12. N. Stone, P. Stavroulaki, C. Kendall, M. Birchall and H. Barr, *Laryngoscope*, 2000, **110**, 1756–1763.
13. P. Lasch, M. Diem, H. Wolfgang and D. Naumann, *J. Chemom.*, 2006, **20**, 209–220.

CHAPTER 11

Single Cell Analysis of TSE-infected Neurons

ARIANE KRETLOW,[a, b] JANINA KNEIPP,[c] PETER
LASCH,[a] MICHAEL BEEKES,[d] LISA MILLER[b] AND
DIETER NAUMANN[a]

[a] P25, Robert Koch-Institut, Nordufer, 20, D-13353, Berlin, Germany;
[b] National Synchrotron Light Source, Brookhaven National Laboratory,
Upton, NY, USA; [c] Federal Institute for Materials Research and Testing, D-
12489, Berlin, Germany; [d] P24, Robert Koch-Institut, Nordufer, 20, D-
13353, Berlin, Germany

11.1 Introduction

Scrapie, a rare neurodegenerative disorder of the central nervous system, is characterized by the accumulation of a β-sheet rich protein, called PrP^{Sc} or scrapie-associated prion protein, primarily in the central and peripheral nervous system. First described in 1732 in sheep, the disease belongs to the family of transmissible spongiform encephalopathies (TSEs) or prion diseases. TSEs in humans, such as Creutzfeldt–Jakob Disease (CJD), can occur sporadically but may also have genetic and infectious origins, which is a unique feature of this group of diseases. According to the prion hypothesis postulated in 1982 by the neurologist and subsequent Nobel Prize winner Stanley B. Prusiner, the causative agent of TSEs is believed to be a proteinaceous infectious particle ("Prion") that lacks agent-specific nucleic acid and consists mainly – if not entirely – of mis-folded and pathologically aggregated prion protein.[1] PrP^{C}, the cellular prion protein, mainly expressed by neuronal cells, is high in α-helical content (42%) and consists of only 3% β-sheet, whereas the misfolded or disease-associated form (PrP^{Sc}) shows a high amount of β-sheet structure (43%) and has less α-helix

RSC Analytical Spectroscopy Monographs No. 11
Biomedical Applications of Synchrotron Infrared Microspectroscopy
Edited by David Moss

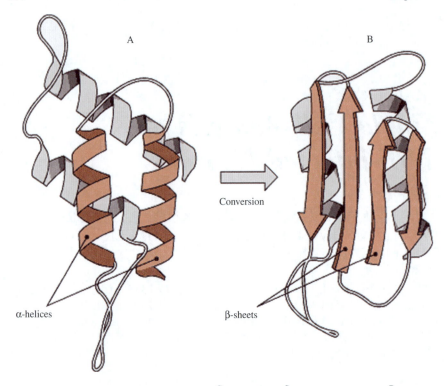

Figure 11.1   Secondary structure of PrP^C (A) and PrP^{Sc} (B). Helices in PrP^C shown in
brown are transformed into β-sheet structure by conversion to PrP^{Sc}.
(From Reference [69]. The structure shown in (B) is purely hypothetical.)

(30%).[2] During the course of the disease, PrP^C gets refolded into the pathological
form, PrP^{Sc} (see Figure 11.1), which then accumulates and causes neuron death.

The nervous system of hamsters infected orally with scrapie has been studied
extensively for the temporal-spatial course of PrP^{Sc} deposition in the brain[3] (for
review see: [4]). After centripetally reaching the spinal cord, subsequent cen-
trifugal spread of PrP^{Sc} deposition proceeds to the corresponding afferent
dorsal root ganglia (DRG), where the protein can first be detected 76 days post-
infection in half of the examined hamsters (Figure 11.2).[5–7] The DRG are
nodules on a dorsal spinal root that contain the cell bodies of afferent nerve
fibers, some of the largest cells in the mammalian body.

11.2   IR-Spectroscopy and the Composition of Complex Biological Material

The following section will give a short introduction to Fourier transform
infrared microspectroscopy (FTIRM) with the emphasis on neuronal tissue.
Please refer to Chapter 2 for more detailed information.

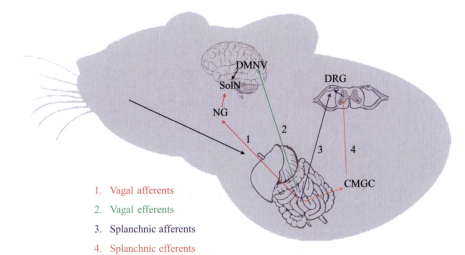

1. Vagal afferents
2. Vagal efferents
3. Splanchnic afferents
4. Splanchnic efferents

Figure 11.2 Initial spread of the 263K agent after oral uptake. Circuitry has been simplified to show major routes only. The infectious agent reaches the CNS by spreading along either efferent (motor) or afferent (sensory) fibers of vagus or splanchnic nerves. Efferent fibers of the vagus nerve have their nerve cell bodies in the dorsal motor nucleus of the vagus nerve (DMNV) and synapse with neurons of the enteric nervous system in ganglia of the submucosal and myenteric plexuses in the wall of the alimentary canal. The nerve cells of vagus nerve afferent fibers are located in the nodosal ganglion (NG) and directly innervate the alimentary canal. These fibers run to the solitary tract nucleus (SolN), where they connect to interneurons projecting to the DMNV. The cell bodies of splanchnic nerve efferent fibers are located in the IML and interrelate to neurons of the celiac mesenteric ganglion complex (CMGC) that, in turn, innervate the gastrointestinal tract. Afferent fibers of the splanchnic nerve originate in the DRG, run through the CMGC, and directly innervate target organs such as the alimentary canal. Enteric and abdominal ganglia (CMCG) have an early involvement in pathogenesis. (Adapted from Reference [7].)

Spectroscopy, in general, is an analytical technique arising from the interaction of a species with electromagnetic radiation, which can either get absorbed (FTIR), scattered (FT-Raman) or emitted (XRF) by the molecules it contains. The electromagnetic spectrum ranges from very low energy radio waves, with large wavelengths (λ) and low frequencies (v), to highly energetic X-rays, which have small wavelengths and high frequencies. The infrared (IR) spectrum is located in the middle, between visible light (400–700 nm wavelength) and microwaves (10^5–10^7 nm wavelength). It is divided into three spectral regions: near: $\lambda = 800$ nm–2.5 µm (wave numbers $= 12500$–4000 cm^{-1}), mid: $\lambda = 2.5$–50 µm (wave numbers $= 4000$–200 cm^{-1}) and far IR: $\lambda = 50$ µm–1 mm (wave numbers $= 200$–10 cm^{-1}). The basic idea behind IR spectroscopy is the fact that the IR region of light covers the same energy range as is necessary to excite the different vibrational modes of molecules (see Figure 11.3

Bond Stretching

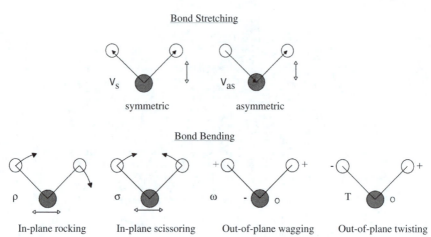

Bond Bending

Figure 11.3   Illustration of some different vibrational modes.

for a selection of vibrational modes). Depending on the atoms in the molecule, their binding forces, masses and orientation, a functional group of a molecule will absorb light at a specific wavelength. This will give rise to a unique spectrum for each molecule. Thus, an absorption spectrum illustrates at what wavelengths the investigated sample absorbed how much energy. For visualization, a typical FTIRM spectrum of a biological sample, *i.e.* hamster peripheral nervous tissue, is given in Figure 11.4. Upon first inspection, the spectrum can be visually divided into two regions: there are usually relatively few peaks on the left half above $2000\,cm^{-1}$, while many peaks of varying intensities are found on the right side, below $2000\,cm^{-1}$. The general rule of thumb is that the stronger the bond, the higher the frequency of the absorbed light. Therefore, some basic rules apply: 1. It is easier to bend a bond than to stretch or compress it, thus bending frequencies are lower than corresponding stretching frequencies. 2. Bonds to hydrogen have higher stretching frequencies than those to heavier atoms. 3. Single bonds have lower frequencies than double bonds, which in turn have lower frequencies than triple bonds (except for bonds to hydrogen). For a more detailed list of functional group assignments, see Table 11.1. As opposed to an IR spectrum of a single substance, FTIRM of cells and tissues provides averaging information on all components in the investigated sample area.

In Figure 11.4, the tallest peak, at around $1650\,cm^{-1}$, represents the so-called amide I band. This band gives information about the secondary structure of a protein in the investigated sample. Assignment of amide I band components to *e.g.* β-pleated sheets or α-helical structures is given in Table 11.2. Its region between 1600 and $1700\,cm^{-1}$ mainly represents the carbonyl stretching vibration of the peptide backbone (C=O). The amide I band is a sensitive marker of protein secondary structures because the vibrational frequency of each C=O bond depends on hydrogen bonding and the interaction between the amide

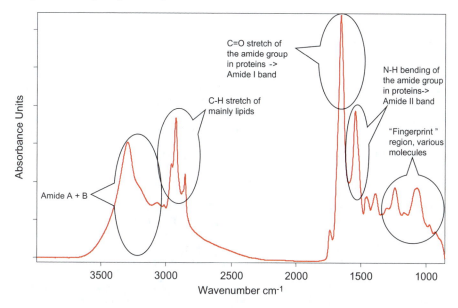

Figure 11.4 Infrared spectrum of hamster peripheral nervous tissue in the spectral range from 4000–900 cm^{-1} measured at a synchrotron source by FTIR microspectroscopy through a $10\times10\,\mu m^2$ aperture.

units, both of which are influenced by the secondary structure. In addition to the usually most prominent amide I band, there are eight other amide bands, called amide A, amide B and amides II–VII.

So, we aim to study the nervous system, *e.g.* neurons of the DRG from scrapie infected animals at different time points during the disease and to compare the obtained results with age-matched control animals.

But why should we choose FTIRM? Well, there are several reasons! First, immunostained sections with the prion specific antibody 3F4 have shown that PrP^{Sc} does not accumulate in all neurons of the DRG, where only certain cells stain strongly. In prion research, conventional techniques such as Western blot analysis are sensitive for examining the total amount of PrP^{Sc} in homogenized tissue. *In situ*, PET blotting (paraffin embedded tissue blotting)[8] labels the misfolded prion protein directly within the tissue, providing spatially resolved information on the location of PrP^{Sc}. However, both techniques are limited to the detection of PrP^{Sc}, and do not provide information on other compositional changes that may occur during development of the disease, especially at early time points in pathogenesis before PrP^{Sc} shows detectable accumulation.

In contrast, FTIRM is an *in situ* technique that is not restricted to the detection of PrP^{Sc}. In fact, we have already learned that an IR spectrum of a biological sample is composed of characteristic absorption bands that originate from *all* tissue components, *e.g.* proteins, lipids, nucleic acids, and carbohydrates. Since the combination of all molecular parameters (structure,

Table 11.1   Infrared band assignment of biological molecules (From references [65], [17], [66], [67], and [68].)

Absorption band (cm⁻¹)	Chemical compound	Assignment
~3300	proteins	N–H stretching vibration (amide A)
~3100	proteins	N–H stretching vibration (amide B)
~3030	lipids	=C–H stretching in alkenes
2955 and 2872	lipids, proteins, carbohydrates, nucleic acids	asymmetric and symmetric C–H stretching vibration from CH_3 groups
2923 and 2853	lipids, proteins, carbohydrates, nucleic acids	asymmetric and symmetric C–H stretching vibration from CH_2 groups
1741	phospholipids, thymine, uracil	C=O stretching vibration of esters
~1550	proteins	N–H bending, C–H stretching, C–O bending, C–C and N–C stretching vibrations (amide II)
~1515	proteins	tyrosine
~1467	lipids, proteins, nucleic acids	symmetric C–H scissoring of –CH_2
~1454	lipids, proteins	asymmetric C–H scissoring of –CH_3
~1400	fatty acids, amino acid side chains	C=O stretching vibrations of –COO^-
~1390	proteins, lipids	symmetric C–H deformation of –CH_3
~1310–1200	proteins	C–H/N–H deformation (amide III)
~1250–1220 (1235)	nucleic acids, phospholipids	asymmetric $P = O$ stretching vibrations in PO_2^-
~1200–900	polysaccharides	C–O–P and C–O–C stretching vibrations, ring vibrations
~1173 and 1154	carbohydrates	symmetric C–O stretching coupled to C–O–H bending
~1085	nucleic acids (DNA), phospholipids	symmetric P=O stretching in PO_2^-, CO–O–C symmetric stretching vibrations
~1070	lipids	symmetric CO–O–C stretching vibration
~1041 and 1055	carbohydrates	symmetric C–O–C stretching vibration
~1023	carbohydrates	symmetric C–O stretching vibration
900–800	nucleotides	C=C, C=N, C–H in ring structure
700	proteins, lipids, nucleic acids	–CH_2 rocking

composition, and/or interactions) in a specific tissue, cell type, or subcellular component is unique, FTIRM is an *in situ* technique suitable for imaging the molecular composition of biological materials.[9] Additionally, no labels, stains or dyes are required for FTIR microspectroscopy and it is a non-destructive, sensitive and fast analytical tool. Thus, coupled with PrP[Sc]-specific imaging techniques, FTIRM is the ideal candidate to be used for determining compositional changes in prion-infected tissue associated with scrapie pathogenesis.

11.3 Why apply Synchrotron FTIR Microspectroscopy (SFTIRM)?

Since the accumulation of PrP^{Sc} in hamster infected DRG is microdisperse,[6] we would need a high spatial resolution down to the diffraction limit in order to get a chance to detect changes in secondary structure based on such tiny deposits. This can only be achieved with the help of synchrotron derived radiation (please refer to Chapter 3 for details about synchrotron radiation). Numerous studies have demonstrated that a synchrotron IR source through a 10 μm pinhole is ~ 100–1000 times brighter than a conventional Globar source. This is mainly because the effective source size is small and the light is emitted into a narrow range of angles. Consequently, signal-to-noise (S/N) ratios with small beam sizes are much higher using IR synchrotron radiation. Thus, only the use of a synchrotron IR beam enables the user to achieve spatial resolutions that are diffraction limited by the optics of the recording system. However, despite its manifold feasibilities, synchrotron FTIRM is not completely flawless. Its major drawback is its expenditure of time. Recently, IR microscopes utilizing focal plane array (FPA) detectors have become popular due to the fact that they provide large data sets in a fraction of both time and costs. However, comparison of spectra from the same sample, obtained with both techniques,[10,11] has shown that S/N levels and spatial resolution are still superior when employing a synchrotron.[10] In the future, coupling FPA based microspectrometers to a synchrotron source might open up a new dimension of SFTIRM (see Chapter 7).

A recent pilot study by Kneipp *et al.* has shown that the microdisperse accumulation of PrP^{Sc} in 263K infected Syrian hamster is only detectable when using small apertures (this phenomenon was termed the "optical dilution effect").[12] In this study, spectra from DRG of terminally diseased 263K hamsters were acquired using such a device, which provides a spatial resolution approaching the diffraction limit of mid-infrared light.[12] Results showed that the peak at around 1657 cm^{-1}, associated with α-helical protein structures (see Table 11.2), decreased and shifted to a lower frequency in spectra from infected animals, and the peak at ~ 1637 cm^{-1}, indicative of β-sheet structure, increased. In some spectra from infected hamsters, an additional peak at

Table 11.2 Assignment of IR absorption bands (cm^{-1}) to different secondary structures of proteins determined by measurements in H$_2$O. (From References [15], [16], [17].)

Amide I secondary structure assignments	
~ 1615–1625	Aggregated strands (intermolecular β-sheets)
~ 1625–1640	β-Sheet
~ 1650–1660	Disordered
~ 1650–1665	α-Helix
various bands between 1650 and 1690	Turns and loops
~ 1680–1695	Anti-parallel β-sheet

around $1631\,cm^{-1}$ appeared. These data suggested a higher β-sheet and a lower α-helical content of the proteins in the investigated areas of the tissue.

For a more detailed insight into disease progression the relative contributions of an increased β-sheet or a decreased α-helix (or both) to the observed phenomenon of scrapie-induced variations in β-sheet to α-helix ratios needed to be elucidated as well. Furthermore, investigations on animals sacrificed at different time points after peroral challenge had to be performed for additional information about the spatial and temporal course of protein composition and prion protein accumulation in scrapie infected nervous tissue during disease progression.

11.4   Materials and Methods

11.4.1   The Study Design

In this study, we wished to image protein content, structure, and distribution during scrapie pathogenesis using synchrotron FTIRM. Syrian hamsters (*Mesocricetus auratus*) perorally infected with the scrapie strain 263K were studied at four different time points: 100 days post infection (dpi), 130 dpi, at first clinical signs (fcs, ∼ 145 dpi) and at the terminal stage (∼ 170 dpi) of the disease. The total protein, the α-helical protein, and β-sheet protein contents and distributions were determined as a function of disease progression and correlated with PrP^{Sc} immunostaining. The aim of this study was to examine changes in protein composition and regulation throughout scrapie pathogenesis, and to correlate these findings with the structural changes associated with the $PrP^{C} \rightarrow PrP^{Sc}$ conversion. Identification of protein changes involved in early scrapie pathogenesis is important for understanding the disease process and for identifying new approaches for early disease detection and treatment.

To summarize, the three main scientific questions we wished to answer were:

1. Are changes in the secondary structure of the proteins detectable by SFTIRM?
2. If yes, are those changes due to the conversion of PrP^{C} to PrP^{Sc}?
3. Where, when and how do they manifest?

11.4.2   Animal Experiments and Sample Preparation

All animal experiments were carried out in accordance with European, German, and USA legal and ethical regulations. Twenty outbred Syrian hamsters at an age of approximately 8 weeks were treated with $1–3 \times 10^{7}$ 50% intracerebral lethal doses of scrapie strain 263K as described elsewhere.[13] Twelve mock-infected hamsters of the same age were similarly treated with normal brain homogenate and served as controls. Five infected and three control hamsters each were sacrificed by euthanasia with CO_2 at four time points: 100

days post infection (dpi), 130 dpi, ∼145 dpi, first clinical signs (fcs), and at the terminal stage of the disease (∼170 dpi). While the first two time points are fixed dates, fcs and terminal stage depend on the progression of the disease in each individual animal. The fcs point is defined as a stage where the animals start to show clinical signs that are specific for infection with 263K scrapie in hamsters. These are most commonly a hypersensitivity to touch and noise, where the animals often twitch, or have difficulties in maintaining balance and rising from a supine position. At the terminal stage, animals show head bobbing, ataxia of gait, and generalized tremor.

After sacrificing the animals, the DRG attached to the thoracic spinal chord were removed and stored at −70 °C. After embedding the samples in Jung tissue freezing medium (Leica Instruments, Germany), 10 μm thick cryo-sections were cut at a temperature of −20 °C and mounted on IR transparent slides (CaF$_2$, 1 mm in thickness; Korth Kristalle GmbH, Altenholz, Germany). Adjacent sections were cut at the same thickness and mounted on standard glass microscope slides for immunostaining. Given that the samples were slightly hygroscopic, the slides were kept in a dry and dark environment until the FTIR microspectroscopic measurements were carried out. One animal was sacrificed before reaching the stage of fcs because of an interfering illness other than scrapie, resulting in a total of 19 infected and 12 mock-infected control animals included in the study.

To gain an idea of the disease progression and where to expect elevated amounts of β-sheet due to the deposition of misfolded prion protein, immunohistochemistry was performed as described elsewhere[14] (see Figure 11.5).

11.4.3 Data Acquisition Techniques

The SFTIRM experiments were conducted at beamline U10B at the National Synchrotron Light Source, Brookhaven National Laboratory (Upton, NY, USA). A Thermo Nicolet Magna 860 FTIR spectrometer, coupled to a Continuum FTIR microscope (ThermoNicolet, Madison, WI, USA), and synchrotron light was used as the IR source. The microscope was equipped with matching 32× Schwarzschild objectives, a motorized x–y mapping stage, an adjustable rectangular aperture, and a mercury cadmium telluride (MCT-A) detector. The aperture of the microscope was set to 10×10 μm. A pre-defined area encompassing approximately 15–20 cells was raster-scanned with a step size of 4 μm using Omnic software 7.3 (ThermoNicolet). At each point, an absorbance spectrum was recorded in transmission mode. Each spectrum was collected in the mid-IR spectral range (800–4000 cm^{-1}) with a spectral resolution of 8 cm^{-1} and 128 scans co-added. Happ–Genzel apodization and a zero-filling of level 2 were applied, resulting in approximately 1 data point per wavenumber. A background spectrum on clean substrate was also recorded by co-adding 512 scans and automatically subtracted from each sample spectrum. For each animal, one or two IR maps were collected, yielding an average of 10500 spectra from 138 cells (∼29 cells per animal) per time point from the

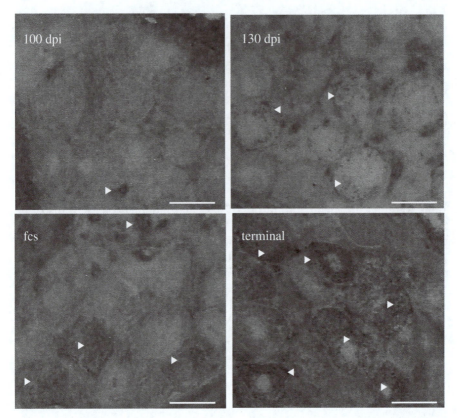

Figure 11.5 Photomicrographs of antibody 3F4 stained adjacent sections. PrP^{Sc} can be detected as granular deposits of immunoreactive material in a number (arrows) but not all of the neuronal and satellite cells at later stages of incubation. Scale bar: 30 µm.

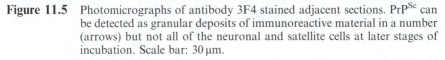

infected animals and 8887 spectra from 110 cells (~ 37 cells per animal) per time point from the controls.

11.4.4 Data Evaluation Techniques

The SFTIRM data were analyzed using Thermo Nicolet's software Omnic 7.3 (Waltham, MA, USA). To examine the protein structure in the tissue, the absorbance spectrum at each pixel was integrated from 1624–1628 cm^{-1} for β-sheet and 1654–1658 cm^{-1} for α-helix determination, respectively (Figure 11.6). The spectral ranges were chosen after curve fitting the amide I band, which was performed using Grams/32 software version 5.1 (Galactic Industries Corporation, USA). To determine the number and position of peaks in the amide I region for the curve-fitting process, a second derivative spectrum was calculated using Thermo Nicolet's software Omnic 7.3, applying the

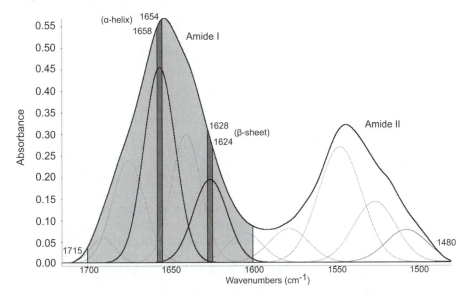

Figure 11.6   Curve-fitting of a spectrum derived from a terminally infected animal. Secondary structure content was calculated by integrating the area between 1624 and 1628 cm^{-1} (β-sheet content) and 1654 and 1658 cm^{-1} (α-helical content), divided by the area under the amide I band (1600–1700 cm^{-1}), using a baseline from 1480–1715 cm^{-1}.

Savitzky–Golay algorithm as the derivative operation and nine points of smoothing in the amide I region (1700–1600 cm^{-1}). The regions 1624–1628 cm^{-1} and 1654–1658 cm^{-1} were chosen because the respective parts of the spectrum had the least overlap with other protein secondary structural components and they had been assigned to β-sheet and α-helical structures earlier as shown elsewhere.[15–18] As the intensity of a peak is directly proportional to the amount of the sample substance tested, a thicker sample would give higher peak intensities and vice versa. Therefore, we needed to normalize the spectra for potential differences in tissue thickness among the samples. This could be achieved by dividing the integral intensity of the α-helix or β-sheet bands by the total area under the amide I band (1600–1700 cm^{-1}) (see Figure 11.6). Chemical images of α-helix or β-sheet content were generated using Transform (Fortner Software, USA). To estimate the relative protein content at each pixel, the protein content (integrated area: 1600–1700 cm^{-1}; baseline: 1480–1715 cm^{-1})[18–20] was divided by the integrated area between 2700 and 3700 cm^{-1}, applying a baseline from 2700 to 3700 cm^{-1}. Since the latter region includes the vibrational modes of C–H, N–H and O–H bonds of all major biomolecules, we could use it in a first approximation as an internal standard to estimate total protein content of the total biomass.

The average values plus standard deviations for α-helical, β-sheet, and total protein content were obtained for each animal at each time point. An unpaired two-sided *t*-test (Student's *t*-test) was performed on all data to test for significant differences ($p < 0.05$) between scrapie and control at each time point, and differences between time points. In the analysis, spectra measured from areas outside the ganglion or from tissue artifacts were excluded from the calculation.

11.5   Results

As already mentioned, the amide I band of pure protein can be deconvolved into six to nine components representative of the different secondary structure motifs in a protein.[21] The calculation of β-sheet and α-helical protein content was performed based on an initial curve-fitting of the amide I band, which is shown in Figure 11.6. To examine changes in the secondary structure of proteins during scrapie pathogenesis, the mean values ± standard deviations for the β-sheet and α-helical content were calculated and are plotted in Figure 11.7. Figure 11.7A depicts the calculated total β-sheet content for infected (red bars) and control (blue bars) hamster DRG at the four investigated time points. As a function of the progress of infection, the results showed that the β-sheet content remained relatively constant in control hamsters, but substantially increased with time in all scrapie-infected animals. The most dramatic difference between the control and infected animals was observed at 100 dpi, where the β-sheet content in scrapie-infected ganglia was significantly lower than in the control ganglia ($p = 0.013$). Then, over the course of the disease, we found a significant increase in β-sheet protein content in infected animals ($p = 0.003$).

In contrast, the α-helical protein content did not change over time, for neither infected nor control ganglia (Figure 11.7B). Significant differences were not observed at any time point.

To determine whether the observed decrease in β-sheet at 100 dpi was due to an increased expression of proteins low in β-sheet or the opposite, the total protein content was determined and is shown in Figure 11.7C. At 100 dpi, infected animals (red bars) showed a significant increase in relative protein expression ($p = 0.020$). While the level in the controls (blue bars) slightly increased over time, that in the scrapie-infected animals gradually declined during pathogenesis and reached a significantly lower level at the terminal stage of the disease ($p = 0.013$).

In addition to measuring the β-sheet content, the spatial distribution of α-sheet proteins within the tissue was examined. Photomicrographs and corresponding SFTIRM images of the β-sheet distribution for representative control (left) and infected (right) ganglia at different time points are shown in Figure 8A. The color scale in the SFTIRM images represents the relative β-sheet content, *i.e.* a value calculated from the ratio of the absorbance area between $1624 \, \text{cm}^{-1}$ and $1628 \, \text{cm}^{-1}$ divided by the total area of the amide I band, ranging from purple (≤ 0.03; very little β-sheet) to red (≥ 0.04; elevated β-sheet).

In all images, the β-sheet content was slightly higher near the cell membranes and extracellular matrix. For the control animals (Figure 11.8A, left columns), the SFTIRM images appeared similar at each time point, indicating no

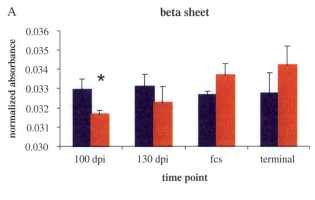

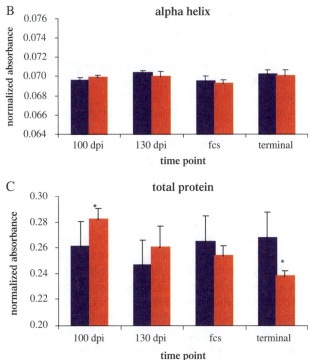

Figure 11.7 Mean values + standard deviation for the (A) β-sheet, (B) α-helical and (C) total protein content for infected (red lines) and control (blue lines) hamster dorsal root ganglia at the four investigated time points. Asterisks mark significant differences ($p < 0.05$) between scrapie and control.

significant change in β-sheet distribution or content over time. For the scrapie-infected animals, it can be seen that with progression of the disease, the β-sheet content in the images shifts to higher values, indicated by more green, yellow, and red areas. At pre-clinical time points, elevated β-sheet was detected only for a few cells and exclusively near the cell membranes. Otherwise, these cells generally exhibited proteins lower in β-sheet content than the controls, as can be seen by the dark blue colored areas. In addition, the cytoplasm of some neurons contained proteins extremely low in β-sheet which seemed to be accumulated in vesicle-like circular structures (arrowheads). To improve visibility, the area shown in the inset is magnified and contrast was improved by exchanging purple with yellow pixels. Some possible explanations for this phenomenon will be presented in the next section. During disease progression, the number of affected cells increased such that, by the terminal stage, the overall β-sheet content increased and was observed throughout a large number of cells, including but not limited to the cell membrane.

Spectra taken from areas representing low ($I_{\beta-sheet} = 0.033$), medium ($I_{\beta-sheet} = 0.038$) and high β-sheet content ($I_{\beta-sheet} = 0.045$) are indicated by the numbers 1–3 in the chemical images (Figure 11.8A) and are shown in Figure 11.8B. The original spectra and second derivatives are displayed. The peak associated with α-helical structures gradually shifted from 1655 to $1652\,cm^{-1}$, while the peaks at 1683 and $1637\,cm^{-1}$ (both attributed to β-sheet) increased and an additional peak appeared around $1624\,cm^{-1}$, suggesting the presence of aggregated protein strands.[12,22–25]

Given that the SFTIRM data showed an increase in β-sheet content as the incubation of infection progressed, which could have resulted from the misfolded prion protein, immunostaining with the 3F4 antibody for the prion protein was performed on adjacent tissue sections (see Figure 11.5). Owing to the fact that the antibody 3F4 stains both the cellular and the misfolded form of the prion protein,[26] a more or less homogeneous brownish staining can be seen in all animals, where the intensity or distribution does not significantly change between time points or infection states. In contrast, it has been previously established by comparative immunohistochemical and PET blot staining that PrP^{Sc} accumulates in neuronal tissue of scrapie-infected hamsters as microdisperse aggregates which are present as characteristic granular deposits of the immunoreactive material.[6] Figure 11.5 shows a clear progression of PrP^{Sc} deposition during pathogenesis by the increased appearance of brown, granular immunostaining, indicating the presence of pathologically aggregated PrP. At 100 dpi, very few neurons and satellite cells showed such immunostaining for PrP^{Sc}, whereas the accumulation of pathological prion protein became more prominent at more advanced stages of incubation as indicated by the increasing number of characteristic granular deposits at 130 dpi, fcs, and at the terminal stage of disease.

To directly correlate elevated β-sheet content with the distribution of aggregated PrP^{Sc}, the SFTIRM samples were subsequently stained with the monoclonal antibody 3F4 (not shown). The results showed that regions which stained positive with immunostaining corresponded well with regions of elevated β-sheet content, indicating the ability to detect aggregated PrP^{Sc} by SFTIRM *in situ*. However, not all regions with elevated β-sheet content stained positive for the misfolded prion protein, suggesting that other proteins might be involved in the disease process as well.

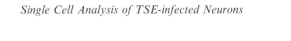

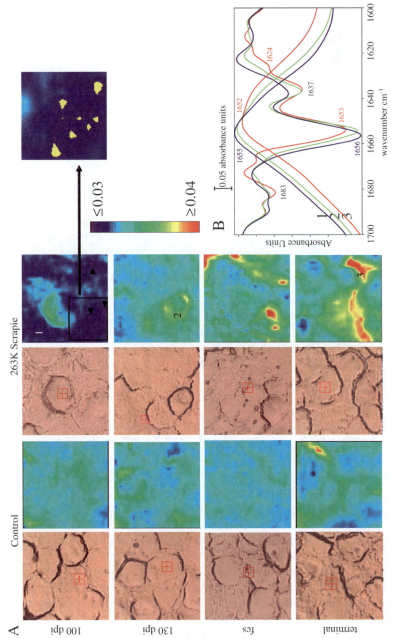

Figure 11.8 (A) Photomicrographs of unstained cryo-sections (first and third columns) and corresponding FTIRM images (second and fourth columns) of the β-sheet distribution for control (left) and infected (right) ganglia at different time points. Areas exhibiting extremely low β-sheet at 100 dpi are indicated by arrowheads. The area shown in the inset is displayed at higher magnification to the right; for better contrast and visibility, purple pixels have been replaced by yellow. (B) Original and second-derivative spectra of areas indicated by numbers in the chemical maps. Red square in photomicrographs: 10×10 μm.

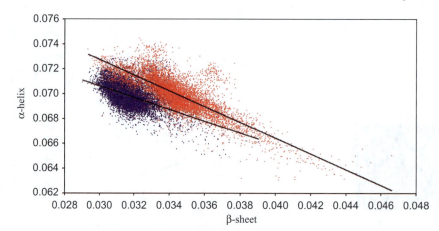

Figure 11.9   SFTIRM pixel correlation of β-sheet *vs.* α-helix content in scrapie-infected dorsal root ganglia at 100 dpi (blue) and terminal stage (red). Each pixel represents the α-helix and β-sheet content of one spectrum. A correlation coefficient of 0.636 (at terminal stage) indicates the presence of proteins mostly high in β-sheet and low in α-helix or *vice versa*, while a correlation coefficient of 0.267 (at 100 dpi) depicts a weaker correlation of secondary structure in the proteins present.

A statistical analysis of the SFTIRM images showed that the pixel correlation, obtained by plotting the α-helical content (*y*-axis value) *vs.* the β-sheet content (*x*-axis value) for each pixel, was low at early time points (100 dpi, $R^2 = 0.267$) but increased during pathogenesis (130 dpi, $R^2 = 0.338$; fcs, $R^2 = 0.345$; terminal, $R^2 = 0.636$) (Figure 11.9). Although the correlation of α-helix to β-sheet increases during pathogenesis, suggesting that α-helical rich proteins were replaced by and/or converted to β-sheet rich proteins, a regression coefficient of about 0.6 also indicates that more complicated changes, probably involving other secondary structures such as turns, loops and unordered structures take place too.

11.6   Assessment, Discussion and Conclusions

As an example of an application of synchrotron FTIRM, thoracic DRG from hamsters perorally infected with 263K scrapie were investigated in order to detect differences in protein secondary structure and distribution between control and infected animals during the progression of the disease. We observed that the β-sheet content did not change with age in the control animals, whereas scrapie-infected animals had a significant increase in the amount of β-sheet protein with disease progression. Immunostained adjacent slides suggested that this increase is at least partly due to the increase of PrP[Sc] in the tissue, supporting the findings of an earlier study on terminally diseased animals.[12] The α-helical protein distribution varied with age, but the trends were similar for control and scrapie-infected animals. In all cases, the α-helical protein content was higher in the control animals, except for 100 dpi where the opposite was found. The total protein content was found to be significantly increased in pre-clinical scrapie and then declined gradually during pathogenesis, ending with significantly less total protein at the terminal stage of the disease.

Only ~0.1% of all proteins in the brain were identified as PrP^{Sc} at the terminal stage.[27] Using the PrP-specific antibody, mAB 3F4, DRG of Syrian hamsters orally infected with 263K scrapie were shown to exhibit PrP^{Sc} accumulation as early as 76 dpi in two of four investigated animals.[6] Our results showed an increase in β-sheet protein content which seems to be well above an amount that can be attributed to PrP^{Sc} only. Moreover, the relative amount of total protein expressed in the tissue at the terminal stage was significantly lower in infected animals than controls. Together, these results suggest that disease progression is associated with (a) an increase in proteins high in β-sheet, including but not limited to the conversion of PrP^{C} into PrP^{Sc}, and (b) a reduced expression of proteins in general, possibly due to degeneration and death of neurons. In addition, not all cells that showed elevated β-sheet content by SFTIRM stained positive for PrP^{Sc}. Thus, we suggest that other proteins in addition to PrP are responsible for the elevated β-sheet at the later stages of the disease, as has also been suggested elsewhere.[28,29] Consulting the literature could give us useful hints about the nature of those proteins. Diedrich and colleagues, for example, reported an increased expression of apolipoprotein E and cathepsin D in activated astrocytes soon after they showed PrP accumulation in scrapie 22L-infected female mice.[28] Further research of the current literature revealed that the latter protein was composed of only 4% α-helix but 50% β-sheet.[30,31] Heat shock protein, transferrin, and γ2-microglobulin were also increased in astrocytes during scrapie,[29] and γ2-microglobulin has been shown to consist of only β-sheet and no α-helix.[32] We therefore concluded that cathepsin D and γ2-microglobulin might be two of the proteins contributing to the detection of increased β-sheet in our study. However, owing to the integrative nature of information provided by FTIRM, assignments to single proteins cannot be achieved.

At the pre-clinical time points, *i.e.* 100 and 130 dpi, the infected animals exhibited a lower β-sheet content than the control animals. These results indicate that changes in protein composition occur long before the onset of clinical signs. These alterations could be due to a downregulation of genes that encode for proteins high in β-sheet and/or an upregulation of genes that encode for proteins low in β-sheet. However, the observed increase in relative protein expression (see Figure 11.7C) argues for the latter possibility, *i.e.* an overexpression of proteins high in α-helix and low in β-sheet. These results suggest that, especially at early stages of the disease, the detected spectral changes may not be due to the misfolded form of the prion protein itself but rather to changes in proteins other than PrP^{Sc}.

When looking at the distribution of protein structure, some cells at 100 dpi showed extremely low values of β-sheet (0.03) paired with high values of α-helix (0.07) in the cytoplasm (see arrowheads and inset in Figure 11.8A), which seemed to be arranged in clusters. Although SFTIRM cannot clarify the nature of these structures, it was shown that prion-infected cells produce transforming growth factor alpha (TGF-α),[33] which in turn increases PrP^C expression in cultured human keratinocytes after exposure.[34] Furthermore, studies of Syrian hamsters intracerebrally infected with 263K showed significantly increased amounts of PrP mRNA in the superior and inferior colliculi at 77 days post-inoculation.[35] PrP^C expression also appeared to be greatly upregulated in B-cells of scrapie-infected sheep,[36] and in senile plaques of Alzheimer's affected brain.[37] After leaving the Golgi apparatus, the cellular prion protein travels to the cell membrane in vesicles, where it becomes attached via a GPI anchor.[38–43] It is possible that the circular structures seen in the chemical images correspond to arrangements of vesicles filled with proteins low in β-sheet, possibly together with PrP^C (which consists of only 3% β-sheet but 42% α-helix) with high local density.

The results presented here show that elevated β-sheet content begins to be detectable near the cell membrane in early stages of the diseases and progresses throughout the cell by the terminal state. Immunohistochemistry studies have previously shown that the deposition of pathological PrP^{Sc} starts in satellite cells and in the cytoplasm and plasmalemma of neurons.[7] As the disease progresses, an increasing number of neurons are affected and more intense staining is also observed in the cytoplasm, primarily in secondary lysosomes,[44–46] aggresomes,[47,48] and in the nucleus.[49] Interestingly, accumulation of PrP in the cytosol has been shown to be neurotoxic.[50] Our finding of a PrP^{Sc} accumulation near the cell membrane of neurons is also consistent with the previously perfomed pilot study reporting the accumulation of large amounts of PrP^{Sc} at the cell membrane of neurons in the DRG of hamsters,[12] as well as in scrapie-infected mouse neuroblastoma cells,[51] in cultured Chinese hamster ovary cells,[52] in the medulla oblongata, pons, and astrocytes of sheep,[53] and in mice with natural scrapie.[54,55] Satellite cells, which are glia cells surrounding neurons in the peripheral nervous system (PNS), were also shown to accumulate the misfolded prion protein.[7,56,57]

The main and characteristic events in prion pathogenesis are the transformation of the cellular prion protein into the pathogenic form followed by an accumulation in the central and peripheral nervous system. However, in scrapie-infected mouse brain, a total of 138 genes were found to be upregulated while 20 were downregulated during pathogenesis,[58] demonstrating the complexity of prion disease pathology beyond transformation of the cellular prion protein.

In summary, this SFTIRM study showed that pronounced protein-related changes occur at the pre-clinical stages of scrapie that precede the transformation of PrP^C to PrP^{Sc}. PrP^{Sc} is known to be involved in the induction of apoptosis in neuronal cells,[59–62] which is accompanied by the upregulation of pro- and the downregulation of anti-apoptotic enzymes.[63] Recently performed SFTIR microspectroscopic studies of single apoptotic cells at different stages of apoptosis revealed a relative increase in α-helical content relative to β-sheet or random coil,[64] at both early and late stages. This is in agreement with our own findings of increased α-helical content and decreased β-sheet content at 100 dpi. However, at the terminal stage, the increase in PrP^{Sc} and other proteins might cause the detected elevation of β-sheet.

Getting back to the questions raised in Section 11.4.3, we can conclude:

1. Yes, alterations in the secondary structure of proteins are detectable by SFTIRM, where significant changes were observed at pre-clinical time points, and
2. the detected changes are partly but not completely due to the conversion of PrP^C to PrP^{Sc}, suggesting that other proteins high in β-sheet are accummulating too, and
3. these changes are manifest throughout scrapie pathogenesis, starting with a significant upregulation of proteins low in β-sheet followed by a gradual increase of β-sheet rich proteins during pathogenesis, while the total protein content significantly declines, likely due to the induction of neuron death.

In conclusion, SFTIRM qualifies as an explorative and rapid analysis technique par excellence, which can be used to diagnose disease or dysfunction via spectral biomarkers that change as indicators of the presence of a particular disease or in response to drug intervention, environmental stress or genetic modification. When nothing or little is known about an observed phenomenon, SFTIRM may provide a first hint for further, possibly more specific investigations. This is particularly the case when changing systems, whether it is a cell suspension of synchronized cells or cells treated with some specific drug, are measured time-dependently. Such experiments can, however, be done with SFTIRM in relatively short time frames compared with serial measurements, using *e.g.* fluorescence labels, testing many genes or separating and analyzing proteins or metabolites from complex mixtures. Therefore, the fundamental fingerprinting nature of the IR spectra of complex biological samples is a big advantage. It is, however, a disadvantage at the same time, since comprehensive understanding of these spectra is desirable but not achievable in many cases.

Acknowledgements

The authors would like to thank Marion Joncic and Kristin Kampf (Robert Koch-Institute) for skillful technical assistance with the animal experiments. We are also grateful to the technical and safety staff at the NSLS, especially to Randy Smith and Andrew Ackerman. This work was funded by the National Institutes of Health Grant R01-GM66873. The National Synchrotron Light Source is supported by the US Department of Energy, Office of Science, Office of Basic Energy Sciences, under Contract No. DE-AC02-98CH10886.

References

1. S. B. Prusiner, Prion diseases and the BSE crisis, *Science*, 1997, **278**(5336), 245–51.

2. K. M. Pan, M. Baldwin, J. Nguyen, M. Gasset, A. Serban, D. Groth, I. Mehlhorn, Z. Huang, R. J. Fletterick and F. E. Cohen *et al.*, Conversion of alpha-helices into beta-sheets features in the formation of the scrapie prion proteins, *Proc. Natl. Acad. Sci. USA*, 1993, **90**(23), 10962–6.

3. M. Beekes, E. Baldauf and H. Diringer, Sequential appearance and accumulation of pathognomonic markers in the central nervous system of hamsters orally infected with scrapie, *J. Gen. Virol.*, 1996, **77**(Pt 8), 1925–34.

4. M. Beekes and P. A. McBride, The spread of prions through the body in naturally acquired transmissible spongiform encephalopathies, *FEBS J.*, 2007, **274**(3), 588–605.

5. M. Beekes, P. A. McBride and E. Baldauf, Cerebral targeting indicates vagal spread of infection in hamsters fed with scrapie, *J. Gen. Virol.*, 1998, **79**(Pt 3), 601–7.

6. P. A. McBride, W. J. Schulz-Schaeffer, M. Donaldson, M. Bruce, H. Diringer, H. A. Kretzschmar and M. Beekes, Early spread of scrapie from the gastrointestinal tract to the central nervous system involves autonomic fibers of the splanchnic and vagus nerves, *J. Virol.*, 2001, **75**(19), 9320–7.

7. P. A. McBride and M. Beekes, Pathological PrP is abundant in sympathetic and sensory ganglia of hamsters fed with scrapie, *Neurosci. Lett.*, 1999, **265**(2), 135–8.

8. W. J. Schulz-Schaeffer, S. Tschoke, N. Kranefuss, W. Drose, D. Hause-Reitner, A. Giese, M. H. Groschup and H. A. Kretzschmar, The paraffin-embedded tissue blot detects PrP(Sc) early in the incubation time in prion diseases, *Am. J. Pathol.*, 2000, **156**(1), 51–6.

9. D. L. Wetzel and S. M. LeVine, Imaging molecular chemistry with infrared microscopy, *Science*, 1999, **285**(5431), 1224–5.

10. P. Dumas, N. Jamin, J. L. Teillaud, L. M. Miller and B. Beccard, Imaging capabilities of synchrotron infrared microspectroscopy, *Faraday Discuss.*, 2004, **126**, 289–302; discussion 303-11.

11. P. Heraud, S. Caine, G. Sanson, R. Gleadow, B. R. Wood and D. McNaughton, Focal plane array infrared imaging: a new way to analyse leaf tissue, *New Phytol.*, 2007, **173**(1), 216–25.

12. J. Kneipp, L. M. Miller, M. Joncic, M. Kittel, P. Lasch, M. Beekes and D. Naumann, *In situ* identification of protein structural changes in prion-infected tissue, *Biochim. Biophys., Acta*, 2003, **1639**(3), 152–8.

13. E. Baldauf, M. Beekes and H. Diringer, Evidence for an alternative direct route of access for the scrapie agent to the brain bypassing the spinal cord, *J. Gen. Virol.*, 1997, **78**(Pt 5), 1187–97.

14. Q. Wang, A. Kretlow, M. Beekes, D. Naumann and L. Miller, *In situ* characterization of prion protein structure and metal accumulation in Scrapie-Infected Cells by Synchrotron Infrared and X-ray Imaging, *Vib. Spectrosc.*, 2005, **38**, 61–69.

15. A. Dong, P. Huang and W. S. Caughey, Protein secondary structures in water from second-derivative amide I infrared spectra, *Biochemistry*, 1990, **29**(13), 3303–8.

16. A. Barth and C. Zscherp, What vibrations tell us about proteins, *Quart. Rev. Biophys.*, 2002, **35**(4), 369–430.
17. H. Fabian and W. Mäntele, Infrared Spectroscopy of Proteins, in: Handbook of Vibrational Spectroscopy. In John Wiley & Sons: 2002, Vol. 5, pp 3399–3425.
18. J. Buijs, W. Norde and J. W. T. Lichtenbelt, Changes in the Secondary Structure of Adsorbed IgG and F(ab')2 studied by FTIR Spectroscopy, *Langmuir*, 1996, **12**, 1605–13.
19. N. Toyran, B. Turan and F. Severcan, Selenium alters the lipid content and protein profile of rat heart: an FTIR microspectroscopic study, *Arch. Biochem. Biophys.*, 2007, **458**(2), 184–93.
20. K. K. Chittur, FTIR/ATR for protein adsorption to biomaterial surfaces, *Biomaterials*, 1998, **19**(4-5), 357–69.
21. A. Troullier, D. Reinstadler, Y. Dupont, D. Naumann and V. Forge, Transient non-native secondary structures during the refolding of alpha-lactalbumin detected by infrared spectroscopy, *Nat. Struct. Biol.*, 2000, **7**(1), 78–86.
22. A. Dong, S. J. Prestrelski, S. D. Allison and J. F. Carpenter, Infrared spectroscopic studies of lyophilization- and temperature-induced protein aggregation, *J. Pharm. Sci.*, 1995, **84**(4), 415–24.
23. A. Martinez, J. Haavik, T. Flatmark, J. L. Arrondo and A. Muga, Conformational properties and stability of tyrosine hydroxylase studied by infrared spectroscopy. Effect of iron/catecholamine binding and phosphorylation, *J. Biol. Chem.*, 1996, **271**(33), 19737–42.
24. G. Damaschun, H. Damaschun, H. Fabian, K. Gast, R. Krober, M. Wieske and D. Zirwer, Conversion of yeast phosphoglycerate kinase into amyloid-like structure, *Proteins*, 2000, **39**(3), 204–11.
25. U. Dornberger, D. Fandrei, J. Backmann, W. Hubner, K. Rahmelow, K. H. Guhrs, M. Hartmann, B. Schlott and H. Fritzsche, A correlation between thermal stability and structural features of staphylokinase and selected mutants: a Fourier-transform infrared study, *Biochim. Biophys. Acta*, 1996, **1294**(2), 168–76.
26. R. J. Kascsak, R. Rubenstein, P. A. Merz, M. Tonna-DeMasi, R. Fersko, R. I. Carp, H. M. Wisniewski and H. Diringer, Mouse polyclonal and monoclonal antibody to scrapie-associated fibril proteins, *J. Virol.*, 1987, **61**(12), 3688–93.
27. M. Beekes, E. Baldauf, S. Cassens, H. Diringer, P. Keyes, A. C. Scott, G. A. Wells, P. Brown, C. J. Gibbs Jr. and D. C. Gajdusek, Western blot mapping of disease-specific amyloid in various animal species and humans with transmissible spongiform encephalopathies using a high-yield purification method, *J. Gen. Virol.*, 1995, **76**(Pt 10), 2567–76.
28. J. F. Diedrich, P. E. Bendheim, Y. S. Kim, R. I. Carp and A. T. Haase, Scrapie-associated prion protein accumulates in astrocytes during scrapie infection, *Proc. Natl. Acad. Sci. USA*, 1991, **88**(2), 375–9.
29. J. F. Diedrich, R. I. Carp and A. T. Haase, Increased expression of heat shock protein, transferrin, and beta 2-microglobulin in astrocytes during scrapie, *Microb. Pathog.*, 1993, **15**(1), 1–6.

30. M. Fusek and V. Vetvicka, Dual role of cathepsin D: ligand and protease, *Biomed. Pap. Med. Fac. Univ. Palacky Olomouc Czech Repub.*, 2005, **149**(1), 43–50.

31. T. Lah, M. Drobnic-Kosorok, V. Turk, R. H. Pain and D. Cathepsin, *Biochem J.*, 1984, **218**(2), 601–608.

32. M. Okon, P. Bray and D. Vucelic, 1H NMR assignments and secondary structure of human beta 2-microglobulin in solution, *Biochemistry*, 1992, **31**(37), 8906–15.

33. E. L. Oleszak, G. Murdoch, L. Manuelidis and E. E. Manuelidis, Growth factor production by Creutzfeldt-Jakob disease cell lines, *J. Virol.*, 1988, **62**(9), 3103–8.

34. J. Pammer, W. Weninger and E. Tschachler, Human keratinocytes express cellular prion-related protein *in vitro* and during inflammatory skin diseases, *Am. J. Pathol.*, 1998, **153**(5), 1353–8.

35. G. Li and D. C. Bolton, A novel hamster prion protein mRNA contains an extra exon: increased expression in scrapie, *Brain Res.*, 1997, **751**(2), 265–74.

36. S. Halliday, F. Houston and N. Hunter, Expression of PrPC on cellular components of sheep blood, *J. Gen. Virol.*, 2005, **86**(Pt 5), 1571–9.

37. I. Ferrer, R. Blanco, M. Carmona, B. Puig, R. Ribera, M. J. Rey and T. Ribalta, Prion protein expression in senile plaques in Alzheimer's disease, *Acta Neuropathol.*, 2001, **101**(1), 49–56.

38. C. Robertson, S. A. Booth, D. R. Beniac, M. B. Coulthart, T. F. Booth and A. McNicol, Cellular prion protein is released on exosomes from activated platelets, *Blood*, 2006, **107**(10), 3907–11.

39. I. Porto-Carreiro, B. Fevrier, S. Paquet, D. Vilette and G. Raposo, Prions and exosomes: from PrPc trafficking to PrPsc propagation, *Blood Cells Mol. Dis.*, 2005, **35**(2), 143–8.

40. B. Fevrier, D. Vilette, F. Archer, D. Loew, W. Faigle, M. Vidal, H. Laude and G. Raposo, Cells release prions in association with exosomes, *Proceedings of the National Academy of Sciences of the United States of America*, 2004, **101**(26), 9683–8.

41. B. Fevrier, D. Vilette, H. Laude and G. Raposo, Exosomes: a bubble ride for prions, *Traffic*, 2005, **6**(1), 10–7.

42. A. Mironov Jr., D. Latawiec, H. Wille, E. Bouzamondo-Bernstein, G. Legname, R. A. Williamson, D. Burton, S. J. DeArmond, S. B. Prusiner and P. J. Peters, Cytosolic prion protein in neurons, *J. Neurosci.*, 2003, **23**(18), 7183–93.

43. M. A. Prado, J. Alves-Silva, A. C. Magalhaes, V. F. Prado, R. Linden, V. R. Martins and R. R. Brentani, PrPc on the road: trafficking of the cellular prion protein, *J. Neurochem.*, 2004, **88**(4), 769–81.

44. A. Taraboulos, D. Serban and S. B. Prusiner, Scrapie prion proteins accumulate in the cytoplasm of persistently infected cultured cells, *J. Cell Biol.*, 1990, **110**(6), 2117–32.

45. M. P. McKinley, A. Taraboulos, L. Kenaga, D. Serban, A. Stieber, S. J. DeArmond, S. B. Prusiner and N. Gonatas, Ultrastructural localization of

scrapie prion proteins in cytoplasmic vesicles of infected cultured cells, *Lab. Invest.*, 1991, **65**(6), 622–30.

46. A. Taraboulos, M. Scott, A. Semenov, D. Avrahami and S. B. Prusiner, Biosynthesis of the prion proteins in scrapie-infected cells in culture, *Braz. J. Med. Biol. Res.*, 1994, **27**(2), 303–7.

47. J. A. Johnston, C. L. Ward and R. R. Kopito, Aggresomes: a cellular response to misfolded proteins, *J. Cell Biol.*, 1998, **143**(7), 1883–98.

48. E. Cohen and A. Taraboulos, Scrapie-like prion protein accumulates in aggresomes of cyclosporin A-treated cells, *EMBO J.*, 2003, **22**(3), 404–17.

49. A. Mange, C. Crozet, S. Lehmann and F. Beranger, Scrapie-like prion protein is translocated to the nuclei of infected cells independently of proteasome inhibition and interacts with chromatin, *J. Cell Sci.*, 2004, **117**(Pt 11), 2411–6.

50. J. Ma, R. Wollmann and S. Lindquist, Neurotoxicity and neurodegeneration when PrP accumulates in the cytosol, *Science*, 2002, **298**(5599), 1781–5.

51. B. Caughey and G. J. Raymond, The scrapie-associated form of PrP is made from a cell surface precursor that is both protease- and phospholipase-sensitive, *J. Biol. Chem.*, 1991, **266**(27), 18217–23.

52. S. Lehmann and D. A. Harris, Mutant and infectious prion proteins display common biochemical properties in cultured cells, *J. Biol. Chem.*, 1996, **271**(3), 1633–7.

53. L. J. van Keulen, B. E. Schreuder, R. H. Meloen, M. Poelen-van den Berg, G. Mooij-Harkes, M. E. Vromans and J. P. Langeveld, Immunohistochemical detection and localization of prion protein in brain tissue of sheep with natural scrapie, *Vet. Pathol.*, 1995, **32**(3), 299–308.

54. M. Jeffrey, C. M. Goodsir, M. E. Bruce, P. A. McBride, N. Fowler and J. R. Scott, Murine scrapie-infected neurons *in vivo* release excess prion protein into the extracellular space, *Neurosci. Lett.*, 1994, **174**(1), 39–42.

55. M. Jeffrey, C. M. Goodsir, M. E. Bruce, P. A. McBride and J. R. Scott, Infection-specific prion protein (PrP) accumulates on neuronal plasmalemma in scrapie-infected mice, *Ann. N. Y. Acad. Sci.*, 1994, **724**, 327–30.

56. F. Archer, C. Bachelin, O. Andreoletti, N. Besnard, G. Perrot, C. Langevin, A. Le Dur, D. Vilette, A. Baron-Van Evercooren, J. L. Vilotte and H. Laude, Cultured peripheral neuroglial cells are highly permissive to sheep prion infection, *J. Virol.*, 2004, **78**(1), 482–90.

57. M. H. Groschup, M. Beekes, P. A. McBride, M. Hardt, J. A. Hainfellner and H. Budka, Deposition of disease-associated prion protein involves the peripheral nervous system in experimental scrapie, *Acta Neuropathol. (Berl.)*, 1999, **98**(5), 453–7.

58. S. Booth, C. Bowman, R. Baumgartner, G. Sorensen, C. Robertson, M. Coulthart, C. Phillipson and R. L. Somorjai, Identification of central nervous system genes involved in the host response to the scrapie agent during preclinical and clinical infection, *J. Gen. Virol.*, 2004, **85**(Pt 11), 3459–71.

59. R. Chiesa, B. Drisaldi, E. Quaglio, A. Migheli, P. Piccardo, B. Ghetti and D. A. Harris, Accumulation of protease-resistant prion protein (PrP) and apoptosis of cerebellar granule cells in transgenic mice expressing a PrP insertional mutation, *Proc. Natl. Acad. Sci. USA*, 2000, **97**(10), 5574–9.

60. Y. Zhang, E. Spiess, M. H. Groschup and A. Burkle, Up-regulation of cathepsin B and cathepsin L activities in scrapie-infected mouse Neuro2a cells, *J. Gen. Virol.*, 2003, **84**(Pt 8), 2279–83.

61. C. Hetz, M. Russelakis-Carneiro, K. Maundrell, J. Castilla and C. Soto, Caspase-12 and endoplasmic reticulum stress mediate neurotoxicity of pathological prion protein, *EMBO J.*, 2003, **22**(20), 5435–45.

62. X. Ye, H. C. Meeker, P. Kozlowski and R. I. Carp, Increased c-Fos protein in the brains of scrapie-infected SAMP8, SAMR1, AKR and C57BL mice, *Neuropathol. Appl. Neurobiol.*, 2002, **28**(5), 358–66.

63. S. K. Park, S. I. Choi, J. K. Jin, E. K. Choi, J. I. Kim, R. I. Carp and Y. S. Kim, Differential expression of Bax and Bcl-2 in the brains of hamsters infected with 263K scrapie agent, *Neuroreport*, 2000, **11**(8), 1677–82.

64. N. Jamin, L. Miller, J. Moncuit, W. H. Fridman, P. Dumas and J. L. Teillaud, Chemical heterogeneity in cell death: combined synchrotron IR and fluorescence microscopy studies of single apoptotic and necrotic cells, *Biopolymers*, 2003, **72**(5), 366–73.

65. H. Fabian and C. Schultz, Fourier Transform Infrared Spectroscopy in Peptide and Protein Analysis, *Encyclop. Anal. Chem.*, 2000, 5779–5803.

66. C. Yu and J. Irudayaraj, Spectroscopic characterization of microorganisms by Fourier transform infrared microspectroscopy, *Biopolymers*, 2005, **77**(6), 368–77.

67. J. M. Chalmers and P. R. Griffith, *Handbook of Vibrational Spectroscopy*, Wiley, 2002, **Vol. 5**.

68. D. Naumann, *Infrared spectroscopy in microbiology*, John Wiley & Sons, Chichester, 2000, p 102-131.

69. S. B. Prusiner, *Prion Biology and Diseases*, 2nd edn., Cold Spring Harbor Laboratory Press, New York, 1999.

CHAPTER 12

Monitoring the Effects of Cisplatin Uptake in Rat Glioma Cells: A Preliminary Study Using Fourier Transform Infrared Synchrotron Microspectroscopy

K. R. BAMBERY,[a] B. R. WOOD,[a] E. SCHÜLTKE,[b,c] B. H. J. JUURLINK,[b] T. MAY[d] AND D. MCNAUGHTON[d]

[a] Centre for Biospectroscopy and School of Chemistry, Monash University, Victoria 3800 Australia; [b] Department of Anatomy and Cell Biology ; [c] Department of Surgery, University of Saskatchewan, Saskatoon, SK Canada; [d] Canadian Light Source, 101 Perimeter Rd, Saskatoon, SK S7N 0X4 Canada

12.1   Introduction

There is a need for a rapid chemosensive test to select the most appropriate drug and optimal dose to cause cytotoxic activity for screening of potential anticancer agents in patients. In this context a number of tests are currently available including the measurement of DNA synthesis by [³H]thymidine incorporation,[1] the measurement of cellular permeability,[2] measurement of cellular metabolism using Alamar Blue,[3] MTT (3-(4,5-dimethylthiazol-2-yl)-2.5-diphenyltetrazolium bromide)[4,5] and the measurement of total cytosolic

RSC Analytical Spectroscopy Monographs No. 11
Biomedical Applications of Synchrotron Infrared Microspectroscopy
Edited by David Moss
Published by the Royal Society of Chemistry, www.rsc.org

poly(A) + mRNA.[5] Most of these assays require prolonged cell culture for up to three days to detect the cytotoxic effects of the anticancer agent, with the exception of the mRNA assay which takes about 12 hours. This, in combination with the fact that some cancer cells are very difficult to culture and the phenotype can change significantly during culturing, results in a lack of correspondence to results reported *in vivo*.[5,6] It has been shown that by measuring poly(A) + mRNA anticancer agent effects can be detected at a much earlier stage for both cytocidal and cytostatic chemosensitivity compared to the more conventional MTT assay.[5] The rapidity of this assay is based on the premise that mRNA synthesis precedes protein synthesis in any repair or apoptotic event. It has been shown that the total amount of mRNA in proliferating cells is significantly higher compared to resting cells,[7] and serum stimulation/ deprivation of cultured leukemic cells resulted in changing levels of total mRNA.[8] However, mRNA tests which involve polymerase chain reaction amplification are labor intensive, time consuming and expensive. In pursuit of a new assay to analyze cellular events in response to anticancer agents we have applied Fourier transform infrared (FTIR) synchrotron spectroscopy to investigate the macromolecular events associated with cisplatin uptake in single rat glioma cells.

Cisplatin [cis-diammine-dichloroplatinum(II)] is an active chemotherapeutic drug used to treat of a variety of human tumors, including cancers of the testis, ovary, cervix, bladder, neck, lung, endometrium, and the malignant brain tumor glioblastoma multiforme.[9] Cisplatin binds preferentially with purine residues in deoxyribonucleic acid (DNA) after hydroxyl groups have displaced cisplatin's chloride ions. Several types of adducts are known to form, the most abundant of which are the 1,2-intrastrand cross-links at the d(GpG) and d(ApG) sites.[10] Cisplatin exerts its antineoplastic effect by inhibiting DNA replication and transcription.

In this study we applied FTIR synchrotron spectroscopy using the Canadian Light Source (CLS), to monitor the effect of cisplatin uptake at the single cell level. The synchrotron provides a brilliant source of infrared (IR) photons that enables one to achieve a higher spatial resolution ($\sim 10\,\mu m$ wavenumber dependent) and signal-to-noise ratio compared to conventional FTIR microscopy using a globar source (where the best spatial resolution is $\sim 20\,\mu m$). The higher spatial resolution available using a synchrotron source enables targeting of single desiccated glioma cells the sizes of which are on the order of 15–20 μm. Only a small fraction of the cellular chemistry will interact with the chemotherapy drug and hence only small changes will be manifest in the recorded spectra, thus, the high signal-to-noise ratio afforded by the synchrotron source is vital for the detection of these small spectral differences. Within the cell population there will also be natural spectroscopic variability that is not directly related to the exposure or lack of exposure to cisplatin. In order to unambiguously identify changes in the spectra as arising from the drug exposure it is necessary to record spectra from a number of cells and to utilize a statistical approach for the analysis. Various methods of multivariate statistical analysis were applied to the analysis of these synchrotron data. Both unsupervised principal component analysis (PCA) and supervised partial least squares (PLS) regression analysis were shown to be successful in classifying the spectra as originating from the control or cisplatin-exposed cell samples. Statistical analysis was also performed using an artificial neural network (ANN) classification analysis. Only the ANN analysis is reported here although essentially the same results were returned from the other two multivariate statistical methods.

We postulate that FTIR spectroscopy in combination with a neural network algorithm shows potential as a rapid technique to monitor the effects of cisplatin uptake and may have potential as a screening technique for a variety of anti-cancer agents.

12.2 Methodology

12.2.1 Cell Culture, Cisplatin Preparation and Treatment

Cells from the rat C6 glioma cell line were cultured in Gibco® D–MEM-Medium and 10% fetal calf serum (FCS) penicillin (40 IU/cm^3) and streptomycin (40 μg/cm^3) at 37 °C in a humidified atmosphere with 5% CO_2 for 24 hours. Cells were seeded at a density of 2×10^6 cells per cm^3 directly onto MirrIR slides and placed in the bottom of two separate 85 mm Petri dishes containing medium as described above. After a further 24 hours, it was found that both groups of cells had expanded to ~60–70% confluence. At this point, cisplatin was added to one of the cultures to a final concentration of 5 μM and both groups were further incubated for another 24 hours, at the end of which the cells were briefly rinsed in distilled water and air dried.

12.2.2 Synchrotron FTIR Microspectroscopy

Spectra were acquired at the CLS on the Mid Infrared Spectromicroscopy beamline using a Bruker IFS66vs FTIR spectrometer and a Hyperion confocal microscope. Spectra were acquired in transmission/reflection or "transflection" mode (6 cm^{-1} resolution, 256 scans co-added, apodized using a Happ-Genzel function and zero filled by a factor of 4) on MirrIR slide substrates. Only one spectrum was collected through a 20×20 μm masking aperture from the center of each individual cell. All the recorded spectra were individually inspected to ensure good quality. Spectra were rejected when they exhibited unacceptably low absorbance (*i.e.* amide I absorbance <0.2) or strong dispersion artifacts (*i.e.* amide II/amide I intensity ratio >0.8). Dispersion can result when there is only an optically thin cell deposit plus some background contributions captured within the masking aperture.[11] Dispersion manifests in the spectrum as a shift in the amide I mode to a lower wavenumber value, an increase in the amide II to amide I ratio and a broadening of the amide features. Sixty spectra for the cisplatin and a further 105 spectra for the control were selected for subsequent analysis.

12.2.3   Neural Network Classification

The ANN analysis was performed using the NeuroDeveloper® 2.5b software.[12] Feed-forward single hidden layer neural networks were trained using the resilient back propagation (Rprop) algorithm. All the networks were trained to solve a two-class classification problem (*i.e.* test for identification of cisplatin-treated cells versus controls). The available spectra were split into three subsets: a training set (50%), a validation set (25%) to monitor and stop the training before network over-training occurs, and a test set (25%) to provide external validation. Network layout was optimized by testing varying numbers of neurons in the hidden layer for the lowest training and validation data set errors. The total number of neurons allowed in each network was constrained to ensure good theoretical generalization.[13,14] The number of available sample spectra was limited, and hence it was necessary to limit the number of weights in the networks. This was achieved by compressing the input spectra with a covariance-based feature selection algorithm in NeuroDeveloper.

While ANNs perform very well in classification and prediction, for analysis they do suffer from their "black-box" aspect. That is, it is difficult to interpret the ANN model to determine which of the input variables (spectral information) are most important for successful class prediction. The ANN analysis was performed on several different spectral windows to investigate which bands within the spectra contained the most important information for distinguishing cisplatin-treated cells from cells in the control group. In this manner it was possible to explore the effect on ANN prediction accuracy of leaving out spectral regions associated with selected functional groups. For each optimized network, the prediction accuracy was determined from classifications performed on the external test set. The NeuroDeveloper software incorporates algorithms for evaluation of the output neuron activations, the "winner-takes-all" (WTA) the "40-20-40" test and an algorithm to detect over-extrapolation. Descriptions of these algorithms have already been published elsewhere.[12,15] If a spectrum fails any of these evaluation tests then its result is marked as unclassified. An unclassified spectrum generally indicates a spectrum with poor signal-to-noise, which NeuroDeveloper has reported as uncertain and, hence, not clearly identifiable. It would be reasonable, therefore, to leave unclassified spectra out of the calculation of accuracy, sensitivity and specificity measures. Here we have instead chosen to leave unclassified spectra in the calculations as "failures" and report more conservative ANN performance measures. The classification accuracy of an ANN's performance was calculated as the number of correctly classified spectra divided by the total number of test spectra, including those marked as unclassified, expressed as a percentage. Sensitivity and specificity binary classification measures were calculated, where an ANN classification of cisplatin-treated cells was taken as a positive test result. For the purpose of calculating ANN sensitivity the unclassified spectra were taken as false negatives, and similarly specificity measures were calculated with unclassified spectra counted as false positives.

To minimize the effects of baseline variations due to Mie scattering, the ANN analysis was performed on second derivative spectra derived from the raw spectra processed using a Savitzky-Golay algorithm (nine points).

12.3   Results

Figure 12.1 shows all the selected second derivative FTIR spectra recorded from the rat C6 glioblastoma cells for both the group exposed to 5 μM of cisplatin for 24 hours (red) and the control group (blue). Figure 12.2 depicts the averages of the second derivative spectra from cisplatin-exposed and control cells from rat C6 glioblastoma cells. The amide I and amide II bands associated with proteins are prominent in the spectra at $1652\,cm^{-1}$ and $1545\,cm^{-1}$ respectively. The ester carbonyl band associated with lipids is apparent at $1740\,cm^{-1}$ and a small peak is visible at $1518\,cm^{-1}$ which may be attributed to tyrosine side chains in proteins.[16] Bands below $\sim 1500\,cm^{-1}$ are more difficult to assign to specific vibrational modes and so only tentative assignments are possible based on previous literature.[17] In biological FTIR spectra the spectral region $1500–1440\,cm^{-1}$ contains bands attributable to scissoring/bending modes of CH_2 and CH_3 asymmetric bending vibrations. These methylene and methyl bands are likely to be dominated by signals from fatty acid chain functional groups of membrane lipids. A small change in the band profile at $1445\,cm^{-1}$ tentatively associated with lipid membranes may be indicative of

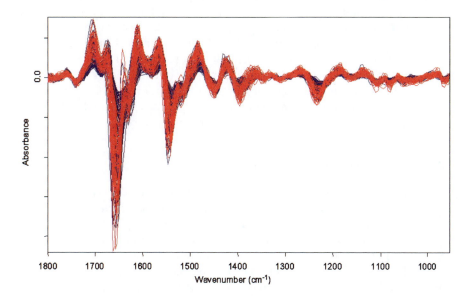

Figure 12.1   Selected second derivative FTIR spectra ($1800–950\,cm^{-1}$) recorded in transflection-mode on Mirr-IR substrates at CLS from rat C6 glioma cells, both the group exposed to cisplatin (red) and the control group (blue).

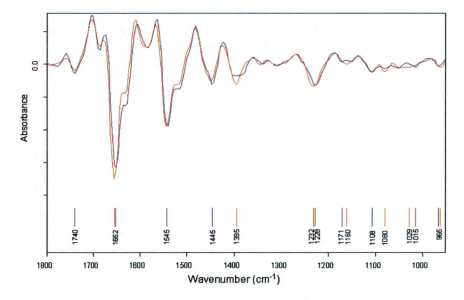

Figure 12.2   Rat cell line control group (blue) and cisplatin exposed group (red)
second derivative average spectra derived from the spectra in Figure 12.1
1 acquired with the infrared beamline at the CLS synchrotron.

lipid composition and content changes in the exposed cells. The band appearing
at $1395 \, \text{cm}^{-1}$ is assigned to the symmetric COO^- stretch of fatty acids and of
amino acid side chains in proteins. Examination of the spectra shows a small
shift in the asymmetric phosphodiester band of nucleic acids from $1228 \, \text{cm}^{-1}$ in
the controls to $1232 \, \text{cm}^{-1}$ in the cisplatin-exposed cells. The region of the
spectrum between 1175 and $1000 \, \text{cm}^{-1}$ contains stretching C–O vibrations as
well as the symmetric phosphate stretching vibration ($\sim 1085 \, \text{cm}^{-1}$). An
increase in the intensity of the band at $1160 \, \text{cm}^{-1}$ assigned to RNA ribose
skeletal modes[18–20] is also observed in cisplatin-exposed cells.

Neural Network Classification

Three spectral windows were chosen to serve as input data for the ANN
analysis to investigate whether each of these regions contained spectral information sufficient for accurately classifying cisplatin-treated and control cells.
The classification accuracy achieved on the external test data set by the ANN
models using the three spectral regions is shown in Table 12.1.

For all three spectral windows trialed there was no significant difference in
the classification accuracy, sensitivity or specificity when the number of cov-
ariant selected IR absorbance values was reduced from 36 to 16 points. This
indicates that the spectra contain some redundant information, which is not
surprising given that the $6 \, \text{cm}^{-1}$ spectral resolution employed for spectral
acquisition is small when compared with the inherent bandwidth of the features

Table 12.1 Summary of ANN trials performed on the external test set of treated and control spectra from the CLS synchrotron beamline.

Spectral window (cm⁻¹)	Number of covariance selected input feature values	Sensitivity (%)	Specificity (%)	Classification accuracy (%)
1800–950	36	82	99	94
1800–950	16	82	90	88
1477–950	36	82	99	94
1477–950	16	85	96	93
1286–1149	36	82	86	85
1286–1149	16	76	94	88

in the IR spectra of cells. The spectral windows 1800–950 cm^{-1} and 1477–950 cm^{-1} both include the absorbance bands due to methyl/methylene, the symmetric COO$^-$ stretch, the symmetric and asymmetric stretching vibrations of phosphodiester bands and the band associated with RNA/DNA ribose, but differ in that the former includes the amide I and amide II bands and the latter does not. When the amide I and amide II bands were included in the ANN training there was no significant improvement in classification accuracy and, in fact, the classification was slightly better without the amide bands included. The amide I band is sensitive to changes in protein secondary structure and, hence, it might be expected that the amide I band should provide useful spectral information for the task of classifying cisplatin-treated versus control cells. However, it appears that amide I band shape distortion produced by the dispersion artifact, which is variably present amongst the training spectra, completely obscures any protein structure or compositional changes that are expected to occur following cisplatin treatment.

Figure 12.3 indicates the position of the 16 absorbance features that were selected by the covariance algorithm over the 1477–950 cm^{-1} spectral window as training inputs for one of the ANNs. The 16 selected features are shown overlaid on an example spectrum. The network trained on these features alone returned a good classification accuracy of 93% on the test set, which suggests that the majority of the spectral changes that occur following cisplatin treatment are represented by these points in the spectra. The relative importance of each feature as determined by the covariance algorithm is shown in the figure. The most significant wavenumber values were found at 1034, 1044, 1122, 1128, 1209, 1253, 1307, 1380 and 1470 cm^{-1}. However, it is important to note that the ANN that was later trained on these features might well have weighted their importance completely differently. The majority of these selected wavenumber values are located on the shoulders of the principal absorbance bands and they presumably provide the ANN model with sensitivity to changes in the band profiles and positions.

The third spectral window covered the wavenumber range 1286–1149 cm^{-1} and so essentially included only the asymmetric phosphodiester stretch from

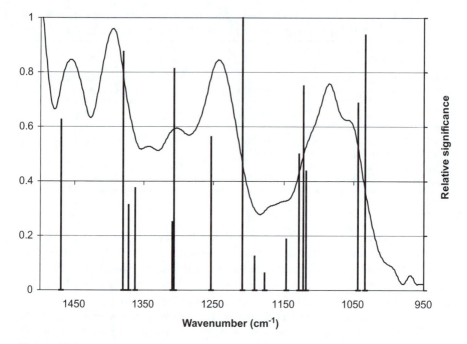

Figure 12.3   Sixteen spectral features selected by the NeuroDeveloper covariance algorithm over the 1477–950 cm^{-1} spectral window as training inputs for one ANN model. The positions of the COVAR selected features are plotted over an example rat cell line control group spectrum.

nucleic acids. This spectral window gave a slightly lower classification accuracy of ~86% but as this was still quite successful it is interpreted to indicate that the exposure to cisplatin is most strongly manifest in the asymmetric phosphodiester vibrational band. The PCA and PLS analyses conducted on the same data set also indicated that the asymmetric phosphodiester band shift was a characteristic spectral marker for cisplatin exposure.

12.4   Discussion

The FTIR synchrotron spectroscopy of rat glioma cells treated with cisplatin revealed changes in the spectra from which the ANN analysis was able to correctly classify cisplatin-exposed or control cells. The spectral features identified as statistically significant for this classification were consistent with features identified by examination of averaged spectra from the two conditions. Moreover the neural network was able to correctly classify an independent test set of spectra based on a model that examined only the profile of the asymmetric phosphodiester stretching vibration band.

   In order to explain why the asymmetric phosphodiester vibration is sensitive to cisplatin uptake it is important to understand the biochemical consequences of cisplatin binding to DNA. All processes of cisplatin intercalation with DNA produce distorted local conformations that unwind the DNA double helix and produce a widening of the minor groove.[10,21] The conformational changes in DNA induced by cisplatin intercalation effectively lower the phosphate density within the cellular nucleus and increase the distance between the phosphate moieties. A number of studies have shown condensation and margination of nuclear chromatin resulting in the breakdown of cells into apoptotic bodies following cisplatin exposure,[22–24] but there are exceptions depending on the cellular model.[25] Studies on the effects of cisplatin on gross morphology and cycling of C6 glioma cells in culture have reported a restructuring of cell processes, an increase in cell size, and a retardation in proliferation resulting in cell death 72 to 96 h after treatment.[26] It was further shown that cisplatin induced alterations in the nuclei of C6 glioma cells, and it was suggested that this may reflect some initial defense- or repair-like reactions leading to apoptotic death.[27]

The observation of IR spectral differences between normal and cancerous human cells and tissue has been reported in the literature.[28–31] These differences are manifested, among other changes, in the intensity of the phosphodiester vibrations (*ca.* 965, 1080 and 1230 cm^{-1}) of nucleic acids.[32] Diem and co-workers established via DNA/RNA digestion studies[33] and spatially resolved IR microspectroscopy[34] that the IR absorption intensities of DNA vary with the state of metabolic and proliferative activity of the cell. One could expect changes in the nucleic acid marker bands of cisplatin-exposed cells because they are usually arrested in the G_2 phase in advance of apoptosis. One could also expect changes in these bands due to condensation and margination of nuclear chromatin in response to cisplatin uptake. Diem and co-workers have demonstrated the sensitivity of the symmetric and asymmetric phosphate bands to DNA condensation.[35,36] Spectra of nucleated chicken cells produce a very similar spectrum to human erythrocytes, which are devoid of a nucleus.[35] However, when the chicken nuclei are swollen in hypertonic solutions the symmetric and asymmetric phosphodiester bands increase dramatically.[35] Recently, Romeo and Diem[11] have argued that the size of the nucleus will influence the overall amplitude of the spectral intensity of an entire cell in two ways: 1) through an averaging process of the nuclear intensity over the entire area of the cell, and 2) via an "optical density" argument employed to explain the absence of DNA signals in some nuclei. It has been demonstrated that in FTIR microspectroscopy studies the nuclear DNA spectral signature is almost entirely absent in condensed nuclei of metabolically inactive cells with coiled DNA tightly packed into nucleosomes, but, in contrast, the DNA spectral signature is detectable whenever DNA is uncoiled (*e.g.* during the S phase).[29,34,35,37,38] It seems likely that in inactive cells the chromosomes are too small to interact with IR radiation, whereas in expanded nuclei the nuclear DNA does absorb some light and, hence, the DNA phosphates become spectroscopically detectable.[29] Based on this hypothesis and the visibility of partially uncoiled DNA it would seem possible that intercalating drugs such as cisplatin could cause an increase in the intensity of nuclear DNA phosphate vibrations because the DNA in this state is more open and able to interact with the IR photons.

Induced DNA lesions can disrupt transcription[39,40] and hence FTIR spectral changes might occur as a result of drug-induced changes in composition and concentration of mRNA transcripts. It is well known that mRNA synthesis precedes protein synthesis in any repair or apoptotic event and therefore one might expect to see changes in spectral features related to RNA. Indeed in the spectra presented of control versus cisplatin, a blue-shift in the asymmetric phosphate stretching vibration is characteristic of RNA but not DNA.[41] This suggests that the changes observed are likely the result of an increase in mRNA synthesis as opposed to changes in nuclei condensation. The formation of cisplatin–DNA adducts can lead to cellular attempts to repair the damaged DNA through expression of damage recognition and repair proteins and p53-regulated pathways or through initiation of a pathway leading to apoptosis,[21] implicating mRNA activity.

Regardless of the mechanism, FTIR spectroscopy shows potential as an independent modality to monitor the effects of cisplatin uptake in cells. The neural network model trained on cells from different cell lines, under different growth and instrumental conditions and tested on a independent data set, provides evidence that this could, in fact, be a tool to supplement conventional total cytosolic poly(A) + mRNA and MTT assays.

12.5   Conclusions

The combination of FTIR synchrotron spectroscopy and multivariate data analysis applied to rat glioma cells exposed to cisplatin indicates mainly a change in the absorbance band profile of the asymmetric phosphodiester stretching vibration (~ 1241–$1228 \, cm^{-1}$ region). Based on the fact that a blue-shift in the asymmetric phosphate stretching vibration, characteristic of RNA, was observed it is hypothesized that the changes observed in the spectra of cisplatin-exposed cells are in response to mRNA activity as opposed to changes in nuclear condensation.

Synchrotron FTIR microspectroscopy enables this assay to be performed at the single-cell level and therefore has potential as a useful research tool for evaluating the effectiveness of various anti-cancer agents. The sensitivity and specificity of the neural network indicate the potential of the technique as an independent assay for cytotoxicity that should be possible to achieve with FTIR instrumentation employing a conventional globar source and still retain clinical value. Future studies will involve investigating cisplatin uptake in living cells recorded in a purpose built flow-through cell.

Acknowledgements

We would like to thank Mrs Connie Wong who prepared the cell cultures for our experiments at the CLS. We would also like to thank Prof. Dieter Naumann of the Robert-Koch Institute for valuable discussions relating to this work. We acknowledge financial support from the Australian Research Council. Dr Wood was supported by an Australian Synchrotron Program Research Fellowship and is currently supported by a Monash Synchrotron Fellowship. Part of the research described in this paper was performed at the Canadian Light Source, which is supported by NSERC, NRC, CIHR, and the University of Saskatchewan.

References

1. D. H. Kern, C. R. Drogemuller, M. C. Kennedy, S. U. Hildebrand-Zanki, N. Tanigawa and V. K. Sondak, *Cancer Res.*, 1985, **45**, 5436–5441.
2. D. D. Ross, C. C. Joneckis, J. V. Ordonez, A. M. Sisk, R. K. Wu, A. W. Hamburger and R. E. Nora, *Cancer Res.*, 1989, **49**, 3776–3782.
3. R. de Fries and M. Mitsuhashi, *J. Clin. Lab. Anal.*, 1995, **9**, 89–95.
4. T. Mosmann, *J. Immunol. Meth.*, 1983, **65**, 55–63.
5. Y. Miura, R. deFries, H. Shimada and M. Mitsuhashi, *Cancer Lett.*, 1997, **116**, 139–144.
6. W. T. Bellamy, *Drugs*, 1992, **44**, 690–708.
7. L. F. Johnson, H. T. Abelson, H. Green and S. Penman, *Cell*, 1974, **1**, 95–100.
8. Y. Mirura, K. Tominaga, T. Arakawa, K. Kobayashi, R. de Fries and M. Mitsuhashi, *Clin. Chem.*, 1996, **42**, 1758–1764.
9. L. Bernhard, *Cisplatin: chemistry and biochemistry of a leading drug*, 1999.
10. J.-M. Malinge and M. Leng, In *Cisplatin: chemistry and biochemistry of a leading anticancer drug*; Bernhard, L., Ed.; Verlag Helvetica Chimica Acta: Zürich, 1999, pp. 563.
11. M. Romeo and M. Diem, *Vib. Spectrosc.*, 2005, **38**, 129–132.
12. SynthonSoftware© **2004**.
13. F. Despagne and D. L. Massart, *Analyst*, 1998, **123**, 157R–178R.
14. D. R. Hush and B. G. Horne, *IEEE Signal Proc. Mag.*, 1993, **1**, 8–39.
15. S. Boydston-White, M. Romeo, T. Chernenko, A. Regina, M. Miljkovic and M. Diem, *Biochim. Biophys. Acta – Biomembr.*, 2006, **1758**, 908–914.
16. K. Range, I. Ayala, D. York and B. A. Barry, *Phys. Chem. B.*, 2006, **110**, 10970–10981.
17. F. S. Parker, *Applications of infrared spectroscopy in biochemistry, biology, and medicine;* Plenum Press, New York: 1971.
18. M. Tsuboi, *Application of infrared spectroscopy to structure studies of nucleic acids*; John Wiley&Sons: New York, 1970.
19. M. Tsuboi, K. Matsuo, T. Shimanouchi and Y. Kyogoku, *Spectrochim. Acta*, 1963, **19**, 1617.
20. E. Benedetti, E. Bramanti, F. Papineschi, I. Rossi and E. Benedetti, *Appl. Spectrosc.*, 1997, **51**, 792–797.
21. D. B. Zamble and S. J. Lippard, In *Cisplatin: Chemistry and biochemistry of a leading anticancer drug*; Lippert B., Ed.; Verlag Helvetica Chimica Acta: Zurich, 1999, pp 73-110.

22. M. Biggiogera, E. Scherini and V. Mares, *Acta Histochem. Cytochem.*, 1990, **23**, 831–839.

23. P. Kopf-Maier and H. J. Merker, *Teratology*, 1983, **28**, 189–199.

24. W. Lieberthal, V. Tiraca and J. Levine, *Am. J. Physiol.*, 1996, **270**, F700–F708.

25. L. J. Muller, C. M. Moorer-van Delft, E. W. Roubus, J. B. Vermorken and H. H. Boer, *Cancer Res.*, 1992, **52**, 963–973.

26. V. Mares, P. A. Giordano, G. Mazzinin, V. Lisa, C. Pellicciari, E. Scherini, G. Bottiroli and J. Drobnik, *Histochem. J.*, 1987, **19**, 187–194.

27. D. Krajci, V. Mares, V. Lisa, A. Spanova and J. Vorlicek, *Eur. Cell Biol.*, 2000, **79**, 365–376.

28. B. R. Wood, D. McNaughton, L. Chirboga, H. Yee and M. Diem, *Gynecol. Oncol.*, 2003, **93**, 59–68.

29. L. Chiriboga, P. Xie, H. Yee, V. Vigorita, D. Zarou, D. Zakim and M. Diem, *Biospectroscopy*, 1998, **4**, 47–53.

30. P. Lasch, W. Haensch, E. N. Lewis, L. H. Kidder and D. Naumann, *Appl. Spectrosc.*, 2002, **56**, 1–9.

31. P. Lasch and D. Naumann, *Cell. Mol. Biol.*, 1998, **44**, 189–202.

32. B. R. Wood, B. Tait and D. McNaughton, *J. Appl. Spectrosc.*, 2000, **54**.

33. L. Chiriboga, H. Yee and M. Diem, *Appl. Spectrosc.*, 2000, **54**, 480–485.

34. P. Lasch, M. Boese, A. Pacifico and M. Diem, *Vib. Spectrosc.*, 2002, **28**, 147–157.

35. B. Mohlenhoff, M. Romeo, M. Diem and B. R. Wood, *Biophys. J.*, 2005, **88**, 3635–3640.

36. M. Romeo, B. Mohlenhoff and M. Diem, *Vib. Spectrosc.*, 2006, **42**, 9–14.

37. M. Diem, S. Boydston-White and L. Chiriboga, *Appl. Spectrosc.*, 1999, **53**, 148A–161A.

38. S. Boydston-White, T. Gopen, S. Houser, J. Bargonetti and M. Diem, *Biospectroscopy*, 1998, **5**, 219–227.

39. C. Cullinane, S. J. Mazur, J. M. Essignamm, D. R. Phillips and V. A. Bohr, *Biochemistry*, 1999, **38**, 6204–6212.

40. E. R. Jamieson and S. J. Lippard, *Chem. Rev.*, 1999, **99**, 2467–2498.

41. B. R. Wood, B. Tait and D. McNaughton, *Appl. Spectrosc.*, 2000, **54**, 353.

CHAPTER 13

Mid-Infrared Reflectivity of Mouse Atheromas: A Case Study

HOI-YING N. HOLMAN[a] AND FRANCIS G.
BLANKENBERG[b]

[a] Lawrence Berkeley National Laboratory, University of California, Berkeley,
CA 94720, USA; [b] Division of Pediatric Radiology/Department of
Radiology, Stanford University Hospital, Stanford, CA 94305, USA

13.1 Introduction

Small transgenic animal models are increasingly being employed to speed the
development of many new diagnostic technologies for cardiovascular disease
and their preclinical testing. Meanwhile the high intensity of light from a
synchrotron-based infrared (IR) source allows for the investigation of the
optical properties of gross specimens of whole tissues. In this chapter, we
describe a case study, combining these two different scientific disciplines, which
may represent a potential opportunity for the development of a new intravascular diagnostic modality.

Cardiovascular disease is the leading cause of death in developed nations and
is rapidly becoming the leading cause of death in developing nations as well.
Every year, more than one million Americans and more than nineteen million
people world-wide experience a sudden cardiac event. In more than half of the
fatal cases, there were no documented coronary events prior to death.[1] One
factor that makes prevention of sudden cardiac death (SCD) particularly
challenging is that those patients with a history of heart disease are the ones
least likely to experience SCD.[2]

RSC Analytical Spectroscopy Monographs No. 11
Biomedical Applications of Synchrotron Infrared Microspectroscopy
Edited by David Moss
© Royal Society of Chemistry 2011
Published by the Royal Society of Chemistry, www.rsc.org

Until the last decade, atherosclerosis was thought of as little more than a plumbing problem in which the progressive build up of fatty deposits (arterial plaques) within the arterial walls resulted in a narrowing (stenosis) or blockage (complete occlusion – thrombosis) of the arteries, often in the heart or brain. The continued growth of these flow-limiting lesions would in time lead to heart attack or stroke, respectively. However, this model failed to explain why so many heart attacks occurred without warning and why commonly employed therapies (such as the stenting of stenotic or occluded arteries) had little impact on long-term survival rates. It is now well documented that the majority of acute myocardial infarctions result instead from the spontaneous rupture of the thin fibrous cap found on top of high-risk/vulnerable plaques (VPs), lesions that in most cases do not limit blood flow before the acute event.[3,4]

Today it is known that atherosclerosis is really a chronic inflammatory disease of arterial blood vessels. Atherosclerosis causes two main problems. The first is stenosis, a narrowing of the blood vessel which results in a restriction in blood flow; the second is the rupture of soft plaque (called vulnerable plaque), which causes the formation of a thrombus that can lead to sudden death. Until recently, efforts to treat atherosclerotic disease tended to focus on the detection and treatment of stenoses. However, more recent research has shown that acute plaque rupture is the far more prevalent and dangerous problem.[5–10]

The current diagnostic challenge is neatly summed up in a recent issue of the journal *Circulation* in which a consensus statement co-authored by more than three dozen leading researchers in the field states: "Despite major advances in treatment of coronary heart disease patients, a large number of victims of the disease who are apparently healthy die suddenly without prior symptoms. Available screening and diagnostic methods are insufficient to identify the victims before the event occurs. The recognition of the role of the vulnerable plaque has opened new avenues of opportunity in the field of cardiovascular medicine."[11] Clearly, conventional diagnostic techniques such as the stress test and the coronary angiogram[12–16] must be supplemented with newer methods to achieve improved diagnosis and treatment of high risk plaques (VPs).

In an attempt to develop a new spectroscopic method to screen asymptomatic patients for unstable plaques we undertook an in-depth examination of the optical properties of the distinguishing histologic characteristics for each grade of atheroma. As we began our studies of atheroma we observed that a fractional amount of mid-IR light originating from a bright synchrotron source was reflected by mouse atheromas *in situ*.[17] In this chapter we describe the details of our investigation of how the pathologic components present within atheromatous plaques, including VPs of ApoE knock-out mice, can reflect significant amounts of mid-IR light. Furthermore, the IR spectra contained a variety of unique spectral signatures, corresponding to biologic features often found in VPs. These data may lay the required foundation for the development of a new intravascular diagnostic modality which can find and fully characterize sites of atherosclerotic disease in the vessels of the heart and brain.

13.2 Existing Diagnostic Methods

Angiography is a well-established medical imaging technique in which an X-ray picture is taken to visualize blood-filled structures, such as arteries and veins. The procedure is performed by inserting a catheter into a peripheral artery, which is then manipulated until it reaches the region of interest. Iodinated contrast medium is then rapidly injected into the blood stream, allowing visualization of the arterial lumen. However, as discussed above, the morphology of the vessel lumen by itself does not provide sufficient information for determining which plaques are at high risk of rupture.

In an effort to characterize the pathophysiologic state of atherosclerotic plaque as opposed to simply imaging the vessel lumen, a number of alternative methods have been tried:

- Contrast enhanced and intravascular magnetic resonance imaging (MRI): Since it is known that standard MR and CT angiography methods underestimate the degree of atherosclerosis, some researchers are experimenting with contrast agents that non-selectively enhance atherosclerotic plaques by taking advantage of the fact that atherosclerotic plaques are more vascular and have increased capillary membrane permeability as compared with normal vessels. While these methods have shown the ability to image the walls of diseased vessels, they fall short with respect to lesion characterization.[18]
- Optical coherence tomography (OCT): OCT obtains images based on light reflection with an exceptional high resolution. However, it cannot obtain images through blood. Furthermore, OCT cannot detect immature stages of atherosclerosis.[19] Recent results are more encouraging, but the sensitivity and selectivity of OCT for fibrous plaques still have room for improvement.[20,21]
- Thermography: Based on the observation that "temperature heterogeneity may be associated with sites of an increased concentration of macrophages and other inflammatory cells that may play a crucial role in heat release within atherosclerotic plaques",[22] vulnerable plaques may be detectable using this method. One possible drawback to this method is that the temperature differences are small, thus the process of scanning an artery is likely to be fairly slow.
- Elastography: The basic concept is that the mechanical properties of a thin fibrous cap (vulnerable plaques) are different from those of a thick fibrous cap, and that these mechanical differences can be detected using elastography. In practice, the approach is difficult to perform owing to motion of the sensor.[23]
- Electron-beam computed tomography (EBCT): While this method has been shown to detect coronary calcium, it cannot identify vulnerable plaques. However, as a screening tool, it may be useful for identifying those patients needing an invasive procedure.[24,25]

- Angioscopy: This method can readily detect ruptured plaques and yellow plaques in their earlier stages. However, the method for determining the subluminal composition of plaques appears to be subjective and is therefore of limited prognostic value.[26]
- Intravascular ultrasound (IVUS): Using a miniature ultrasound transducer, this method is quite useful for assessing stenosis and surface morphology of a diseased vessel. IVUS is, however, unable to characterize the composition of most plaques.[27] Furthermore, the safety of this practice has been called into question.[28]

One newer method, which bears some resemblance to the currently proposed method, requires discussion. Specifically, we are referring to the method by Moreno which involves the use of near-IR light for the identification of vulnerable plaque.[29] In a manner analogous to the proposed method, Moreno detects spectral patterns in the reflected light. Despite the strong results published, there are several concerns regarding the use of only the near IR. As described in the original paper, the light source is a tungsten lamp, which limits signal-to-noise ratio (SNR). The paper also notes that the system appears to have little specificity: "Because lipid-rich and non-lipid-rich samples contain many similar constituents (including water, which absorbs strongly in these wavelengths), the gross appearance of reflectance spectra is similar for most samples. Precise identification of subtle chemical differences is achieved only through a highly developed mathematical method . . . ". The observed features are spectrally broad – from 1600–1800 nm is a cholesterol peak and at 1500 and 1800 nm are broad water peaks. Therefore, it appears that the method hinges on small changes in biological water content and increased cholesterol in lipids. While this approach did work in five *ex vivo* samples, it is not clear that this differentiation can be robustly observed *in vivo* (owing to inter-patient differences in hydration level, and other differences that may tend to obsfucate this signal).

In summary, the need for better methods of detection of VPs is well supported, both in the literature and in clinical practice. Despite the large number of methods proposed for the detection of VPs none has been able to characterize the distinguishing biochemical and cellular features fully within each histologic grade (stage) of atherosclerotic disease. There is therefore a great unmet need for a new diagnostic imaging modality that can specifically identify VP in patients at risk of sudden cardiac death with or without clinical symptoms.

13.3 Pathologic and Biochemical Features of Vulnerable Plaques

Approximately 75% of acute coronary events and 60% of recently symptomatic carotid artery disease are caused by disruption of an atheromatous plaque.[30,31] Vulnerable plaques are metabolically active and histologically complex atheromas with thin fibrous caps that are prone to spontaneous rupture, often leading to sudden death. The lesion characteristics of VPs include the lack of any severe stenosis, thin fibrous caps (less than 100 μm thick), large lipid pools just deep to the luminal surface of an artery (subintimal), the infiltration of the subintimal layer with numerous highly active and proliferating and rapidly dying inflammatory cells such as macrophages and T-lymphocytes, the loss of vascular smooth muscle cells (SMCs) via apoptosis, decreased amounts of collagen and other structural extracellular proteins due to the local secretion of proteolytic enzymes such as the matrix metalloproteinases (MMPs) from intralesional macrophages, increased amounts of tissue factor, cell adhesion molecules (I-CAM, integrins, RGD-peptides), increased amounts of oxidized low density lipoprotein (LDL)-cholesterol (from diet and/or hereditary factors) and reactive oxygen species such as superoxide anion (O_2^-), hydrogen peroxide (H_2O_2), and peroxynitrite ($ONOO^-$).[32]

How all these features combine and generate the pathophysiology seen in VPs is still a matter of intense debate; however, most experts agree that the primary insult is the pathologic accumulation of oxidized LDLs and reactive oxygen species in the subintimal layer of the arterial wall. This oxidative stress adversely affects the intimal layer and induces endothelial cell activation/proliferation and dysfunction with pathologic intimal thickening (PIT) coupled to the local release of chemokines and cell adhesion molecules. In response to the release of these factors, macrophages and T-lymphocytes enter the subintimal layer in an attempt to remove the oxidative damage to the arterial wall. As macrophages ingest oxidized materials, they accumulate large amounts of lipid and become "foam cells". Macrophages along with T-lymphocytes elaborate a number of proteolytic enzymes, including several MMPs, that serve to weaken the vessel wall by degrading the collagen-basement membrane. The vascular smooth muscle cells of the media react to the pathologic changes in the subintimal layer by undergoing highly regulated and organized cell death (apoptosis), proliferation, migration, and differentiation into fibroblasts in an attempt to repair the weakened arterial wall. As oxidized LDLs cannot be degraded by macrophages and foam cells the amount of lipid in these cells simply increases until they reach their limits and undergo apoptosis (becoming disintegrating foam cells). The accumulated lipid is released in the surrounding extracellular space, forming a lipid pool. Eventually the lipid pool becomes hypoxic and evolves into a large necrotic lipid core surrounded by inflammatory cells with reduced amounts of collagen and vascular smooth muscle cells. The necrotic core is highly thrombogenic and quickly forms large clots when in direct contact with blood; a situation seen with disruption of the thin fibrous cap of a VP. The progressive build up of lipid and inflammatory cells in the subintimal layer, along with weakening of the supportive smooth muscular cell layer (media), tends to create eccentrically shaped atheromas that, owing to continual remodeling, cause little to no narrowing of the vessel lumen.

A competing but complementary school of thought on the pathogenesis of VPs focuses on the neoangiogenic response of the microvessels observed

histologically during autopsies of the outer wall of major arteries, especially those containing ruptured fibrous plaques in patients who died from
SCD.[33,34] Although the underlying mechanisms for the conversion of an
asymptomatic fibroatheroma to a lesion vulnerable to rupture has not been
established, the significance of intraplaque hemorrhages in lesion stability has
recently been proposed as an important contributor. In the neoangiogenic
theory the pathologic thickening (PIT) of the intima leads to local tissue
hypoxia, and over time this induces the formation of new microvessels of the
arterial wall (*i.e.* the vasa vasorum, which normally supply the adventitial or
outer layer of a large artery or coronary vessel). As these new vessels are
immature they lack supportive pericytes, basement membranes, and vascular
smooth muscle cells; they are therefore disorganized, leaky and prone to
rupture.

Neoangiogenesis is greatly enhanced by activated macrophages and T-lymphocytes which release large amounts of vascular endothelial growth factor
(VEGF) and other growth factors, as well as proteolytic enzymes. This
disorganized tangle of tumor-like vessels spreads into the media and intima and
is greatest around the necrotic–hypoxic lipid core. Ultimately all these pathologic changes result in a leaky, hemorrhage-prone intima and necrotic
lipid core. Over time red blood cells leak out of these fragile vessels (micro-
hemorrhages) and decompose in the arterial wall. As red blood cells have the
highest cholesterol content of any cell in the body, the accumulation of lipids
(including oxidized LDLs) from these degraded cell membranes expands the
necrotic lipid core and its cholesterol content. At a certain point the necrotic
core either comes into direct contact with the circulation, forming an occlusive
thrombus (with rupture of the thin fibrous cap) or experiences a massive
intraplaque hemorrhage (neoangiogenic vessel rupture) leading to SCD or
stroke.

13.4   Concept of Mid-infrared Reflectivity of Atherosclerotic Aorta

For decades, mid-IR spectroscopy has been used to characterize the pathologic
biochemical changes within thin sections of atheromas, including VPs,
ex vivo.[35–38] A variety of absorption peaks and bands within the mid-IR
transmission spectra have been specifically linked to the vibrations of atoms
within functional groups of many characteristic molecules that compose an
atherosclerotic lesion, such as atherogenic lipoproteins, phospholipid particles,
cholesterol ester and fatty acids. Unfortunately, the high native water
content within the arterial wall and its strong interfering absorption has
severely limited the use of mid-IR spectroscopy as an *in vivo* imaging modality
for atherosclerosis.

There are, however, alternative ways to use IR light to interrogate tissue
samples that can circumvent these difficulties, including spectral analyses of the
absorption patterns from "reflected" as opposed to "transmitted" IR light,

coupled to a bright but harmless light emitter such as a synchrotron.[39] We have previously measured the reflected IR (FTIR; Fourier transform IR measurement) spectra from individual living pulmonary fibroblasts in cell culture on highly reflective substrates interrogated with a highly focused (1–10 μm) beam of IR light from an on site synchrotron.[40] This approach is essentially a double-pass transmission–absorption method, and is often referred to as a "reflection–absorption" technique. We postulated that this spectroscopic technique could also be applied to whole tissues in which a subsurface change in refractive index in effect creates a reflective interface among the pathologic molecules of interest. Our prior work measuring the reflection–absorption FTIR spectra of microbial communities located on geological materials (which contain multiple reflective interfaces of different optical properties) gave us confidence that the reflection–absorption technique could be applied to the investigation of atherosclerotic vessels; biologic structures that contain multiple potential reflective interfaces.[39] Specifically, atheromas contain high amounts of non-native lipid (with an average refractive index of 1.45 to 1.50, depending on the composition and sizes of lipid particles)[41] and calcium deposits such as hydroxyapatite and calcium phosphate (with an average refractive index of at least 1.63).[42] In contrast, normal (non-diseased) intimal tissues are ~ 80% water, relatively uniform in optical properties, and generally have an average refractive index of approximately 1.35–1.38 (by extrapolation from the literature).[43,44]

Conceptually, mid-IR light that enters an atheroma would encounter multiple refractive boundaries, wherein some light might be reflected and re-emerge under certain conditions (as shown in Figure 13.1). From first principles, if the surface roughness at the refractive boundary is larger than the wavelength of the incoming light, the light that is not absorbed would undergo multiple diffuse reflections before re-emerging. If the surface roughness is smaller than the wavelength (*i.e.* the surface is nearly smooth), unabsorbed light can experience a quasi-specular reflection before re-emerging. The lack of accurate information on the wavelength-dependent refractive index, absorption coefficients and the surface roughness of atherosclerotic interfaces makes it difficult to numerically predict the detectable reflectivity. As an approximation, we assume a near normal incident light source. For diffuse reflection, we turn to Fresnel's equation which states that, at a normal incidence, the reflection is given by $(n2 - n1)^2/(n2 + n1)^2$; where $n1$ is the average refractive index of a water-rich native tissue material and $n2$ is the average refractive index of lipid-rich non-native materials such as lipid particles and foam cells. This yields an average of less than 1% of the incident light that would reflect back via diffuse reflection. Given the morphologic complexity of most atheromas, the final detectable signal would likely be the sum of these two modes of reflectivity. It is also likely that the features of the reflected IR spectra would contain the absorption patterns generated by the major pathologic components of an atheroma; namely, atherogenic lipoproteins, extracellular phospholipid particles, smooth muscle cells, foam cells, and disintegrating foam cells as they undergo apoptotic cell death.[45,46]

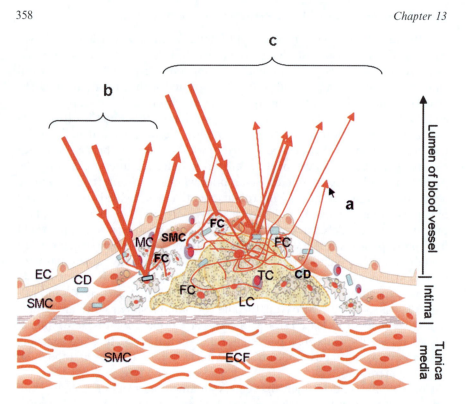

Figure 13.1 Mid-infrared reflectivity of an aortic atheroma. Depending on the refractive index, absorption coefficients and the surface roughness of components in an atheroma, incident IR light may encounter a refractive index boundary and undergo (a) diffuse reflection, (b) quasi-specular reflection, or (c) their combination. CD, calcium deposits; EC, endothelial cells; ECF, extracellular collagen fiber; FC, foam cells; MC, macrophages; LC, lipid particles and lipid core; SMC, smooth muscle cells; TC, T-cells; Arrows and lines, IR light paths.

13.5   Mid-infrared Reflectivity of Experimental Atherosclerosis

To test whether such reflectivity exists, we made IR measurements on the explanted aorta of a 9-month-old female adult ApoE ($-/-$) mouse fed a high-fat diet. The ApoE ($-/-$) mice (*i.e.* homozygous for the Apoe^{tm1Une} mutation; C57BL/6×129SvJ) and wild-type controls were obtained from Jackson Laboratories (Bar Harbor, Me). Apolipoprotein E (ApoE) is a ligand for receptors that clear remnants of chylomicrons and very low density lipoproteins (VLDLs). Lack of the ApoE lipoprotein causes accumulation in plasma of cholesterol-rich remnants, which accumulate in the circulation promoting atherogenesis. ApoE deficient mice ($-/-$) have been generated for use as a murine model of spontaneous atherosclerosis. A high fat/cholesterol diet greatly accelerates the development of large macroscopic plaques in these knockout mice. Foam cell-rich deposition is usually noted by the age of 3 months followed by severe occlusion with repeated intramural plaque hemorrhages, particularly within the brachiocephalic vessels, by the age of 8 months, with a 50% incidence of sudden death 28 weeks after initiation of a high fat/cholesterol diet. Most importantly, ApoE knockout mice develop all phases of lesions found in humans in a time dependent manner and are recognized as an appropriate model for human atherosclerosis.[47,48]

The ApoE $(-/-)$ mice in the current study were placed on a high fat/cholesterol (1–2%) diet for 3 months until sacrifice. We employed Beamline 1.4.3 at the Advanced Light Source (ALS) synchrotron at LBNL (http://infrared.als.lbl.gov/) as our mid-IR source. The synchrotron IR beam has brightness in the range of approximately 10^{11} to 7×10^{12} photons/sec-mm^2 mrad-0.1%BW when focused to a 10 μm spot. Each aortic specimen was placed luminal side up beneath an IR-transparent ZnSe window (International Crystal Labs, NJ) inside a custom-built environmental chamber where the temperature was kept at $\sim 5\,^{\circ}\mathrm{C}$ to maintain the "freshness" of the specimens during reflectivity measurements. All FTIR measurements were recorded in the standard reflection mode in the mid-IR 4000 to 650 cm^{-1} region, and consisted of 128 co-added spectra at a spectral resolution of 4 cm^{-1}.

Figure 13.2 shows examples of the FTIR measurements collected along a cluster of different types of atheromas including histologically advanced lesions (VPs; confirmed later by routine histologic examination) within the aortic arch. A significant number of mid-IR photons (Figure 13.2a) that entered any given atheroma were reflected and emerged from the tissue sample (Figure 13.2b). In contrast, few were reflected from the non-diseased portions of aorta (Figure 13.2c). A striking feature was that the reflected light contained the spectral features (Figure 13.2d) known to be associated with atherosclerotic disease.[36-38] For instance, the 3100 to 2800 cm^{-1} region revealed similar spectral features known to arise from the carbon hydrogen bond stretching vibrations of fatty acids and cholesterol esters, including the –HC=CH– moiety of their unsaturated hydrocarbon chains at ~ 3010 cm^{-1}, or from their acyl CH$_2$ groups at ~ 2925 and ~ 2852 cm^{-1}.[37,38] The strong absorption at ~ 1745 cm^{-1} also matched the spectral profile of the C=O stretching vibrations of ester carbonyl ($>$C=O) groups that are found in atherogenic phospholipid. The broader absorptions in the 1700 to 1500 cm^{-1} region had comparable features to those generated by the $>$C=O stretching of the amide I and of the N–H bending of the amide II modes present in the peptide groups of proteins. In the 1500–1000 cm^{-1} fingerprint region, absorption characteristics at ~ 1465 and ~ 1375 cm^{-1} were similar in nature to the vibrations of the lipid acyl CH$_2$ and to the symmetric bending of the lipid methyl CH$_3$ groups, respectively. The absorptions at ~ 1240 and ~ 1090 cm^{-1} matched the spectral patterns known to arise from the asymmetric and symmetric stretching modes of PO$_2^-$ in the phosphodiester groups in phospholipids, whereas absorptions centered at ~ 1165 cm^{-1} and ~ 1060 cm^{-1} matched the ester C–O–C vibrations of phospholipid, cholesterol ester and fatty acid.[36-38]

A dendrographic analysis of all FTIR measurements at sites of atheromas (employing the *d*-value distances measure, Ward's algorithm, together with the

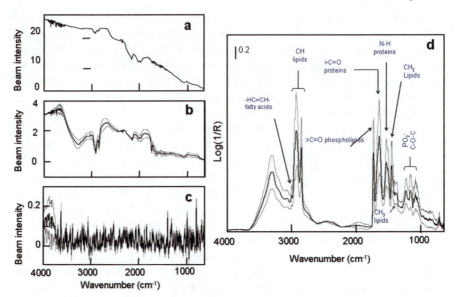

Figure 13.2 Examples of infrared reflectivity from aortic atheromas. Typical intensity profiles of (a) the incoming synchrotron IR beam, (b) reflected signals from atheromas, and (c) from non-atherosclerotic sites. (d) The corresponding reflection–absorption spectra of (b) shared spectral characteristics that were consistent with known excitation effects by IR photons on atoms of molecules that are known to characterize atherosclerotic plaques. Each plot shows the averaged spectrum (black trace) $\pm$ 1.0 standard deviation (gray trace); $n = 26$.

distribution of the z-values, Figure 13.3) suggested that the FTIR spectra of atherosclerotic aorta could be grouped into four categories (types I–IV) (Figures 13.4a to 13.4d; heavy lines), based on their spectral similarity. To determine whether the observed spectral patterns of reflected light could be linked to particular pathologic features, we conducted additional FTIR measurements on several model systems; including atherogenic lipoproteins, phospholipid particles, activated macrophages, late-stage or disintegrating foam cells, and smooth muscle cells (Figure 13.5). These biologic materials and cell lines are characteristic of atheromatous disease and its progression.[45,46] Although the exact hydrogen-bond environments of molecules in these model systems are different from those in actual tissue samples, they do mirror the different compositions of pathologic material found within atherosclerotic plaques, and form the basis of our subsequent comparative analyses (Figure 13.4a to 13.4d; light lines).

Using the data from the model systems above we found that the type-I spectrum (Figure 13.4a, heavy lines), based on its spectral shape and the sharply defined high absorptions (at ~ 2925, ~ 2852, and ~ 1745 ~ 1465, ~ 1375, ~ 1240, ~ 1165, ~ 1090, and $\sim 1060 \, \text{cm}^{-1}$), indicates the dominant presence of lipoproteins and phospholipids (Figure 13.4a; light line). For the type-II

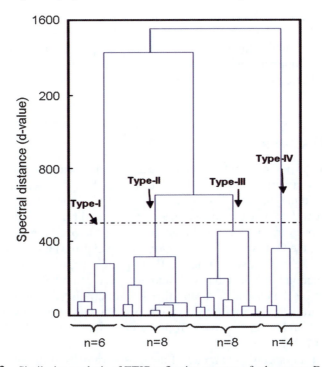

Figure 13.3 Similarity analysis of FTIR reflection spectra of atheromas. Dendrogram
of FTIR reflection–absorption spectra measured at 26 randomly chosen
locations within the cluster of atheromas (including vulnerable plaques)
(see text and Figure 2b). The dendrogram was based on the spectral
information contained and on employing the d-value distances measure
and Ward's algorithm. An analysis of the dendrograms together with the
distribution of the z-values (not shown) suggests that a d-value of ~ 500
(dotted lines) would be a reasonable spectral criterion for dividing the
reflection spectra into four types (see text and Figure 13.4).

spectrum (Figure 13.4b; heavy line), its overall spectral shape indicates the
presence of foam cells (Figure 13.4b; light line) in addition to lipoprotein and
phospholipid; a characteristic of VPs. The type-III spectrum exhibits a fine
structure of a triplet of peaks at ~ 1275, ~ 1235, and $\sim 1205 \, cm^{-1}$, and a
doublet of peaks at ~ 1083 and $\sim 1033 \, cm^{-1}$ (Figure 13.4c; heavy line), which
are indicative of type-I collagen,[36,38] and possibly smooth muscle cells (Figure
13.4c; light line). Smooth muscle cells in VPs have been reported to orchestrate
the assembly of type-I collagen.[49] Finally, the type-IV spectrum (Figure 13.4d;
heavy line) exhibits spectral features in the region $> 1400 \, cm^{-1}$ typical of the
spectral character of disintegrating foam cells (Figure 13.4d; light line), especially the spectral region between 1200 and $1000 \, cm^{-1}$ in which there was a
conspicuous absence of the PO_2^- group absorption peaks that are characteristic
of nucleic acids and various oligo- and polysaccharides, most of which were
probably degraded during apoptotic cell death.

We then repeated FTIR measurements on the diseased aortas from three additional ApoE $(-/-)$ mice (3-month-old adult females) and two wild-type controls, and found similar results. These data support the hypothesis that the pathologic components within atheromas cause both the reflection of IR light and the characteristic absorption patterns observed in experimental atherosclerosis.

13.6 Discussion

The diagnostic potential of the spectral information depends on the locations of the spectral character in the mid-IR spectrum. The broad reflection–absorption peak in all four types of VP spectra centered at $\sim 3300\,cm^{-1}$ and a shoulder near $3450\,cm^{-1}$ match the mixed spectrum of our library of atherosclerotic features. However, they are predominantly from the overlapping absorptions arising from the O–H stretching vibrations of the hydroxyl group in water, and the O–H and the N–H stretching mostly in protein peptides. These are molecules of common biological materials and not specific to atherosclerotic disease. Their effect on the character of an atherosclerotic reflection–absorption spectrum standard was not important as their spectrum was centered a significantly higher wavenumber further away from the other more specific components of the atherosclerotic spectra.

An important spectral character of diagnostic value is in the spectral region between 3100 and $2700\,cm^{-1}$, which is sensitive to the vibrations of C–H in the polyunsaturated acyl chains of cholesterol esters and fatty acids. Given that high levels of polyunsaturated cholesteryl ester have been linked to deposition of cholesteryl ester from plasma,[50] we think this could be a specific indicator of the transport of non-native cholesterol materials into the vessel wall. The very narrow but prominent absorption maximum at $\sim 1745\,cm^{-1}$ matches the

---

Figure 13.4   Comparing spectral variations with the pathologic components of atheromas. (a) Spectral variations in the type-I spectrum of atheromas (heavy line) can be explained (see text) by the two-component model spectrum of mixed AL + PP (light line). (b) Spectral variations in the type-II spectrum of atheromas (heavy line) can be explained (see text) by the three-component model spectrum of mixed AL + PP + FC (light line). (c) Spectral variations in the type-III spectrum of atheromas (heavy line) can be explained (see text) by the four-component model spectrum of AL + PP + FC + SMC (light line). Insert: a close-up comparison of the type-III spectrum of atheromas (heavy line) with the composite spectrum of AL + PP + FC + SMC (light line). The triplet of peaks at 1300–1200 cm^{-1} and the doublet of peaks at 1100–1000 cm^{-1} closely match the typical spectral pattern of type-I collagen (see text). (d) Spectral variations in the type-IV spectrum of atheromas (heavy line) can be explained (see text) by the spectrum of disintegrating foam cells (light line). R, reflectance; AL, atherogenic lipoproteins; PP, phospholipid particles; AM, activated macrophages; FC, late-stage foam cells; SMC, smooth muscle cells. Spectral markers of key functional groups in the solid-state atherosclerotic components are highlighted.

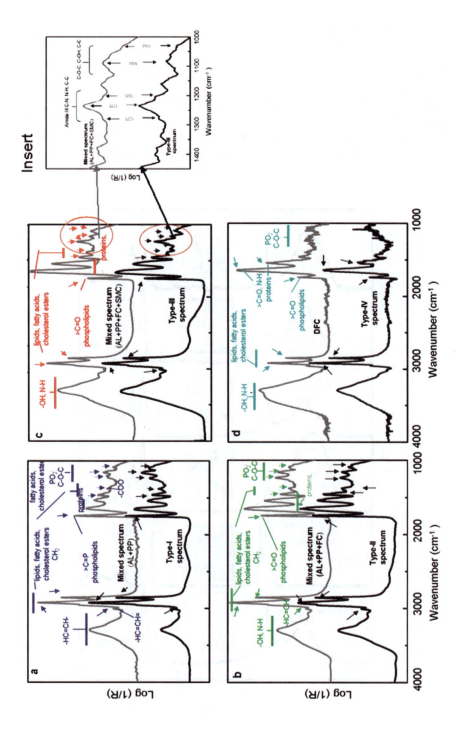

increased absorption feature of the mixed atherosclerotic component spectra with the added lipid particle spectral components. The spectral position of the absorption between ~ 1730 and $\sim 1750 \, cm^{-1}$ is very sensitive to the hydrogen-bonding environment of the ester carbonyl groups and the redox conditions of the lipids. For example, the absorption maximum for a lipid component can shift from a frequency less than $1730 \, cm^{-1}$ to $\sim 1745 \, cm^{-1}$ upon oxidation.[51] This blue-shift upon lipid oxidation has been reported to be associated with a more hydrophobic environment where the lipid carbonyl groups are free from forming complexes with other neighboring biomolecules via hydrogen bonding interactions.[52] We found that type-I, -II, and -III VP have the absorption maximum at $\sim 1745 \, cm^{-1}$ and higher wavenumbers. This blue-shift of

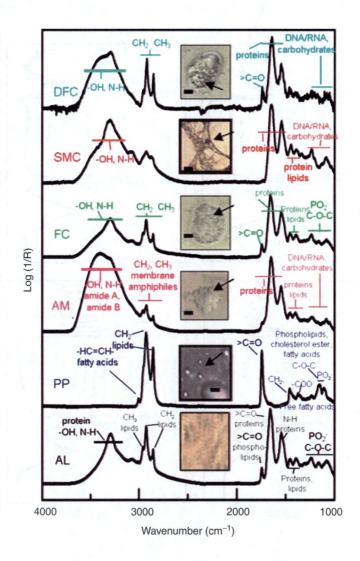

absorption maximum is a useful diagnostic feature for detecting the presence of less hydrogen-bounded and more hydrophobic lipid components of atherosclerotic disease.

In both the type-III VP spectra and the EP spectra, we observed a triplet of peaks at ~ 1275, ~ 1235, and $\sim 1205 \, \text{cm}^{-1}$, and a doublet of peaks at ~ 1083, and $\sim 1033 \, \text{cm}^{-1}$. Given that these fine spectral features are known to be unique to type-I collagen,[38,53] and since type-I collagen is reported to comprise approximately two-thirds of the total collagen during atherosclerosis process,[54] we believe this is another potentially valuable spectral structure for detecting sites of atherosclerotic disease.

Further studies are required to determine whether these new findings can be translated to the diseased vessels of other animal models, or in humans where the dimensions of atherosclerotic plaques and their position beneath the endothelial cell layer are much larger than that seen in the diseased aortas of ApoE $(-/-)$ mice. This question has recently been partially answered by the application of a mid-IR laser to image a patterned reflective surface through 150 mm thick films of blood.[55] Our results show that one can detect the presence of atherosclerotic risk factors beneath a layer of endothelial cells, at least in the mouse model system. If this can be applied to humans *in vivo*, and with continuing physics and engineering innovations, there will be an opportunity for the development of a new intravascular diagnostic modality which can

---

Figure 13.5   Mid-infrared FTIR reflection spectra of individual biologic risk factors. Photomicrographs and reflection–absorption FTIR spectra of athrogenic lipoprotein (AL), phospholipid particles (PP), activated macrophage (AM), late-stage foam cells (FC), smooth muscle cells (SMC), and disintegrating foam cells (DFC) are displayed. Each spectrum is an average of at least 20 samples. Purified chicken egg yolk and bacterial lipopolysaccharides (from Salmonella minnesota) were purchased from Sigma-Aldrich (MO, USA). The J774.1 murine macrophage cells were obtained from the American Tissue Culture Collection (ATCC). Thin films of egg yolk were prepared from the powder of the purified egg yolk which was suspended in distilled sterile water and dried under a stream of nitrogen gas. We recorded spectra of extracellular phospholipid particles and smooth muscle cells that were isolated from transverse micro-dissections of frozen ApoE$(-/-)$ mouse tissues. To obtain spectra of macrophages and macrophage-derived foam cells, we activated mouse macrophages (J774.1) in culture with bacterial molecules, and obtained foam cells by feeding the activated macrophages cholesterol-containing liposomes. Spectra of different stages of foam cells were measured at different time points: 24 hours (early-stage foam cells), 48 hours (mature-stage foam cells), and 72 hours (late-stage foam cells). To obtain spectra of disintegrating foam cells, we fed activated macrophages excess cholesterol-containing liposomes for 96 hours. Except for the smooth muscle cells that were recorded on a low-E slide, all other atherosclerotic component spectra were recorded with the samples placed on a gold-coated glass microscope pieces (0.5 cm × 1.0 cm) inside a custom-built environmental chamber where the temperature was kept at $\sim 5 \, ^\circ\text{C}$. Scale bars for photomicrographs: PP, 5 μm; AM, FC, DFC, 10 μm; SMC, 15 μm. Spectral markers of key functional groups are highlighted.

detect and characterize sites of atherosclerosis discovered at the time of screening coronary angiography.

Acknowledgements

This work was performed with support by the Directors of the Office of Science, Office of Biological and Environmental Research, Medical Science Division, the Office of National Petroleum Technology Program, the Office of Science, Basic Energy Sciences, Materials Science Division, of the United States Department of Energy under Contract No. DE-AC03-76SF00098.

References

1. R. J. Myerburg, A. Interian Jr, R. M. Mitrani, K. M. Kessler and A. Castellanos, *Am. J. Cardiol.*, 1997, **80**, 10F.
2. W. Rosamond, K. Flegal, G. Friday, K. Furie, A. Go, K. Greenlund, N. Haase, M. Ho, V. Howard, B. Kissela, S. Kittner, D. Lloyd-Jones, M. McDermott, J. Meigs, C. Moy, G. Nichol, C. J. O'Donnell, V. Roger,
   J. Rumsfeld, P. Sorlie, J. Steinberger, T. Thom, S. Wasserthiel-Smoller and Y. Hong, Heart disease and stroke statistics 2007 update a report from the American Heart Association Statistics Committee and Stroke Statistics Subcommittee, *Circulation*, **115**, 169e.
3. P. K. Cheruvu, A. V. Finn, C. Gardner, J. Caplan, J. Goldstein, G. W. Stone, R. Virmani and J. E. Muller, *J. Am. Coll. Cardiol.*, 2007, **50**, 940.
4. D. G. Katritsis, J. Pantos and E. Efstathopoulos, *Coron. Artery Dis.*, 2007, **18**, 229.
5. A. Burke, A. Farb, G. T. Malcolm, Y. H. Liang, J. Smialek and R. Virmani, *N. Engl. J. Med.*, 1997, **336**, 1276.
6. R. Virmani, F. D. Kolodgie, A. P. Burke, A. Farb and S. M. Schwartz, *Arterioscler. Thromb. Vasc. Biol.*, 2000, **20**, 1262.
7. C. L. Lendon, M. J. Davies and G. V. Born, *Atherosclerosis*, 1991, **87**, 87.
8. R. Ross, *N. Engl. J. Med.*, 1999, **340**, 115.
9. J. Narula, R. Virmani and B. L. Zaret, *in Atlas of nuclear cardiology*, E. Braunwald, V. Dilsizian, J. Narula (eds.), Philadelphia, Current Medicine, 2003, pp. 217.
10. J. R. Davies, J. H. Rudd, P. L. Weissberg and J. Narula, *J. Am. Coll. Cardiol.*, 2006, **47**, C57.
11. M. Naghavi, P. Libby, E. Falk, S. W. Casscells, S. Litovsky, J. Rumberger, J. J. Badimon, C. Stefanadis, P. Moreno, G. Pasterkamp, Z. Fayad, P. H. Stone, S. Waxman, P. Raggi, M. Madjid, A. Zarrabi, A. Burke, C. Yuan, P. J. Fitzgerald, D. S. Siscovick, C. L. de Korte, M. Aikawa, K. E. J. Airaksinen, G. Assmann, C. R. Becker, J. H. Chesebro, A. Farb, Z. S. Galis, C. Jackson, I. K. Jang, W. Koenig, R. A. Lodder, K. March, J. Demirovic, M. Navab, S. G. Priori, M. D. Rekhter, R. Bahr, S. M.

Grundy, R. Mehran, A. Colombo, E. Boerwinkle, C. Ballantyne, W. Insull, Jr, R. S. Schwartz, R. Vogel, P. W. Serruys, G. K. Hansson, D. P. Faxon, S. Kaul, H. Drexler, P. Greenland, J. E. Muller, R. Virmani, P. M. Ridker, D. P. Zipes, P. K. Shah and J. T. Willerson, *Circulation*, 2003, **108**, 1664.

12. P. Libby, *Am. J. Cardiol.*, 2001, **88**(Suppl B), 3.

13. Z. A. Fayad and V. Fuster, *Circ. Res.*, 2001, **89**, 305.

14. J. E. Muller, G. S. Abela, R. W. Nesto and G. H. Tofler, *J. Am. Coll. Cardiol.*, 1994, **23**, 809.

15. E. Falk, P. K. Shah and V. Fuster, *Circulation*, 1995, **92**, 657.

16. P. Libby, *Sci. Am.*, 2002, 47.

17. H.-Y. N. Holman. C. Bjornstad, C. Rosenberg, M. C. Martin, W. R. McKinney, E. A. Blakely, and F. G. Blankenberg, *J. of Biomedical Optics*, 2008, **13**, 030503-1.

18. J. Barkhausen, W. Ebert, C. Heyer, J. F. Debatin and H. J. Weinmann, *Circulation*, 2003, **108**, 605.

19. M. Zimarino, F. Prati, E. Stabile, J. Pizzicannella, T. Fouad, A. Filippini, R. Rabozzi, O. Trubiani, G. Pizzicannella and R. De Caterina, *Atherosclerosis*, 2007, **193**, 94.

20. H. Yabushita, B. E. Bouma, S. L. Houser, H. T. Aretz, I.-K Jang, K. H. Schlendorf, C. R. Kauffman, M. Shishkov, D.-H Kang, E. F. Halpern and G. J. Tearney, *Circulation*, 2002, **106**, 1640.

21. J. Schmitt, D. Kolstad and C. Petersen, Coherence Tomography Opens a Window onto Coronary Artery Disease, *Optics & Photonics News*, February 2004, 20.

22. S. Verheye, G. R. Y. de Meyer, G. Van Langenhove, M. W. M. Knaapen and M. M. Kockx, *Circulation*, 2002, **105**, 1596.

23. C. L. De Korte, S. G. Carlier, F. Mastik, M. M. Doyley, A. F. W. van der Steen, P. W. Serruys and N. Bom, *Eur. Heart J.*, 2002, **23**, 405.

24. R. C. Detrano, *Br. Heart J.*, 1994 October, **72**, 313.

25. R. Erbel, A. Schmermund, S. Möhlenkamp, S. Sack and D. Baumgart, *Eur. Heart J.*, 2000, 720.

26. Y. Ueda, A. Hirayama and K. Kodama, *Herz*, 2003, **28**, 501.

27. R. J. G. Peters, W. E. M. Kok, M. G. Havenith, H. Rijsterborgh, A. C. van der Wall and C. A. Visser, *J. Am. Soc. Echocardiogr.*, 1994, **7**, 230.

28. E. H. Yang and A. Lerman, *Nature Clin. Pract. Cardiovasc. Med.*, 2005, **2**, 392.

29. P. R. Moreno, R. A. Lodder, K. R. Purushothaman, W. E. Charash, W. N. O'Connor and J. E. Muller, *Circulation*, 2002, **105**, 923.

30. P. K. Cheruvu, A. V. Finn, C. Gardner, J. Caplan, J. Goldstein, G. W. Stone, R. Virmani and J. E. Muller, *J. Am. Coll. Cardiol.*, 2007, **50**, 940.

31. D. G. Katritsis, J. Pantosa and E. Efstathopoulos, *Coronary Artery Dis.*, 2007, **18**, 229.

32. M. Aikawa and P. Libby, *Cardiovasc. Pathol.*, 2004, **13**, 125.

33. F. D. Kolodgie, H. K. Gold, A. P. Burke, D. R. Fowler, H. S. Kruth, D. K. Weber, A. Farb, L. J. Guerrero, M. Hayase, R. Kutys, J. Narula, A. V. Finn and R. M.D. Virmani, *N. Engl. J. Med.*, 2003, **349**, 2316.

34. R. Virmani, F. D. Kolodgie, A. P. Burke, A. V. Finn, H. K. Gold, T. N. Tulenko, S. P. Wrenn and J. Narula, *Arterioscler. Thromb. Vasc. Biol.*, 2005, **25**, 2054–2061.

35. J. M. Gentner, E. Wentrup-Byrne, P. J. Walker and M. D. Walsh, *Cell. Molec. Biol.*, 1998, **44**, 251.

36. K. Z. Liu, I. M. C. Dixon and H. H. Mantsch, *Cardiovasc. Pathol.*, 1999, **8**, 41.

37. A. Becker, M. Epple, K. M. Muller and I. Schmitz, *J. Inorg. Biochem.*, 2004, **98**, 2032.

38. D. L. Wetzel, G. R. Post and R. A. Lodder, *Vib. Spectrosc.*, 2005, **38**, 53.

39. H.-Y. N. Holman and M. C. Martin, *Advances in Agronomy*, D. Sparks (Ed.), Elsevier, New York, 2006, 79.

40. H.-Y. N. Holman, M. C. Martin, E. A. Blakely, K. Bjornstad and W. R. McKinney, *Biopolymers: Biospectroscopy*, 2000, **57**, 329.

41. V. P. Maltsev, A. V. Chernyshev, K. A. Semyanov and E. Soini, *Meas. Sci. Technol.*, 1997, **8**, 1023.

42. B. J. Tarasevich, C. C. Chusuei and D. L. Allara, *J. Phys. Chem. B*, 2003, **107**, 10367.

43. L. J. J. Dirckx, L. C. Kuypers and W. F. Decraemer, *J. Biomed. Optics*, 2005, **10**, 044014–1.

44. H. D. Downing and D. Williams, *J. Geophys. Res.*, 1975, **80**, 1656.

45. P. Libby and P. Theroux, *Circulation*, 2005, **111**, 3481.

46. R. Virmani, S. Malik, A. Burke, K. Skorija, N. Wong, F. Kolodgie and J. Narula, *Circulation*, 2006, **114**, 385.

47. J. Neuzil, J. K. Christison, E. Iheanacho, J.-C. Fragonas, V. Zammit, N. H. Hunt and R. Stocker, *J. Lipid Res.*, 1998, **39**, 354.

48. R. L. Raffaï, L. M. Dong and R. V. Farese, *Proc. Natl. Acad. Sci. USA*, 2001, **98**, 11587.

49. M. M. Kockx, G. Y. De Meyer, N. Buyssens, M. W. M. Knaapen, H. Bult and A. G. Herman, *Circ. Res.*, 1998, **83**, 378.

50. J. H. Rapp, W. E. Connor, D. S. Lin, T. Inahara and J. M. Porter, *J. Lipid Res.*, 1983, **24**, 1329.

51. R. Hielscher, T. Wenz, S. Stople, C. Hunte, T. Friedrich, P. Hellwig, *Biopolymers*, 2006, **82**, 291.

52. R. Lewis, R. N. McElhaney, W. Pohle and H. H. Mantsch, *Biophys. J.*, 1994, **67**, 2367.

53. S. Mordechai, R. K. Sahu, Z. Hammody, S. Mark, K. Kantarovich, H. Guterman, A. Podshyvalov, J. Goldstein and S. Argov, *J. Microsc. – Oxford.*, 2004, **215**, 86–91.

54. K. Murata, T. Motayama and C. Kotake, *Atherosclerosis*, 1986, **60**, 251.

55. B. Guo, Y. Wang, C. Peng, H. Zhang, G. Luo, H. Le, C. Gmachl, D. Sivco, M. Peabody and A. Cho, *Optics Express*, 2004, **12**, 208.

Subject Index

Note: page numbers in *italic* refer to figures and tables.